MATERIALS FOR THE 21ST CENTURY

Dedicated to my brother, Jeffrey Segal

Materials for the 21st Century

David Segal

OXFORD
UNIVERSITY PRESS

OXFORD
UNIVERSITY PRESS

Great Clarendon Street, Oxford, OX2 6DP,
United Kingdom

Oxford University Press is a department of the University of Oxford.
It furthers the University's objective of excellence in research, scholarship,
and education by publishing worldwide. Oxford is a registered trade mark of
Oxford University Press in the UK and in certain other countries

First Edition published in 2017

Impression: 1

Published in the United States of America by Oxford University Press
198 Madison Avenue, New York, NY 10016, United States of America

British Library Cataloguing in Publication Data
Data available

Library of Congress Control Number: 2017933535

ISBN 978–0–19–880407–9 (hbk)
978–0–19–880408–6 (pbk)

DOI: 10.1093/oso/9780198804079.001.0001

Printed and bound by
CPI Group (UK) Ltd, Croydon, CR0 4YY

Preface

There are many pressing issues facing nations in the 21st century and materials have a role in finding solutions to these issues. For example, there is much concern among governments with established healthcare systems with the potential growth of Alzheimer's disease among ageing populations and the associated costs to these systems. Alzheimer's disease is associated with protein fibres or fibrils that associate to form plaque in the brain. There is much interest in the 21st century in the development of personalised medicine to improve the health of individuals while the emergence of diseases such as Ebola in West Africa in 2014 can have devastating consequences for populations. There is an increasing search for antimicrobial agents in the 21st century to combat drug-resistant infections. While the modern pharmaceutical industry dates to the end of the 19th century novel techniques based on recombinant DNA technology offer potential for new classes of drugs known as biopharmaceuticals. A growing world population in the 21st century requires sources of pure water that may be obtained from brackish and salty waters.

The requirement for reliable sources of energy to mitigate the effects of global warming has highlighted the role of nuclear power and renewable energy sources such as wind, solar, geothermal, hydroelectric and tidal. Throughout the 20th century many materials in particular synthetic polymers have been derived from petrochemical sources. There has been increasing emphasis in 21st century on the development of materials from renewable sources such as bioethanol from crops for blending with petrol and on the development of biodegradable plastics in order to limit the amount of plastics that litter the environment, as well as reducing the reliance on petrochemical sources for production of synthetic polymers. Biomass, namely natural materials such as plants, agricultural wastes and wood chips, are sources of lignocellulose, which is composed of cellulose, hemicelluloses and lignins that are potentially valuable sources of materials and biofuels, thus reducing a dependency on petrochemical sources.

The 21st century is one of digital communication and in everyday life smartphones and computers such as desktop, laptop and tablets are ubiquitous and all of these devices are characterised by displays while new forms of energy-efficient lighting are increasingly used in the home, in offices and in the environment. Increasing ranges of consumer products are available to the general public, whether televisions, games consoles, digital cameras, ink-jet printers or in-vehicle navigation systems as examples. Compact lithium-ion batteries are widely used in smartphones and portable devices while there is continuing development of fuel cells as a clean source of electricity for applications including transport. Medical diagnostic techniques include magnetic resonance imaging (MRI) and computerised tomography while self-testing for measuring blood glucose concentrations is widely used. New ways of communication such as social media proliferate, allowing people from across the world to interact with each other. Personal products such as cosmetics are widely advertised and the vast range of cheap and mass-produced foods are visible to all in advanced nations. Nickel-based superalloys for blades in gas turbines, nickel-titanium alloys for use in implantable coronary stents, Zircaloy alloys as fuel rods for nuclear fuels in thermal nuclear reactors and superconducting metallic alloys for generating magnetic fields in MRI are examples of speciality materials. Textiles, whether based on natural products or on synthetic fibre, are produced in a wide range of colours and are widely available. Surfaces and their modification affect material properties and throughout the text are examples of coating processes to modify surfaces. Surfaces are also important in the biological area, for example in the interaction of viruses with cells.

The threat of international terrorism using homemade explosives is an ever-present threat in the 21st century while personal protection is worn by police and members of the armed forces.

It may come as a surprise to members of the general public as well as to specialists that a common theme to the pressing issues in the 21st century involves materials that have the potential to contribute to the issues' solutions.

The purpose of this monograph is to show how materials can contribute to solve problems facing nations in the 21st century. In order to do this a glossary of 500 materials is included that gives descriptions of the materials, some historical details on their development, their uses or potential uses, and a range of references that the interested reader may wish to follow up. The choice of materials reflects in part my interests in ceramic materials but I have attempted to widen the range of materials to cover diverse applications and potential applications. In addition to this Glossary I have prepared a set of thirteen chapters on what I consider to be key topics that are important to everyday life, healthcare and the economies of nations in the 21st century.

The technical level of this book is aimed at the interval between school and university. The primary readership will be students at secondary school who are concentrating on examinations such as 'A' levels in the United Kingdom or equivalent examinations elsewhere with a view to seeking a career in the physical sciences, life sciences or engineering, or indeed students who have completed their 'A' levels. The first part of the book can be read as a whole by them to gain an overview of materials that will be important throughout their lives. They may also want to refer to selected entries in the glossary for clarification. In this book I have included examples from many disciplines and the entries can complement more detailed descriptions in specific reference books. Although a list of references are supplied in each chapter I have, where appropriate, included simplified descriptions of technical terms for clarity.

However, the book will also appeal to other groups of people. Undergraduates who want to continue their careers through higher degrees will find the entries in the glossary of interest as these entries may help them to crystallise their thoughts on which direction of research to follow. The book will also be of interest to postgraduate students who may want to follow up on entries they may not be familiar with through the references. Finally, in an age when information has never been more available because of access to the Internet, members of the general public will find the book a useful reference work to clarify areas of science and technology they may have encountered in their lives.

David Segal
Abingdon, UK

About the Author

David Segal has maintained an interest in materials chemistry and patent literature throughout his working life. He has researched and worked on a variety of materials and their applications, including high-temperature ceramic superconductors, gas-to-liquid technology for the synthesis of synthetic diesel, the treatment of diesel emissions from vehicles by use of non-thermal plasmas, activity transport in pressurized water reactors, sol-gel chemistry for the synthesis of a wide range of ceramic materials in the form of powders, coatings and fibres, ceramic nanofiltration membranes and inorganic phosphors. He has analysed patent portfolios in diverse technologies, including three-dimensional printing, light-emitting diodes, medical devices and electronic devices. David is the author of *Exploring Materials through Patent Information* (Royal Society of Chemistry, 2014) and *Chemical Synthesis of Advanced Ceramic Materials* (Cambridge University Press, 1989). David read natural sciences at Trinity Hall, Cambridge University, and completed a PhD on foaming in lubricating oils at the University of Bristol. He has worked for the UK Atomic Energy Authority at the Harwell Laboratory, AEA Technology and Coller IP Management. He is an author or co-author of over forty scientific papers and an inventor or co-inventor of over twenty-five patent families.

Contents

Approximate Timeline for Developments in Materials

Note that in many cases teams of people are involved in materials development and omission of people's names in this list should not be interpreted as meaning that their contribution was unimportant.

---

600 AD—Chinese soft porcelain

1709—Meissen porcelain produced for the first time in Europe (Johann Böttger)

1745—Bone china produced for the first time in England

1824—Portland cement (Joseph Aspdin)

1824—Waterproof clothing using natural latex rubber (Charles Mackintosh)

1844—Vulcanisation of rubber (Charles Goodyear)

1846—Cellulose nitrate (Frederick Schönbein)

1856—Bessemer process for low-carbon steel (Sir Henry Bessemer)

1856—Parkesine (Alexander Parkes)

1856—Purple dye mauveine (Henry Perkin)

1860s—Vulcanite (ebonite)

1861—Identification of colloidal systems (Thomas Graham)

1869—Celluloid (John Hyatt Jr)

1884—Acetate rayon (Hilaire de Chardonnet)

1886—Extraction of yellow textile dye from eucalyptus leaves (J. H. Maiden)

1887—Initial theoretical study of photonic materials (Lord Rayleigh)

1888—Liquid crystals (Friedrich Reinitzer)

1892—Viscose rayon (Edward Bevan and Charles Cross)

1890s—Cellulose acetate

1897—Diesel compressor engine using biodiesel (peanut oil) (Rudolf Diesel)

1897—Synthetic indigo (BASF)

1899—Patent for a machine to produce candy floss (J. C. Wharton and W. J. Morrison)

1900—Acetylsalicylic acid (Felix Hoffmann)

1907—Bakelite (Leo Baekeland)

1907—First observation of electroluminescence (Henry Round)

1910—Ammonia synthesis in the Haber–Bosch process (Fritz Haber)

1911—Detection of superconductivity in mercury (H. Kamerlingh Onnes)

1912—Stainless steel invented (Krupp Steelworks, Germany)

1913—Development of the plastic laminate Formica (US Formica)

1915—Synthetic diesel in Fischer–Tropsch synthesis

1916—Production of biobutanol (Chaim Weizmann)

1918—Cellophane (Jacques Brandenberger)

1921—Isolation of insulin (Frederick Banting and Charles Best)

1921—Polysilanes (F. S. Kipping)

1931—Synthetic rubber (neoprene) (Julius Nieuwland)

1932—Silica aerogels (Samuel Kistler)

1933—Polyvinylchloride (Waldo Semon)

1933—First preparation of low-density polyethylene (Reginald Gibson and Eric Fawcett)

1933—Poly (vinylidene chloride) (Ralph Wiley)

1935—Nylon (Wallace Carothers)

1937—Polyurethane (Otto Bayer)

1938—Polytetrafluoroethylene (Roy Plunkett)

1938—Polystyrene

1938—Xerographic process (Charles Carlson)

1939—Polymethylmethacrylate (Otto Rohm)

1940s—Ionic liquids

1940s—Superalloys

1947—Acrylic fibres

1947—Transistors (John Bardeen, Walter Brattain and William Schockley)

1948—Polyesters Terylene (Rex Whitfield and James Dickson, UK Calico Printers Association) and Dacron (Du Pont)

1950s—Styrene-acrylonitrile-butadiene

1950s—Float glass process (Sir Alistair Pilkington and Kenneth Bickerstaff)

1952—Concept of integrated circuits proposed (Geoffrey Dummer)

1952—X-ray crystallography of DNA fibres results in photograph 51 (Rosalind Franklin and Raymond Gosling)

1953—Structure of DNA (Francis Crick and James Watson)

1953—Polycarbonate

1953—High-density polyethylene

1954—Ziegler–Natta catalysts (Karl Ziegler and Guillo Natta).

1956—Living polymers

1956—Reactive textile dyes (ICI)

1958—Integrated circuits developed on a practical scale (Jack Kilby)

1960—Ruby laser (Theodore Maiman)

1960—Phase-change chalcogenide memories (Stanford Ovshinsky)

1961—Velcro (George de Mestral)

1960s—Red light-emitting diodes (N. Holonyak)

1964—Kevlar (Stephanie Kwolek)

1965—Moore's Law (Gordon Moore)

1965—Nitinol shape memory alloys for stents (W. Buehler)

1967—Polyphenylene oxide (Allan Hay)

1967—Poly (p-phenylene sulphide)

1968—Initial theoretical studies on metamaterials (V. Veselago)

1969—First commercial quartz wristwatch (Seiko)

1970s—Recombinant DNA technology (Stanley Cohen and Herbert Boyer)

1970s—Three-way automotive catalysts

1972—Adhesive for Post-it notes (Spencer Silver)

1973—Superconductivity in niobium-germanium alloys, Nb_3Ge (Gavaler)

1974—Twisted nematic liquid crystals (W. Helfrich and M. Schadt)

1975—Monoclonal antibodies (Cesar Milstein)

1975—Silicon carbide fibre

1976—Cyanobiphenyl liquid crystals (George Gray)

1976—Gore–Tex (Robert W. Gore)

1976—Discovery of sucralose artificial sweetener (Leslie Hough)

1977—PEEK, polyether ether ketone (Imperial Chemical Industries)

1978—Sialons (Kenneth H. Jack)

1979—Development of lithium-ion batteries (John Goodenough)

1980—Polyacetylene (Alan Heeger)

1980s—Macro-defect free cement

1980s—MRI equipment with superconducting magnetic metal coils for medical imaging

1982—Quasicrystals (Daniel Schectman)

1984—Stereolithography (Charles Hull)

1985—Fullerenes (Harold Kroto)

1986—High-temperature superconducting oxide ceramics (J. Bednorz and K. Muller)

1987—Polymerase chain reaction (Kary Mullis)

1989—First reference to the phrase 'three-dimensional printing' (Emanuel Sachs)

1989—Biodegradable starch-based plastics (Ferruzi spA)

1991—Carbon nanotubes (Sumio Iijima)

1991—Commercial launch of lithium-ion batteries (Sony Corporation)

1990s—Bio-based chemicals, e.g. biodiesel, bioethanol, bio-succinic acid

1990s—Blue light-emitting diodes (Shuki Nakamura)

1990s—Auxetic materials

1990s—Metamaterials

1993—Poly-(p-phenylenevinylene) (Richard Friend)

1997—First reference to the 'Lotus effect' of surfaces (Wilhelm Barthlott)

2000s—Biopharmaceuticals e.g. Humira

2004—Graphene (Andre Geim and Konstantin Novoselov)

2010s—Bioprinting

2012—Development of gene editing technology (CRISPR)

2014—ZMapp for treatment of Ebola virus

2015—Continuing development of lithium-air batteries

2015—Continuing development of two-dimensional materials

2015—Continuing development of synthetic biology

2015—Continuing development of self-healing materials

2015—Continuing development of scaffolds for tissue engineering

2015—Continuing worldwide development of gene-editing technology (CRISPR)

1

Introduction

The Importance of Materials for 21st-Century Economies

1.1 Introduction

At the start of *The Graduate,* a film directed by the late Mike Nichols and released in 1967, the character Benjamin Braddock played by Dustin Hoffman is advised by a family friend to seek a career in 'plastics' in order to secure financial and professional success. If this film was to be remade in the 21st century then Benjamin may well be advised to seek a career in 'materials', although whether financial and professional success would follow is debatable. Materials are often taken for granted in everyday life. For example, millions of non-degradable plastic bags have been issued annually to shoppers in supermarkets and other shops although since 5 October 2015 a charge of 5 pence per bag has been levied in the United Kingdom in order to cut down on the disposal of the bags. These bags are often made out of the synthetic polymer polythene (polyethylene) but there was a time before polythene and it is unlikely that the general public will be aware that polythene was first invented in 1933. Indeed it was prepared accidentally and was not the intended reaction product. In fact, polythene had a very important role in the Second World War when it was used as an electrical insulator for cables in fighter aircraft of the Royal Air Force for the newly developed compact radar systems based on magnetron cavities. There is no rigid definition of the word materials and while materials are frequently associated with objects made out of metal, ceramic or plastic the range of materials goes far beyond these narrow categories. If Benjamin had sought a career in materials in the 21st century he would soon realise that they underpin many industries, are critical for the development of consumer goods, are key components of medical diagnostic techniques and offer hope for the treatment of currently incurable diseases. In this introductory chapter an overview of the range of materials important for 21st century economies of nations is given. The study of materials is interdisciplinary. An appreciation of materials can be made by having an open mind and looking beyond the specific interests of an individual whether in chemistry, physics or engineering while retaining an awareness that basic concepts such as thermodynamics, chemical

Materials for the 21st Century. David Segal.
© David Segal 2017. Published 2017 by Oxford University Press.
DOI: 10.1093/oso/9780198804079.001.0001

kinetics and solid-state phenomena are often relevant to many different types of materials. The reader may want to peruse the glossary for more detailed descriptions of selected materials.

1.2   The significance of surfaces on material properties

Entries in the glossary do not highlight in an obvious way the important role surfaces have in material properties independent of the material composition. For example, superhydrophobicity is a property of a surface and superhydrophobic materials are not wetted easily by water. In fact, they have contact angles greater than 140°. The leaves of certain plants, in particular the lotus plant, are superhydrophobic and this property arises because their surfaces have a hierarchical structure with protrusions on the microscale and nanoscale which prevent contact of water droplets with the surface so that droplets roll off the plant carrying dirt particles with them. Hence the leaves are self-cleaning. There is much interest in developing materials that mimic the surface structure of plant leaves to produce self-cleaning structures such as self-cleaning glass, textiles and buildings. Heterogeneous catalysis is also an area where chemical reactions take place on the surfaces of catalytic particles and examples of these reactions include the Fischer–Tropsch reaction for the production of synthetic diesel from synthesis gas, the treatment of vehicle emissions of unburnt hydrocarbons, nitrogen oxides and carbon monoxide in so-called three-way automobile catalysts and the treatment of diesel emission that include nitrogen oxides and particulate carbon. As the particle size of materials decreases, an increasing percentage of atoms reside in the surface and when systems have a dimension around 1 micrometre or less they are referred to as colloidal. Nowadays the phrase nanotechnology is often used to describe systems that have dimensions similar to that of colloidal systems and nanomaterials or nanoparticles are words associated with the term nanotechnology. The properties of nanoparticles, in particular their optical properties, are a function of their size and this effect is utilised in stained glass windows that incorporate small gold particles. Also the optical properties of quantum dots, in particular the emitted wavelengths from these phosphors, are a function of their particle size. Surfaces are important in adhesion processes, for example the attachment of Post-it notes to noticeboards and the use of Super Glue. But not all adhesives are applied from a container as illustrated by the ability of the gecko lizard to climb walls and hang upside down from ceilings. The feet of this lizard contain millions of extremely thin protruding protein fibres and during contact between the feet and the wall or ceiling, an attractive **van der Waals force** between the fibres and wall or ceiling hold the lizard in place. Material properties can be affected by coatings, for example coatings that act as thermal barriers in gas turbines that enable the turbine blades to operate at temperatures above their melting points, antireflectance coatings on spectacle lenses and yellow coatings of titanium nitride on cutlery.

Van der Waals force is an attractive force between atoms and molecules. It arises from electrostatic interactions between two molecules with permanent dipole moments and from dipole-induced dipole interactions in which the dipole of a molecule induces a dipole in a neighbouring molecule. The force can also be considered to arise from fluctuations within dipoles.

1.3 The 20th century: A golden age for synthetic polymers

Polymers are macromolecules in which monomers are joined together chemically by polymerisation to form larger molecules. **Homopolymers** are derived from one type of monomer while heteropolymers or **copolymers** are made from two or more different monomers.

*If a molecule is designated by the letter A, then a **homopolymer** formed from this monomer is represented by . . . A-A-A- . . . . If two different molecules are represented by the letters A and B then a **copolymer** can be represented by . . . A-B-A-B-A . . . .*

Synthetic polymers are derived from petrochemical sources and throughout the 20th century oil has been a source of intermediate chemicals for conversion to polymers. Synthetic polymers are frequently referred to as plastics. Some synthetic polymers have been discovered accidentally, for example polyethylene and polytetrafluoroethylene, although a common misconception is that the latter was developed during the 'Space Race' in the 1960s, whereas it was first produced in 1938. Nylon had spectacular success as a substitute for silk in women's stockings when it was introduced as a commercial product while Kevlar, poly (p-phenylene terephthalamide), is associated with bullet-proof vests for anti-terrorist activities. Many synthetic polymers are commodity products that are produced on the industrial scale and common usage includes packaging (e.g. polystyrene beads, polythene), foamed products for furniture (polyurethane) and textiles (e.g. polyester and acrylic fibres). However, there are niche markets, thus for Kevlar, while poly (vinylidene fluoride) is used as a porous separator in lithium-ion batteries to prevent electrical shorting of the electrodes. In reverse osmosis, a process for purification of salty or brackish waters that is thermodynamically driven and not kinetically driven, polysulphone membranes have been used in the separation process. Synthetic polymers, with the exception of polyvinyl chloride, are potentially flammable and when used in consumer products flame retardants are often incorporated. The polymers do not degrade easily when disposed and there are concerns with pollution of land and oceans by plastic bags and other plastics. This concern has led to the search for **biodegradable** polymers and one such polymer is polylactic acid that can be produced from petrochemical sources or by fermentation of corn starch. Biodegradable polymers also have medical applications, for example as sutures in surgical operations.

Biodegradable materials will decompose in the environment over time to carbon dioxide and water.

1.4   The 21st century: A golden age for natural polymers?

There is increasing interest in healthcare delivering personalised medicine to people in which the genome of an individual would be sequenced; that is, the sequence of nucleic acids in an individual's DNA (deoxyribonucleic acid) would be determined. DNA can be considered to be a polymer, in particular a heteropolymer. Proteins are also **natural polymers** and they are made up of sequences of amino acids.

Natural polymers are found in nature and not prepared by chemical synthesis in a laboratory.

A pressing medical condition in the 21st century is the potentially increasing number of elderly people developing Alzheimer's disease, which is associated with protein fibres or fibrils that associate to form plaque in the brain. Other natural polymers include collagen, elastin, carbohydrates, starches and cellulose. The latter is a major component of biomass that includes plants and wood chips and which is a source of biofuels and materials with a reduced dependency on **petrochemical sources**. For example, cellulose can be converted to a textile fibre that has a silk-like texture and is marketed under the name of Tencel or Lyocell by first dissolving it in a solvent, followed by extruding the solution from a rotating disc known as a spinerette, after which the solvent can be recovered.

Petrochemical sources are derived from oil.

1.5   From candy floss (cotton candy) to composites

Many materials are produced in the form of fibres. Children and their parents have enjoyed candy floss while at a funfair or at the seaside. Candy floss consists of fibres of a sugar glass and is made by extruding molten granulated sugar (sucrose) through a spinning disc with holes in it known as a spinerette or through holes in the base of a rotating drum. This method of making fibres has been applied to the manufacture of short glass fibres from a melt and for the production of continuous Kevlar fibres. Glass fibres have been used to reinforce polymer **matrices** such as those based on polyester and the product known as fibreglass has been used as panels in vehicles, for the hulls of small boats and for bathtubs.

*The **matrix** is the material, for example the polymer, into which fibres are dispersed.*

Reinforcement of a matrix with fibres increases the toughness of the composite, that is, its resistance to fracture. Oxide fibres have also been manufactured by solution processes rather than from a melt and involve sol-gel precursors; an example of such a fibre is Nextel, a boric oxide-doped aluminosilicate material. Oxide fibres have applications as thermal insulation in buildings and for fibre-reinforced metals such as aluminium as combustor liners in gas turbine engines. Polymer matrices can be reinforced with flakes rather than with fibres, for example with graphene flakes for use in sporting equipment such as tennis rackets. Carbon-fibre-reinforced composites have applications where lightweight, strong and tough structural components are required particularly in space, military and civilian aircraft.

1.6 Biopharmaceuticals: From small to large molecules

The modern pharmaceutical industry dates from the late 19th century with the successful synthesis of acetylsalicylic acid that was sold and continues to be marketed under the trade name aspirin. Throughout the 20th century pharmaceutical compounds were usually synthesised in chemical laboratories and they are characterised by having a low molecular weight. For example, acetylsalicylic acid has a molecular weight of 180. Biopharmaceuticals (or biologics) are a new class of pharmaceutical compounds that are not prepared by conventional chemical synthesis. They are monoclonal antibodies, they contain sequences of amino acids in their structure and their synthesis involves recombinant DNA technology. An example of a biologic drug is Humira for the treatment of rheumatoid arthritis. It contains over 1,000 amino acids and has a molecular weight around 40,000. Antibodies are a class of proteins known as immunoglobulins. They are produced *in vivo* and bind to **antigens** in order to destroy them.

> **In vivo** *means within the body in contrast with* in vitro, *which means outside the body, for example a cell culture in a Petri dish.*
>
> *An* **antigen** *is an organism, for example a virus or bacterium that attacks another organism.*

A key feature of pioneering work carried out by Cesar Milstein and co-workers was to show how specific antibodies could be produced in large quantities. These specific antibodies are known as monoclonal antibodies so that all antibodies produced are identical. There is much interest in industrial processes such as those in the chemical and pharmaceutical industries in reducing the generation of waste products as well as in recovering and recycling solvents. This approach is sometimes referred to as utilising 'green chemical' processes. One class of material that has potential applications in green chemistry are ionic liquids which are salts that are liquid at room temperature. They have low vapour pressures, a useful property when recycling solvent vapours is required, and have applications including scrubbing gases.

1.7   Chips with everything

The 21st century is an age of digital communication. Smartphones and computers, whether desktop, laptop or tablet, proliferate, business processes migrate to the digital economy and are carried out over the Internet while social media allow people to communicate with each other and share information across the world. None of these devices or processes would be possible without silicon chips, which are key components of computers. A silicon chip is a piece of semiconducting silicon that has been doped with impurity atoms and contains electronic circuits that carry out logical operations in an algorithm. Silicon chips contain millions of transistors, resistors and capacitors and the transistors act as switches for carrying out the logical operations. Photolithography and curable organic resins known as photoresists, combined with the controlled addition of dopant or impurity atoms and etching, are used to fabricate the transistors and other components. The development of silicon chips is a triumph for materials research, requiring contributions from chemistry, physics, engineering and mathematics.

1.8   Energetic materials and terrorism

Cellulose, a natural polymer made up of glucose molecules that are joined together, is a major component of biomass, that is, natural materials such as plants, agricultural wastes and wood chips that are nowadays potentially valuable sources of materials and biofuels, thus reducing a dependency on petrochemical sources. However, in the 19th century cellulose was viewed as a chemically inert material until it was found that it would react with nitric acid to form nitrocellulose, or guncotton, which was an energetic material, that is, an explosive when dry. When the ratio of nitric acid to cellulose was reduced, useful reaction products that were not explosive were obtained. These products included celluloid, which had applications in the expanding film industry and also as buttons, combs and dentures as examples; collodion, which was used as a wound dressing; and Parkesine, which could be shaped into ornamental products. Developments with explosives have taken place since nitrocellulose was discovered but an emerging threat to societies in the 21st century is the use of unstable explosives that can be made from household materials. An example of such an explosive is triacetone triperoxide (TATP). Protective equipment worn by police and army personnel often contains Kevlar in the form of bullet-proof vests. Kevlar is a synthetic polymer, hence a plastic, and it is not obvious why it should be such a strong material. This plastic contains polymer molecules which form liquid crystals that have a higher degree of molecular ordering than the liquid state. Qualitatively the entropy of the system decreases as the liquid crystals form but close packing of polymer chains and hydrogen bonding between chains lowers the free energy of the system, favouring formation of the liquid crystalline phase. This combination of close packing and **hydrogen bonding** contributes to the strength of the plastic. Governments are also concerned that terrorists may acquire highly radioactive fission products and construct a 'dirty bomb' or even obtain fissile material such as ^{239}Pu.

*A **hydrogen bond** is an interaction between molecules that have hydrogen atoms bound to electronegative atoms such as oxygen, nitrogen and fluorine. Electronegative atoms can be viewed as atoms of elements that tend to gain electrons and form negative ions. Hydrogen bonding represents an attractive force between atoms.*

1.9   Speciality metallic alloys

People encounter synthetic polymers, that is, plastics, in all aspects of their lives, for example in clothing and in packaging, in furniture and in consumer goods, and may conclude that plastics are the most important materials in 21st century economies. But many plastics are commodity items and there are a number of speciality metallic **alloys** that have important roles but remain hidden from everyday life.

***Alloys** are materials that consist of two or more metals.*

Manufacturers of aircraft and airlines seek more efficient aircraft with improved fuel economy. Nickel-based superalloys are used to fabricate turbine blades in gas turbines and the blades are often coated with a thermal barrier of a ceramic material that increases the operating temperature of the turbine, increasing its efficiency. The blades operate at about 200°C above their melting point by use of a combination of a thermal barrier coating and internal cooling passages. The use of single-crystal casting technology maximises **creep** resistance and the removal of grain boundaries removes sources of weakness in the blades.

***Creep** is the continuous deformation of a material, usually a metal, by a constant stress while grain boundaries refer to the interface between crystals in a solid body, such as a metal.*

The single-crystal casting technology aids the design and fabrication of the cooling passages which allows cooler air from the compressor to be fed into the turbine blades during operation. Alloys of zirconium (with iron, chromium and tin) known as Zircaloy are used as fuel rods in thermal nuclear reactors while alloys are used in superconducting coils to generate large magnetic fields in magnetic resonance imaging. Nickel–titanium compositions are used as shape memory alloys in implantable coronary stents while light-emitting diodes contain metallic **alloys**, for example those formed from gallium and arsenic.

*Note that **superalloys** contain a number of metallic additives to nickel, in particular aluminium and chromium. The additives have atoms with a different size to nickel and they disrupt the ordered crystalline structure of nickel, a process that can reduce deformation of the component.*

1.10   Renewable energy

There is worldwide research to develop technologies that can reduce carbon dioxide emissions in order to limit the effects of global warming and to use renewable energy sources such as solar, hydroelectric, wind, geothermal and tidal. Materials have an important role in developing these renewable sources. For example, components used in wind turbines require high strength and fracture toughness to avoid catastrophic failure. Overhead cables on pylons that transmit electricity need to be strong and electrically conducting. The conventional material that is used in solar or photovoltaic cells is **polycrystalline silicon** in the form of panels that are mounted in modules.

* **Polycrystalline silicon** is distinct from a silicon component that is a single crystal with no grain boundaries.*

Dye-sensitised materials have the potential to increase the efficiency of the cells. The dyes are often based on ruthenium complexes. An alternative approach for fabrication of solar cells is to deposit a thin film of a photosensitive material such as copper indium gallium diselenide onto a low-cost substrate such as stainless steel or glass. This latter approach is often referred to as using thin film photovoltaics. Nuclear power is strictly not a source of renewable energy but as emissions of carbon dioxide from thermal nuclear reactors are minimal and do not contribute to global warming, this source of electricity is under active consideration worldwide. Materials have an important role in the construction and operation of thermal nuclear reactors. Thus, materials for structural components of the reactor, the pressure vessel and the reactor shell as well as materials for the fuel rods and the **enriched** uranium fuel. While not a renewable energy source it is worth mentioning here that compact lithium-ion batteries are widely used in smartphones and portable devices while there is continuing development of fuel cells as a clean source of electricity for applications including transport.

* **Enriched** means a higher concentration of fissionable ^{235}U in the uranium oxide fuel than occurs in uranium ores. Fissionable means that thermal neutrons will split this atom into smaller atoms with the release of energy.*

1.11   Renewable materials

There is a desire to move away from petrochemical sources as routes to materials and derive useful products from renewable materials such as biomass, which refers to natural materials such as plants, agricultural wastes and wood chips. Biomass is a source of lignocellulose which is composed of cellulose, hemicelluloses and lignins, which are potentially valuable sources of materials. Cellulose is a natural polymer consisting of glucose molecules that are joined together chemically. An example of a process using renewable materials is the production of bioethanol from **fermentation** of sugar extracted from sugarcane and the bioethanol is then blended with conventional fuels. Bioplastics such as polylactic acid can be prepared from glucose in a process that involves a fermentation step with genetically engineered bacteria. During the First World War biobutanol was obtained by fermentation of molasses, the residue left over from sugar refining, and acetone was obtained by fermentation on the industrial scale and converted to cordite, an explosive used by the British Army. Biodiesel can be obtained from vegetable oils or animal fat and is associated with its production from rapeseed oil. Bioethanol and biodiesel are known as biofuels.

*In **fermentation**, enzymes, for example in yeast, catalytically convert sugar to alcohol, a process used for brewing beer.*

1.12 The importance of microstructure

Wings of a butterfly may appear to be blue but in fact they are colourless and do not contain pigments. The colours arise because there is a periodic structure of particles in the wing made out of proteins and other components of the wing and this structure acts as a diffraction grating as light falls onto the wing. Similarly, the colours in the mineral opal arise from a periodic structure of close-packed silica particles. These periodic structures are examples of microstructures and such periodic structures, whether natural or artificial, are referred to as photonic crystals. The latter have the potential to direct light in specific directions. Traditional ceramics such as earthenware or porcelain have microstructures consisting of glassy phases, crystalline regions and pores that are produced during the calcination process in a kiln. Composite materials can consist of fibres dispersed in a matrix of polymer, ceramic or metal. The fibres increase the toughness of the material as they deflect cracks and prevent crack propagation and hence composites have a specific microstructure. Many organisms including molluscs have shells made out of composite materials that can withstand attack from a predator. DNA has a double-helical structure in which hydrogen bonding occurs between pairs of four bases but the pairing of bases is not random and this structure would not arise if the pairing was random. Thus, the helical structure is the microstructure that is characteristic of DNA and these examples highlight how material properties can depend on microstructure. Conversely microstructural control can be used to modify material properties.

1.13   Lighting technology

There have been rapid changes to lighting technology in recent years. Conventional **incandescent light bulbs** and fluorescent lights have been replaced by solid-state lighting, which is underpinned by developments in materials.

> **Incandescent light bulbs** *are light bulbs with a tungsten filament.*

For example, light-emitting diodes are increasingly used in the home and in business environments as they are energy-efficient. These devices rely on the semiconducting properties of materials, for example gallium arsenide and gallium nitride. Organic light-emitting diodes (OLEDs) are based on semiconducting organic molecules, either small molecules or polymers, and semiconducting quantum dots, which are small nanoparticles which behave as phosphors, are also being used in lighting technology. The increasing use of portable devices such as smartphones and tablet computers requires compact light sources for their displays and light-emitting diodes and quantum dots are used in these displays. OLEDs have the potential to be used as light sources in flexible electronics. It is worth stating here that while liquid crystals are used in displays such as televisions, the liquid crystals do not emit light and require a backlight in use.

1.14   Sugars and foods

There is much debate in countries with established health services such as the United Kingdom on rising levels of obesity and its role in the increase of Type 2 diabetes among the populations with the resulting additional costs to the health services for treatment of this disease. This debate often refers to sugars, carbohydrates, starches and foods generally and it is often overlooked that foods can be considered to be a class of materials. Thus, carbohydrates are polymers made up of glucose monomer units joined together chemically while starches are also polymers based on glucose molecules but can contain linear and branched molecules, whereas glucose is a component of sucrose that is widely used in everyday life as granulated sugar. While carbohydrates are broken down *in vivo* into smaller molecules including glucose they are often described as sugars, which spreads confusion among people concerned over issues relating to their weight. Consideration of foods consisting of materials such as natural polymers will aid the debate on obesity. Artificial sweeteners, either synthetic, such as sucralose, or natural products including those from the stevia plant, are widely used.

1.15   Intellectual property and materials

Companies, universities and other organisations that carry out research and development on materials will want to obtain a return on their investments. Patent coverage is often taken out and this confers a monopoly to the patent owner for a limited period of time in exchange for disclosing the invention in the public domain. The effect of the expiry of a patent, often twenty years after the filing date, can be observed in the pharmaceutical industry, as an example, when generic drugs can be produced by a third party and sold to governments at much lower prices than the patented version. Expert advice from qualified practitioners is required for anyone interested in obtaining patent protection and litigation relating to patent infringement is a frequent occurrence. Litigation can involve non-practicing entities (NPEs), which are individuals or firms who own patents but who do not use them to produce goods or services. Instead NPEs assert their patents against companies that do produce goods or services. Non-practicing entities are sometimes referred to as patent trolls although this phrase has derogatory overtones when used in this way. Trademarks that are words and logos are used to distinguish goods or services of one business from another and can continue to be used even when a patent has expired, as is the case for the material Lycra, which is also known by the names spandex and elastane.

1.16 Summary

Materials have an important role in 21st century economies and have the potential to contribute to solving issues in those economies. Examples where materials can make a positive contribution include renewable energy, renewable materials, lighting technology, biopharmaceuticals, semiconducting materials and the fabrication of silicon chips, synthetic and natural polymers, speciality metallic alloys for specialist applications, foods and their role in obesity, medical diagnostics, composite materials based on polymers, ceramic and metal and control of surface structure for tailored properties, as in superhydrophobic materials.

1.17 Further reading

Akhaven, J. *The Chemistry of Explosives*, 3rd edn. Cambridge: Royal Society of Chemistry, 2011.

Askeland, D. R., and Wright, W. J. *The Science and Engineering of Materials*, 7th edn. Boston, MA: Cengage Learning, 2011.

Atkins, P. *Atkins' Molecules*, 2nd edn. Cambridge: Cambridge University Press, 2003.

Bansal, N. P., and J. Lamon (eds). *Ceramic Matrix Composites: Materials, Modelling and Technology*. London: Wiley, 2015.

Basile, A., and C. Charcosset (eds). *Integrated Membrane Systems and Processes*. London: Wiley, 2016.

Brook, R. J. (ed.). *Concise Encyclopaedia of Advanced Ceramic Materials*. Oxford: Pergamon Press, 1991.

Burrows, J. Turbine technology. *Materials World* (2015), 23(5), 40–2.

Christie, R. M. *Colour Chemistry*, 2nd edn. Cambridge: Royal Society of Chemistry, 2015.

Felice, M. Materials for aeroplane engines. *Materials World* (2013), 52–3.

Fried, J. R. *Polymer Science & Technology*, 3rd edn. Upper Saddle River, NJ: Prentice Hall, 2014.

Ginley, D. S., and D. Cahen (eds). *Fundamentals of Materials for Energy and Environmental Sustainability*. Cambridge: Cambridge University Press, 2012.

Goldstein, D. (ed.). *The Oxford Companion to Sugar and Sweets*. Oxford: Oxford University Press, 2015.

Jones, R. A. L. *Soft Condensed Matter*. Oxford: Oxford University Press, 2013.

Kerton, F., and Marriot, R. *Alternative Solvents for Green Chemistry*, 2nd edn. Cambridge: Royal Society of Chemistry, 2013.

Lecce, L., and A. Concilio (eds). *Shape Memory Alloy Engineering for Aerospace, Structural and Biomedical Applications*. Oxford: Butterworth-Heinemann, 2015.

Ngo, C., and van de Voorde, M. *Nanotechnology in a Nutshell: From Simple to Complex Systems*. Amsterdam: Atlantis Press, 2014.

Novotny, L., and B. Hecht. *Principles of Nano-optics*, 2nd edn. Cambridge: Cambridge University Press, 2006.

Papachristodoulou, D., A. Snape, W. H. Elliott, and D. C. Elliott. *Biochemistry and Molecular Biology*, 5th edn. Oxford: Oxford University Press, 2014.

Price, G. *Thermodynamics of Chemical Processes*. Oxford: Oxford University Press, 2009.

Sarin, A. *Biodiesel: Production and Properties*. Cambridge: Royal Society of Chemistry, 2012.

Schneider, S. J. (gen. ed.), updated by I. H. Smith. *1000 Movies You Must See before You Die*, 460–1. London: Cassell Illustrations, 2015.

Solymar, L., D. Walsh and R. A. Syms. *Electrical Properties of Materials*, 9th edn. Oxford: Oxford University Press, 2014.

Tilley, R. J. D. *Understanding Solids: The Science of Materials*. London: Wiley, 2013.

Walton, D., and P. Lorimer. *Polymers*. Oxford: Oxford University Press, 2005.

2

Candy Floss, Cellulose, Sugars and Foods

2.1 Introduction

Many children and adults have eaten candy floss (cotton candy) at a funfair or at the seaside. Candy floss has a combination of properties. It is made in fibrous form, it is a glass, namely a sugar glass and glucose is a component of granulated sugar used in its preparation. Its method of preparation is relevant to other materials and as a food it can be enjoyable to eat. Glucose has a central role in the structure of some natural products in particular cellulose and starches and initial studies in the 19th century showed how cellulose could be converted into useful products including plastics. Although the 20th century saw the development of polymers and plastics from petrochemical sources, cellulose has an increasing role in the 21st century as a renewable material and a source of useful products. This chapter describes some material aspects of cellulose and other natural products with particular reference to the role of glucose in their chemical structures and the uses for the materials.

2.2 Cellulose

Cellulose is the most abundant biorenewable material on earth and consists of linear polymeric chains formed by repeated connection of β-D-glucose building blocks through a 1-4 glycoside linkage.

*A simple definition of **cellulose** is that it a natural polymer of glucose molecules, where glucose is the monomer.*

Together with hemicellulose and lignin, cellulose is the third component of ligno-cellulose, often referred to as biomass. Lignocellulose is the major structural component of plants and is found in sawdust, wood chips, straw and bagasse (sugar cane residue). Cellulose is crystalline and polymer chains are held tightly

Materials for the 21st Century. David Segal.
© David Segal 2017. Published 2017 by Oxford University Press.
DOI: 10.1093/oso/9780198804079.001.0001

together by hydrogen bonding and the van der Waals force and this structure makes cellulose insoluble in water and common solvents. This tightly bound structure cannot be digested, that is broken down into smaller molecules by humans. However, it was found around the middle of the 19th century that cellulose (e.g. wood pulp) could be nitrated with concentrated nitric acid to form nitrocellulose, also known as guncotton. Strictly speaking nitrocellulose, also known as cellulose nitrate, is not an organic nitrate but an ester and was an unstable material that exploded when dry. A number of people had been involved in the development of nitrocellulose at that time but it has been reported that a leading researcher, Frederick Schönbein, first produced nitrocellulose when he spilled nitric acid onto one of his wife's cotton aprons which he was wearing while carrying out experiments in his kitchen. The apron promptly burst into flames. Nitrocellulose could be handled in a more stable form when soaked in boiling water and then converted to a pulp and was relatively safe to handle when wet. Naval vessels were then able to carry 'guncotton' torpedoes quite safely.

Useful materials besides explosives were obtained by reducing the amount of nitric acid used to react with cellulose, especially when plasticisers such as camphor were added to the reaction product. For example, collodion was a liquid used as a wound dressing that left a thin film of cellulose nitrate over the wound. Parkesine, which can be considered to be a plastic, was a mixture of cellulose nitrate and a plasticiser and was used for ornamental objects and also for making combs and buttons, as examples. However, successful exploitation of celluloid on the commercial scale was made by John Hyatt, who referred to a mixture of cellulose nitrate and camphor as celluloid. One reason for the commercial success of celluloid was because John Hyatt and his brother developed an **injection moulding** machine which allowed bars and sheets of celluloid to be produced that could then be worked into products. Celluloid found a niche market in the nascent film industry as the substrate and reels for films that were shown on projectors in cinemas but its flammability resulted in many fires in projector rooms and it was eventually phased out in this application. Celluloid was also used to produce billiard balls as a substitute for balls made out of ivory because hunting threatened the survival of elephant populations. Spectacle frames, fountain pens, dentures and dolls were all manufactured from celluloid. It has been reported that some billiard balls made out of celluloid exploded on collision during games of billiards because too much nitric acid had been used in the preparation of cellulose nitrate.

*In **injection molding**, a process often associated with the shaping of plastics, a sheet or paste of material is subject to pressure so that the material takes the shape of a die. Injection molding is used to produce plastics bowls in kitchenware.*

Other plastics and fibres could be obtained from cellulose following experiments carried out towards the end of the 19th century. Thus, cellulose sources such as wood pulp were dissolved in glacial acetic acid to form solutions of cellulose acetate. The solutions were spun into the textile fibre known as acetate rayon and formed into sheets of a thermoplastic that could be shaped by pressing a hot sheet in a mold. Applications for cellulose acetate besides textile fabrics included lacquers and shatterproof glass for windscreens. Another type of rayon known as viscose rayon was prepared by first dissolving, for example, wood pulp in alkaline carbon disulphide, a process yielding cellulose xanthate solution. The latter was converted to continuous regenerated cellulose fibres of viscose rayon by extrusion through fine nozzles into sulphuric acid. The fibres of viscose rayon could be woven into fabrics. Extrusion of cellulose xanthate solution through a narrow slit produced sheets of cellophane, which is used as a wrapping material although it is flammable. Plastics and fibres derived from cellulose are known as cellulosics. The preparation of synthetic fibres from petrochemical sources in the 20th century has to some extent overshadowed the early work on cellulosics but a new cellulosic fibre was developed by the company Courtaulds in the 1970s. Here, wood pulp could be dissolved in a specific solvent and the solution was spun into a water bath that regenerated the cellulose as continuous fibre. The fibre has a silk-like texture and is known as Lyocell and by its trade name of Tencel.

2.3 Candy floss (cotton candy)

Candy floss is made by dropping molten granulated sugar onto a spinning disc or rotating drum where the disc or base of the drum contains small holes so that continuous fibres or threads of molten sugar are extruded from the holes. The rotating disc is known as a spinerette. The streams of molten sugar solidify to a sugar glass in the colder air. This general method of making fibres was applied to glass fibres in the 1930s and glass-fibre-reinforced polymers such as polyester resins, known as fibreglass, have been used in lightweight structures, such as bathtubs and the hulls of small boats. Continuous fibres of Kevlar are also made by using a spinerette.

2.4 Sugars and carbohydrates

There is much debate in the public domain, often acrimonious, on the causes of obesity. The debate frequently highlights sugars as the culprit and during the discussion reference is made to carbohydrates. Articles on food and diets also refer to sugars and carbohydrates as well as what foods should or should not be avoided but very rarely are definitions given to the materials of interest, such as glucose and sugars generally. Simple sugars such as glucose and fructose are also known as monosaccharides and cannot be chemically broken down into simpler molecules (Table 2.1). Glucose forms the polymer backbone of cellulose but humans are unable to digest cellulose as they do not possess the enzymes for saccharification, the process by which cellulose is hydrolysed to smaller molecular

Table 2.1 *Common sugars and their classification*

Sugar	Classification
Deoxyribose	Monosaccharide, a component of DNA
Fructose	Monosaccharide
Galactose	Monosaccharide, stereoisomeric with glucose
Glucose	Monosaccharide
Lactose	Disaccharide between glucose and galactose
Maltose	Disaccharide between two glucose molecules
Mannose	Monosaccharide, stereoisomeric with glucose
Ribose	Monosaccharide, a component of RNA
Sucrose	Disaccharide between glucose and fructose

species, such as monosaccharides. Sucrose, which is used as granulated sugar in everyday life, consists of a molecule of glucose chemically joined to a molecule of fructose. The latter can be commercially prepared by enzymatic action on glucose from corn syrup. Other simple sugars include ribose and deoxyribose that form the backbone of ribonucleic acid (RNA) and deoxyribonucleic acid (DNA). Monosaccharides are carbohydrates as are polysaccharides that have a larger molecular weight than simple sugars and are polymers consisting of long chains of monosaccharides. Polysaccharides have molecular weights up to several million and include starches and cellulose. In contrast with these very high molecular weights, oligosaccharides are produced as intermediates during the digestion of polysaccharides and are carbohydrates containing up to twenty monosaccharide units; hence they are polymers. Raffinose is an example of an oligosaccharide. However, carbohydrates are not sugars even if they are built up from glucose molecules, as is the case for cellulose. So anyone eating a bread roll that contains carbohydrates is not eating sugar as such.

Starches are a mixture of the linear polysaccharide amylose, which is soluble in hot water, and the insoluble branched polysaccharide amylopectin. Amylase enzymes can cleave the polysaccharide chains in starch, producing glucose and the sugar maltose that is important for the manufacture of beer and malt whisky. The absence of tightly bound structures in cooked foods containing starch allows starch to be digested in contrast with the inability of humans to digest cellulose. Thus, the absence or presence of hydrogen bonding between neighbouring polymer chains can have an important role on whether polysaccharides are useful products for foods. Chitin is a naturally occurring polysaccharide found in the **exoskeletons** of shellfish such as crabs and insects where it strengthens these structures and has a chemical structure similar to cellulose. Chitosan is a derivative of chitin, is soluble in acidic media and has applications in pharmaceuticals and cosmetics.

Exoskeletons *are their shells.*

Bioethanol, a biofuel, is produced in commercial quantities by the enzymatic fermentation of sugars derived from crops and blended with conventional fuels. Animal fats are mixtures of lipids, mainly triglycerides that are esters of glycerol (also called glycerine). Biodiesel, also a biofuel, refers to the mono-alkyl esters of long-chain fatty acids derived from vegetable oils or animal fat, for example the methyl ester of rapeseed oil. It can be derived from waste household cooking oils and added to petroleum-based diesel to produce a biodiesel blend.

Many foods contain artificial sweeteners, for sugar-free products including carbonated soft (fizzy) diet drinks, chewing gum, desserts, yoghurt and cough mixtures. Saccharin, the first artificial sweetener, was identified in 1879. Sodium cyclamate was discovered in the 1930s. Both cyclamate and saccharin were discovered accidentally because a researcher in a laboratory either licked his finger and noticed a sweet taste or lit a cigarette that was contaminated with cyclamate, giving the cigarette a sweet taste. Aspartame that is marketed as Nutrasweet is a combination of two naturally occurring amino acids, aspartic acid and phenyl-alanine, and was discovered in 1965. Another synthetic artificial sweetener is sucralose, which is a chlorinated sucrose and sold under the name of Splenda, which is not heat sensitive and can be used in cooking, baking and frying. There are trends for using sweeteners that are natural products such as those derived from the stevia plant and sold under the name Truvia.

2.5   Storage of sugar in the body

When sugar is consumed, the sugars are broken down into monosaccharides such as glucose, fructose and galactose (stereoisomeric with glucose) by a combination of stomach acids and **enzymes**.

*An **enzyme** is a biological catalyst, usually a protein, but some enzymes are made of RNA.*

For example, the enzyme lactase breaks down lactose in milk into glucose and galactose, while the enzyme sucrase breaks down table sugar. Monosaccharides are converted into glucose in the body. However, free glucose is not stored directly in cells. This is because the osmotic pressure of a solution is proportional to the number of molecules and the osmotic pressure of a solution of the monosaccharide glucose in the body would be dangerously high and result in cell damage and possibly cell rupture as water would migrate from regions of low-sugar to high-sugar concentrations, that is from regions of low **osmotic pressure** to regions of high osmotic pressure. Instead glucose is stored as the polysaccharide glycogen, a highly branched starch with a chemical structure similar to amylopectin. Glycogen is stored in the liver and also in muscle cells while some is metabolised immediately to produce a source of energy. The high molecular weight of glycogen compared to glucose produces a low osmotic pressure.

Note that when two solutions of different concentrations are separated by a membrane that allows molecules of solvent but not solute to pass through, that is a semipermeable membrane. Then the solvent, for example water, will pass from the dilute to the more concentrated solution. This process is known as osmosis and is thermodynamically driven. Osmosis stops when the concentrations of solute (e.g. sugar) are equal or by applying a hydrostatic pressure, the **osmotic pressure***.*

The hormone insulin, which is a protein, is produced in the pancreas and promotes the uptake of glucose by cells in the liver and muscles and controls its concentration in the blood. A shortage of insulin will increase blood sugar levels and can lead to Type 2 diabetes.

Many foods, for example ice cream and mayonnaise, fall into the category of soft matter (also known as soft material) or more formally soft condensed matter (or soft condensed material). Soft matter is a general name for non-crystalline condensed matter and is a state that is neither a simple liquid or crystalline solid. Examples of soft matter include colloidal dispersions, liquid crystals and polymers that are not in a crystalline state, such as polymer melts. Soaps, paints and glues are also examples of soft matter. Common features of soft matter are length scales between atomic sizes and macroscopic scales (e.g. colloidal dispersions), particles that exhibit **Brownian motion** and the ability to self-assemble as in block copolymers. Soft matter represents **disordered systems**. Examples of soft matter are described throughout this book (e.g. colloidal dispersions, liquid crystals) but these examples have not specifically been referred to as soft matter as such.

Brownian motion *is the continuous random movement of microscopic solid particles around 1 μm in diameter when suspended in a fluid medium.*

Disorder systems *are systems that are non-crystalline or amorphous.*

2.6 Summary

Carbohydrates and polysaccharides are polymers and include cellulose, the most abundant biorenewable material and a component of lignocellulose, or biomass, which is the major structural component of plants. Glucose monomer has an important role in the formation of these polymers, particularly cellulose. Carbohydrates also include starches. There is much debate on the causes of obesity and sugars, which are also carbohydrates or monosaccharides, are considered to have a role in this condition; sugars are also a source of biofuels such as bioethanol. There is much interest in the use of artificial sweeteners, either natural products or those derived from chemical synthesis. When consumed, sugars are broken down into the monosaccharide glucose that is stored in the body, in particular in the liver and muscle cells. The protein hormone insulin controls the blood sugar concentration and a lack of insulin can result in Type 2 diabetes. Many foods, for example ice cream and mayonnaise, fall into the category of soft matter.

2.7 Further reading

Akhaven, J. *The Chemistry of Explosives*, 3rd edn. Cambridge: Royal Society of Chemistry, 2011.

Atkins, P. *Atkins' Molecules*, 2nd edn. Cambridge: Cambridge University Press, 2014.

Brazil, R. The sweet and the low. *Chemistry World* (2015) 11, 50–63.

Coultate, T. Food: The Chemistry of its Components, 6th edn. Cambridge: Royal Society of Chemistry, 2016.

Emsley, J. *Molecules at an Exhibition*. Oxford: Oxford University Press, 1999.

Fried, J. R. *Polymer Science and Technology*, 3rd edn. Upper Sadddle River, NJ: Prentice Hall, 2014.

Goldstein, D. (ed.). *The Oxford Companion to Sugar and Sweets*. Oxford: Oxford University Press, 2015.

Jones, R. A. L. *Soft Condensed Matter*. Oxford: Oxford University Press, 2014.

Morrison, W. J., and J. C. Wharton. Candy Machine. United States Patent 618428, 1899.

Papachristodoulou, D., A. Snape, W. H. Elliott and D. C. Elliott. Biochemistry and Molecular Biochemistry, 5th edn. Oxford: Oxford University Press, 2014.

Rennie, R. (ed.). *Oxford Dictionary of Chemistry*, 7th edn. Oxford: Oxford University Press, 2016.

Sarin, A. Biodiesel: *Production and Properties*. Cambridge: Royal Society of Chemistry, 2012.

3

Chips with Everything

3.1 Introduction

The development of the Internet in recent years has brought about a revolution in business practices and the way people communicate. For example, social media has allowed people across the world to communicate with each other and exchange information in the form of text or photographs or video. Online banking, online shopping, music downloads, video-on-demand, streaming services, online dating agencies and estate agents are services that were not available just a few years ago. There has been a large increase in the range of consumer goods in recent years, thus smartphones, desktop, laptop and tablet computers, digital cameras, in-vehicle navigation systems and games consoles, as examples. Schoolchildren use the Internet to help with their schoolwork and governments transfer information on their services to the Internet. Unfortunately, novel ways of carrying out fraudulent activities have been invented and used. However, none of these goods, services or activities would be possible without the use of silicon chips that are at the heart of computer systems. In this chapter aspects of the fabrication of silicon chips and how they came to be developed are described with particular reference to the role of materials in their manufacture.

3.2 Cat's whisker, vacuum tubes and relays

In the first two decades of the 20th century amateur radio enthusiasts used a fine wire known as a 'cat's whisker' in contact with a specific point on the surface of a crystalline material, a semiconductor. The cat's whisker acted as a rectifier that changed an alternating current radio signal into a direct current, allowing it to be converted into sound. Note that Henry Round who discovered electroluminescence in 1907 had applied a small voltage to a piece of silicon carbide, a semiconductor. Vacuum tubes (also known as thermionic valves) could do the same function as the cat's whisker and vacuum tube-based radio sets became common throughout the 1920s and later years. Vacuum tubes resemble small incandescent light bulbs. Electrons flow from a glowing wire that passes through a vacuum to the other electrode. Vacuum tubes could amplify and switch on and off the signals they

Materials for the 21st Century. David Segal.
© David Segal 2017. Published 2017 by Oxford University Press.
DOI: 10.1093/oso/9780198804079.001.0001

received. It was shown that a zero could be represented by an off-state and a one by an on-state. This ability to have a switch that operates in two states is crucial to how digital computers work. There is a historical link in the development of digital computers between vacuum valves and switching mechanisms used in telephone exchanges. Throughout the 1930s and 1940s exchanges used electromechanical devices known as relays to connect telephone calls but they were slower switches than vacuum tubes. Thus, electromechanical relays operated as switches that could have an off position and an on position.

Some further details of the historical role vacuum tubes had in the development of the electronic age are given here. The production, distribution and use of electricity throughout the United States took place in the last two decades of the 19th century and the creation of electric power is associated with Thomas Edison of the General Electric Company and George Westinghouse of Westinghouse Electric. Nikola Tesla had an important role in the development of motors and generators for the distribution of alternating current supplies over large distances. The earlier industrial age was transformed into an electrically driven era. The novel technology of wireless communication developed at the turn of the 20th century relied on electric power and allowed information to be sent through the air. Thus, wireless telegraph messages from shore to ship and back, two-way voice transmissions and one-way transmission by radio broadcasting. It was the vacuum tube that characterised the new era of electronics. As previously stated they act as on–off switches, producing electrical signals that can be amplified.

Lee de Forest, an employee of the Federal Telegraph Corporation in Palo Alto, California, discovered by accident that if the outgoing current from a vacuum tube was fed back into the input then the signal was significantly amplified. It has been stated that 'if one dropped a handkerchief a few inches from a telephone transmitter, there was a loud thud in the earphones'. This discovery that was made around 1906 had a big effect on the development of communications, because the ability to amplify enabled devices to be made and used for long-distance telephone and wireless communications. The work of de Forest helped to usher in the age of electronics and technologies dependent on vacuum tubes, namely telephone transmission, telegraphy and radio.

Production of vacuum tubes in the United States grew from the 1920s to 1940s, from 1 million to nearly 100 million per year. Ownership of telephones spread throughout the nation and radio captivated the public. By 1940 nearly 40 per cent of American homes had a telephone and almost three-quarters of households had a radio compared to the presence of a radio in 5000 homes in 1920.

3.3   Computability and the foundations of digital computers

The mathematical description of computability was described in the 1930s by Alan Turing in the United Kingdom and Alonso Church in the USA and it underpins the development of digital computers in the 1940s and later years. The subject of computability is beyond the scope of this book but some aspects of digital computers are summarised. A modern computer carries out logical operations on an input of information in the form of a binary sequence, that is a string of ones and zeros. The information can be text, numbers, pictures, music, speech, video or any quantity as long as it can be represented by a binary sequence and the information can be transmitted by computers, mobile phone and over the Internet, as examples. Logic circuits carry out the operation and are based on Boolean algebra. The key feature of the computer that allows the operations to be carried out is that it must be able to distinguish between a one and a zero in the binary sequences and this can be achieved by using components that act as switches. Early computers in the 1940s and 1950s were bulky and contained many valves that could be switched electronically from a zero (off) to a one (on) by sending a signal to the valve. These early mainframe computers were housed in their own rooms and inputs of data were made by the use of punched cards.

Digital computers consist of three parts: (i) logic, (ii) main memory and (iii) storage. In early computers, the main memory consisted of thousands of doughnut-shaped magnetic ferrite beads that could be magnetised or not magnetised so that these two states correspond to binary one or binary zero. Logic refers to the digital circuitry (central processing unit, or CPU) that carries out logical operations on binary inputs. Information from the CPU was stored in the main memory as binary ones or binary zeros depending on the magnetised state of the cores. Storage holds data for long periods and transfers it to the main memory when required.

However, the invention of transistors as electronic switches in the late 1940s eventually resulted in transistors replacing vacuum tubes in digital computers and gave rise to the development of the silicon chip, or microchip.

3.4 Transistors

Transistors are semiconductors that can carry out all of the operations that valves and electromechanical relays do, that is, amplify and switch on and off signals sent to them. They are much smaller than valves and operate at ambient temperatures without the necessity to use hot filaments. They are solid-state semiconductors whose electrical conductivity can be modified by the incorporation of elemental dopants from the gas phase into specific regions so that each region or junction exhibits their own electrical behaviour. The development of transistors led to the miniaturisation of electronic logic circuits. The first type of transistor that was invented in 1947 and known as a point-contact transistor consisted of applying a voltage across two wires in contact with a doped semiconductor made out of germanium. Dopants can change the electrical properties of the crystal from conductor to insulator. Point-contact transistors are discrete devices with wire contacts but have been superseded by transistors based on doped semiconducting silicon in silicon chips.

A more detailed description of transistors is given here as they have such a crucial role as electronic switches for carrying out logic operations in digital circuits. (The references by Hummel and Nixon in this chapter's Further Reading section are recommended to the interested reader.) Transistors have three terminals. In a bipolar transistor two p-n semiconductor diodes are effectively joined together back-to-back so that, for example, a region of p-type semiconductor is connected to n-type regions to give an npn transistor. In practice dopants are introduced into a substrate such as silicon by techniques including metal–organic chemical vapour deposition (MOCVD) to produce the p-type and n-type regions. The base terminal is connected to the p-type region and the emitter and collector terminals are connected separately to n-type regions. If the diode configuration consisting of the base and emitter is forward biased and the diode configuration of base and collector is reversed biased, then electrons entering the emitter region can diffuse through the base area and are then accelerated into the collector region. The acceleration causes amplification of the input alternating current signal. However, the electron flow, hence current from emitter to collector, can be switched on or off by changes in the base-emitter voltage. This role of the transistor as a switch is used for logic and memory functions in computers. The name bipolar transistor arises because the current passes through both n-type and p-type semiconductor materials so that conduction involves both holes and electrons as charge carriers.

A disadvantage of bipolar transistors is their relatively high power consumption and nowadays the dominant transistor used is the field effect transistor (FET), which has one type of charge carrier, electrons or holes controlled by the doping process. The method used for construction is known as metal-oxide-semiconductor (MOS) and the transistors are known as MOSFETs. Compared to the bipolar transistor, the terminals on a MOSFET are the drain (replaces the collector), the source (replaces the emitter) and the gate (replaces the base). The gate is insulated from the p-type substrate and the gate–substrate configuration acts as a capacitor. The aim of the device is to control the current flow from source to drain and this is achieved by varying the voltage applied to the gate. No current flows when the voltage is absent. However, when a voltage is applied between the gate and source, electrons flow into a channel between the drain and source, the channel conductivity increases, current flows between source and drain and the device is switched on. MOSFETs are unipolar transistors because the electron path from source to drain through the channel involves just one type of semiconductor.

There are two types of MOSFET. The first is a depletion-type, or NMOSFET, which has highly doped source and drain regions and a lightly doped channel of the same polarity (e.g. n-type) deposited on a p-type substrate. The high doping lowers the resistance for electrical connections. The second type, enhancement-type MOSFETs, or positive-type MOSFETS (PMOSFETS), depends on holes as the majority charge carrier. PMOSFETS do not have a built-in channel and when a voltage is applied to the gate, holes below the gate are repelled into the substrate, allowing electrons to flow into a so-called inversion channel and flow between the source and drain. A key feature of PMOSFETS is their high input impedance that reduces Joule heating in use and they are widely used in the integrated circuit industry. Integration of NMOSFETs and PMOSFETs on one chip is the main approach used in manufacturing integrated circuits and the technology is known as CMOSFET, or CMOS, namely complementary MOSFET. CMOS technology produces devices with low operating voltages (0.1 volts), low-power consumption (less heat) and short channels that give a higher speed for processing. Details of other transistors are given in the Further Reading section.

3.5 Silicon chips

An outline description of the stages used to manufacture silicon chips may help to understand how they operate:

(a) A single crystal of semiconducting silicon is grown from a melt using a seed crystal in what is known as the Czochralski method. Circular slices or wafers are cut from the cylinder and polished to obtain a smooth surface on the silicon substrate (Fig. 3.1).

(b) The wafer is heated in an oven to deposit a layer of silica (SiO_2) onto the wafer.

(c) A photocurable resin known as a photoresist is coated onto the wafer and photolithography is used to project a shrunken image of the desired digital circuit onto the wafer using a photomask. In photolithography, ultraviolet light from a laser, usually an argon-fluoride (ArF) excimer gas laser with a wavelength of 193 nm is used. Areas of the photoresist that are exposed to radiation form solid regions.

(d) Solid regions of cured photoresist are removed with a solvent and exposed regions of silica are also removed by etching using gaseous reactants so that areas of silicon are exposed. The remaining resist is then removed from the wafer by use of a solvent.

(e) Dopants can be introduced into the exposed regions of silicon to modify the electrical conductivity. Both n-type and p-type dopants are used and these are deposited by techniques including diffusion from the gas phase, ion implantation, molecular beam epitaxy and MOCVD.

(f) Exposed regions of silicon are made conductive by depositing metallic 'wires' from gas phase reactants. These wires are known as interconnects.

(g) A layer of insulating silica is then deposited onto the wafer and further layers of circuitry are then fabricated. The final layer is of silica.

(h) Holes known as vias are etched onto the substrate that allow electrical connections to be made between different layers.

(i) The wafer is packaged for protection and the silicon chip, after testing for reliability, is ready for use. The plastic or ceramic or package has the size of a postage stamp.

The smallest feature that can be obtained by current photolithographic processes is about 32 nm, which corresponds to the width of the deposited lines on the wafer. Reduction of the minimum feature size allows more transistors to be placed on a chip, hence increasing the computing power. About 2 billion (2×10^9) transistors can be packed onto a silicon wafer with a size of 1 cm^2.

An integrated circuit is a silicon chip that has been doped with elemental impurities and which has a surface pattern or surface topography of interconnected components, thus transistors, resistors, diodes and capacitors that carry out the logic operations in digital computers. In contrast with point-contact transistors, integrated circuits are made out of a solid block of material, here silicon with no connecting wires. It is the development of integrated circuits that has led to the miniaturisation of electronic circuits so essential for the manufacture of portable

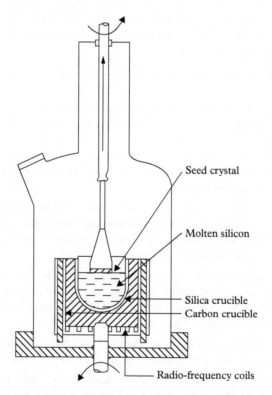

Fig. 3.1 *Schematic diagram for growth of single crystals by the Czochralski method.*

(*R. E. Hummel*. Electronic Properties of Materials, *4th edn, p. 161. Berlin: Springer, 2013 and Wacker Siltronic AG.*)

electronic devices such as mobile phones. The development of the silicon chip is a triumph for interdisciplinary research across areas of chemistry, physics, materials science and engineering.

3.6   Moore's Law

Gordon Moore, the founder and president of Intel Corporation, suggested in 1965 that the number of components such as transistors on a silicon chip in an integrated circuit would increase exponentially over time; in particular, they would double every two years. This suggestion is known as Moore's Law and since the suggestion was made the law has been followed and the number of transistors that have been fabricated on a silicon chip has doubled nearly every two years.

3.7   Directed self-assembly of block copolymers

Reduction of the minimum feature size allows more transistors to be placed on a chip, hence increasing the computing power. Directed self-assembly of block copolymers has the potential for increasing resolution of existing lithographic tools for semiconductor fabrication. Phase separation can occur in a block copolymer, for example in polystyrene-polymethylmethacrylate copolymers that result in spheres or cylinders of a colloidal dimension that can align into ordered arrays or domains. These domains can be formed on a silicon wafer and can be selectively removed. In this way, there is the potential for increasing the line density on the surface of wafers.

3.8   Quantum computing

Digital computers encode data as binary sequences using binary digits of one or zero. Quantum computers are computers that carry out processing using quantum effects, in particular **quantum entanglement**. Quantum computers would store data as quantum bits or qubits. These computers, if constructed in the future, would be much faster than digital computers currently in use and would rely on properties of atoms, ions or photons, as examples, rather than the semiconducting properties of silicon. One area where quantum computers would have a large impact is cryptography. Encryption of information sent over the Internet is widely used but it is difficult to break the encryption codes in a realistic time scale. A quantum computer may have the capability of breaking computer codes used for encryption that currently appear to be unbreakable.

Quantum entanglement refers to a particle or system that does not have a definite state but exists as an intermediate form of two entangled states.

A bit or binary digit is the smallest unit of information and can only take one of the values zero or one. The qubit is the smallest unit of information in quantum computing, has two quantum mechanical states and as such can, because of **superposition**, have values of zero and one. An example of a quantum mechanical state is spin angular momentum or spin that can have a spin-up or spin-down state. Spin is associated with protons in the technique of magnetic resonance imaging and also with electron spin although it should be stressed that electrons do not spin in the way an ice skater spins. Coloured diamonds are mentioned in the Glossary. In these materials, for example pink diamonds, the insertion of nitrogen atoms into the carbon lattice results in a vacancy and a single trapped electron. The latter interacts with the unpaired electrons in three neighbouring carbon atoms and the two unpaired electrons from the nitrogen atom. The resulting six-electron system has an electronic ground state that has split energy levels associated with two different values of electron spin. Coloured diamonds are examples of materials that have a colour centre and this class of materials has potential and exciting applications in the development of practical quantum computers as they have two quantum states (spin), a requirement for quantum computing.

*Note that **superposition** refers to the effect of interacting waves at a point and the effect is obtained by adding the amplitudes of the waves at a point. In quantum mechanics the interacting waves correspond to wave functions.*

3.9   DNA computer

The four bases in DNA, abbreviated to A, T, C and G, encode the information in the human genome. In a DNA computer, at present only a theoretical concept, strands of DNA represent the data for solving problems and these would be codified by sequences of the four bases just as binary sequences are used in digital computers. Just as logic gates based on transistors are used in current computers, biological logic gates based on genetic material may be available. The area of biological computing illustrates how developments in one technical area can have potential applications in a totally unrelated area.

3.10   Summary

The development of digital computers has developed from the use of electromechanical relays and vacuum valves that act as switches to the use of silicon chips. The latter are fabricated by use of photolithography combined with photoresists. Dopants are introduced into the silicon substrates from the gas phase in order to modify the semiconducting properties. The suggestion known as Moore's Law that the number of transistors on a silicon chip would increase exponentially was made in 1965 and it has been followed since that time. Directed self-assembly of block copolymers has the potential for increasing the line density on the surface of wafers. Quantum computers and biological computers based on DNA are under consideration as future computing systems.

3.11 Further reading

Askeland, D. K., and W. J. Wright. *The Science and Engineering of Materials*, 7th edn. Boston, MA: Cengage Learning, 2011.

Church, A. An unsolvable problem of elementary number theory. *American Journal of Mathematics* (1936) 58, 345–63.

Cockshott, P., L. M. Mackenzie and G. Michaelson. *Computation and its Limits*. Oxford: Oxford University Press, 2015.

Copeland, B. J. *Turing: Pioneer of the Information Age*. Oxford: Oxford University Press, 2014.

Fox, M. *Optical Properties of Solids*, 2nd edn. Oxford: Oxford University Press, 2010.

Hummel, R. E. *Electronic Properties of Materials*, 4th edn. New York: Springer, 2012.

Ince, D. *The Computer: A Very Short Introduction*. Oxford: Oxford University Press, 2011.

Isaacson, W. *The Innovators*. London: Simon and Schuster, 2015.

Lee, J.-H., J-H. Cho, J.-S. Cho and D.-J. Lee. Spin-on glass composition and method of forming silicon oxide layer in semiconductor manufacturing process using the same. United States Patent 7,270,886, 2007.

Nixon, M. *Digital Electronics: A Primer*. London: Imperial College Press, 2015.

Russell, T. P., S. Park, D. H. Lee and T. Xu. Self-assembly of block copolymers on topographically patterned polymeric substrates. United States Patent 8,247,033, 2012.

Segal, D. *Exploring Materials through Patent Information*. Cambridge: Royal Society of Chemistry, 2014.

Thackray, A., D. C. Brock and R. Jones. *Moore's Law*. New York: Basic Books, 2015.

Turing, A. M. On computable numbers with an application to the Entscheidungs problem. *Proceedings of the London Mathematical Society* (1936–7) 42, 230–65.

Whitaker, A. *The New Quantum Age*. Oxford: Oxford University Press, 2015.

Further reading

4

Polymers

4.1 Introduction

The successful preparation of polymers and plastics known as cellulosics and derived from cellulose in, for example, wood pulp was achieved around the middle of the 19th century. At around the same time success was achieved on the vulcanisation of natural rubber that greatly aided the development of the transport industry with the use of vulcanised rubber in tyres. Throughout the 20th century a series of synthetic polymers were prepared from petrochemical sources and plastics and polymers are widely used in everyday life, for example for packaging, furniture and textile fabrics. The growth of molecular biology since the second half of the 20th century and the sequencing of deoxyribonucleic acid (DNA) as well as structure determination for proteins highlighted that genetic material such as DNA as well as proteins can be viewed to be polymers. There is much interest in the 21st century on developing products such as plastics from renewable sources as well as biodegradable plastics. Details of the preparation of polymers are given in the Glossary and they are not repeated here. Instead some salient features for selected polymers are described in this chapter.

4.2 Polymers

Polymers are macromolecules in which monomers are chemically joined together by polymerisation to form larger polymer molecules.

> *For example, a molecule designated as A, a monomer can form a **polymer** designated as ...A-A-A.... Another way to view a polymer is to make an analogy with a necklace. The latter consists of small identical rings that are joined together. If each ring is described as a monomer than the necklace can be described as a polymer. If there are two types of ring than the necklace can be described as a copolymer.*

They include biopolymers that are naturally occurring as well as synthetic polymers derived from petrochemical sources. **Synthetic** polymers are often referred to as plastics.

Materials for the 21st Century. David Segal.
© David Segal 2017. Published 2017 by Oxford University Press.
DOI: 10.1093/oso/9780198804079.001.0001

Synthetic means not occurring naturally.

Bioplastics are derived from renewable sources and some are biodegradable. In homopolymers only one type of monomer is used, for example ethylene monomer in polyethylene. Heteropolymers or copolymers are made from two or more different monomers. Nylon, styrene-acrylonitrile-butadiene and Kevlar are examples of heteropolymers. DNA is a heteropolymer as there are sequences of four bases, adenine, cytosine, guanine and thymine, along the backbone so that these bases can be considered to be the monomers. **Block copolymers** consist of two polymer chains or blocks that are covalently bonded to each other and which are chemically different. For example, polystyrene-polymethylmethacrylate, which is derived from styrene and methylmethacrylate monomers, contains polystyrene and polymethylmethacrylate.

*If two polymers are designated as C and D, then a **block copolymer** is designated as . . . C-D-C-D-C . . . . .*

The stereochemistry of polymers, that is, the relative spatial positions of side groups on the polymer chains affects the material properties. For example, natural rubber for use in tyres and gutta percha are both isomers of polyisoprene; that is, they have the same chemical composition. However, rubber is in the *cis* form while gutta percha is in the *trans* form and while rubber behaves as an elastomer, gutta percha is a hard material. Thermoplastics are polymers that can be re-shaped by heating while thermosetting plastics are shaped initially by a heat treatment that causes a polymerisation process in a resin but the resulting cross-linked chains in the solid plastic are resistant to softening and deformation. In 2011 about 1 million tonnes of bioplastics were produced compared to a total of 280 million tonnes for all plastics, corresponding to about 5 per cent of global oil production. Annual production of polyethylene was about 5×10^7 tonnes, polystyrene about 2×10^7 tonnes and polyester about 10^7 tonnes. These figures compare with about 3×10^7 tonnes for natural fibres (e.g. wool).

4.3   Vulcanisation of rubber

Charles Goodyear showed in 1844 that natural rubber latex from trees could be converted into solid products known as elastomers, that is, materials which can be deformed but retain their original shape when deformation forces are removed. The process by which solid material was produced is known as vulcanisation and involved heating the latex suspension of polymeric isoprene molecules with about 3 weight per cent sulphur. The addition of sulphur **crosslinked** polyisoprene molecules and it has been reported that Goodyear discovered the process accidentally when a mixture of latex and sulphur was accidentally spilled onto a hot stove. The vulcanisation process opened up the possibility of using solid rubber for making tyres for automobiles at the end of the 19th century. Use of higher sulphur contents produced a hard non-deformable solid known as vulcanite that was used to produce ornaments.

Crosslinked *means joining together different polymer chains.*

Another latex liquid with the same chemical structure as polyisoprene but with a different isomeric form could be cured to a hard non-deformable solid known as gutta percha. The latter was exploited commercially for use as a durable insulator for long-distance underwater telegraph cables and promoted communication technologies in the latter half of the 19th century. Neoprene, a synthetic rubber first prepared in 1931, is a polychloroprene. Nowadays the main synthetic rubber used in automotive tyres is based on styrene-butadiene copolymers and blends of these elastomers with natural rubber.

4.4   Polymers in the 20th century

Leo Baekeland invented Bakelite at the beginning of the 20th century and was one of the first commercially successful synthetic plastics. Note that celluloid was a commercially successful plastic in the 19th century. It was obtained through the condensation reaction between phenol and formaldehyde. Similar products were obtained as condensation products from the reaction between urea and formaldehyde and melamine and formaldehyde. Whereas phenol-formaldehyde resins were only available in dark colours, the other condensation products could be obtained in a range of colours, a desirable feature for consumer products such as laminates for table tops.

Synthetic polymers tend to be flammable and when used in consumer products flame retardants are blended into the polymer before forming the product such as electrical plugs (Table 4.1). They have relatively low melting points and molten plastics can cause serious burns. Many flame retardants are based on halogen-containing compounds but polyvinyl chloride is one plastic that has a 'built-in' flame resistance.

Plastics are taken for granted in everyday life, for example polythene bags and food packaging. Many plastics do not degrade in the environment and to reduce the number of bags that are thrown away annually and litter the land and the oceans, governments such as that in the United Kingdom have introduced charges for polythene bags in supermarkets and shops. Polythene, formally polyethylene, was discovered in the early 1930s by Reginald Gibson and Eric Fawcett. They reacted the monomer ethylene with benzaldehyde at 170°C and 1900 atm. The intention was to produce ethyl phenyl ketone but in fact the ethylene unexpectedly underwent polymerisation to form polyethylene. Polythene obtained under high pressure is known as low-density polyethylene (LDPE). A more economic route was developed in the early 1950s based on the use of so-called Ziegler–Natta catalysts and the key feature of this route is that it is a low-temperature,

Table 4.1 *Examples of polymeric materials*

Class of polymer	Example
Cellulosics	Acetate rayon
	Cellulose nitrate
	Celluloid
	Gun cotton
	Parkesine
	Tencel
	Viscose rayon
Synthetic polymers	Acrylic
	Kevlar
	Nylon
	Phenol-formaldehyde
	Polycarbonate
	Polyester
	Polyethylene
	Polymethylmethacrylate
	Polypropylene
	Polytetrafluoroethylene
	Polycarbosilanes
	Polyphosphazenes
	Polysilanes
	Polysilazanes
	Polystyrene
	Polyurethane
	Urea-formaldehyde
Biodegradable polymers	Polylactic acid
Natural polymers	DNA
	Proteins
	Ribonucleic acid (RNA)

low-pressure route that yielded high-density polyethylene. Ziegler–Natta catalysts can be used for the synthesis of other polymers, for example polypropylene. An important application for polythene in the Second World War was as electrical insulation for cables used in compact radar systems installed in aircraft of the Royal Air Force. While thin low-density polythene film is used to wrap sandwiches and other foods and is sometimes referred to as Clingfilm, the original Clingfilm was based on poly (vinylidene chloride) that was invented by Ralph Wiley in 1933 and has also been called Saran Wrap.

A well-known polymer is nylon, invented in the 1930s by Wallace Carothers, and nylon is known as an aliphatic polyamide. The polymer is a condensation product between two different monomers that react at near ambient temperatures in an interfacial reaction. The polymer could be spun into fibres but unlike Kevlar, nylon does not form liquid crystalline phases. An early application that was a spectacular success was as a substitute for silk in women's stockings at the beginning of the Second World War but it has many uses including textile fabrics, toothbrushes and parachute cords.

Polytetrafluoroethylene (PTFE) is a polymer derived from the monomer tetra-fluoroethylene and is associated with its use as a non-stick coating for cooking utensils such as frying pans. These coatings have been marketed under the name Teflon. It is sometimes assumed among the general public that PTFE was developed during the 'Space Race' of the 1960s but in fact it was first prepared, unexpectedly, by Roy Plunkett in the late 1930s (Table 4.2). A gas cylinder containing the monomer under pressure was left overnight and the next morning it was noticed that the pressure in the cylinder had fallen to zero. When the cylinder was cut open a white waxy solid, PTFE, was observed. The applications for PTFE are very dependent on its surface properties. Hence, it has a low coefficient of friction, a useful property for bearings and chemical inertness due to the C–F chemical bond. In fact, its first application was in the Manhattan Project for the separation of isotopes of uranium hexafluoride in the development of the atomic bomb as it was chemically resistant to hydrogen fluoride and vapours of uranium hexafluoride.

Table 4.2 *Timeline for the development of synthetic polymers*

Year	Synthetic polymer
1846	Cellulose nitrate
1856	Parkesine
1907	Bakelite (phenol-formaldehyde)
1918	Cellophane
1931	Neoprene (synthetic rubber)
1933	Polyvinyl chloride
1933	Polyvinylidene chloride
1933	Low-density polyethylene
1935	Nylon
1937	Polyurethane
1938	Polytetrafluoroethylene (PTFE)
1938	Polystyrene
1939	Polymethylmethacrylate
1953	High-density polyethylene
1953	Polycarbonate
1956	Living polymers
1964	Kevlar
1977	Polyether ether ketone (PEEK)

Acrylic textiles are based on polyacrylonitrile fibres that are spun from solutions of acrylonitrile and acrylic fibres can be blended with natural or synthetic fibres such as cotton. Polyacrylonitrile fibres can undergo a controlled pyrolysis to yield carbon fibres for the production of composites.

Kevlar is an aramid fibre also referred to as an aromatic polyamide which was invented by Stephanie Kwolek in 1964 and as with many other synthetic polymers it is produced as a condensation product from two or more monomers. In everyday language Kevlar can be described as a plastic. The polymer can be spun using a spinerette into continuous fibre but Kevlar has a property that is not observed in all polymers; namely, the polymer molecules aggregate in the liquid phase to form liquid crystals during the spinning process. This means that the polymer molecules align themselves in a parallel direction and hydrogen bonding between polymer chains forms a compact and strong structure. It is reasonable to enquire why such association structures form. Qualitatively it can be said that when a liquid crystal forms, the entropy of the system decreases and this decrease will tend to make the Gibbs free energy positive. In order to maintain a thermodynamically stable system this free energy needs to be negative. This can be achieved if close packing of polymer chains and hydrogen bonding between adjacent chains are favourable configurations. The strength of Kevlar is essential for its best-known application in bullet-proof body armour. There is considerable concern among governments in the 21st century about international terrorism and the use by suicide bombers of homemade, unstable but powerful explosives such as triacetone triperoxide (TATP). Bullet-proof vests consist of layers of woven Kevlar and larger protective structures can be made by using alternate layers of steel and Kevlar. In use the polymer fibres absorb energy from a projectile, and woven sheets deform in order to dissipate the energy from a bullet. Another aramid fibre known as Nomex is used in fire-resistant clothing and gloves and has a chemical structure similar but not identical to Kevlar.

While many polymers are commodity products such as polythene, polystyrene, polyester, polymethylmethacrylate and polycarbonate, others are speciality materials with niche markets. For example, polysulphones for use as membranes in reverse osmosis for purification of sea water and brackish water, polyvinylidene fluoride (PVdF) for use as porous separators in lithium-ion batteries to prevent electrical shorting of the electrodes and the semi-conductor polymer poly-(p-phenylenevinylene) for potential use as organic light-emitting diodes (OLEDs). The latter are particularly attractive for applications of flexible electronics as coatings of them do not crack and lose adhesion with the polymer substrate.

Injection-molded styrene-acrylonitrile-butadiene (ABS) thermoplastic resins have many applications, including the body shells of office equipment such as computers, home appliances and toys. Acrylonitrile imparts strength, butadiene imparts impact resistance and styrene aids the processing of the polymer.

The latter part of the 20th century and the first part of the 21st century have seen the growth of studies in molecular biology and genomic research. There is much interest in personalised medicine in which an individual's genome would be sequenced; that is, the order of the four bases in DNA would be identified. The 100,000 Genomes Project in the United Kingdom aims to sequence the genome for 100,000 people and combine the data with information from health records. In this way, it is hoped that individuals will receive targeted medical treatment based on specific drugs. Proteins that are expressed by DNA are made up of combinations of up to twenty amino acids and hence the amino acids can be considered to be monomers as can the four bases in DNA. Thus, DNA and proteins are considered to be naturally occurring polymers. The shape and folding of proteins are important for understanding diseases. For example, Alzheimer's disease is associated with protein fibres or fibrils that are mis-folded and form what are referred to as amyloid plaques in the brain with devastating effects on health. The sequencing of bases in DNA, the order of amino acids in proteins and the factors affecting the shape of proteins can be considered to fall within the area of polymer chemistry (Table 4.3).

Some more details of protein structure are given here as there are analogies with other polymers. Amino acids have the general formula $H_2N–CHR–COOH$ where R represents the side chain and they are linked together by a condensation reaction with the elimination of water by a peptide bond, thus $–CO–NH–$, although the actual mechanism of how linkages are formed *in vivo* is complicated. Note that the preparation of nylon and Kevlar involves condensation reactions. The primary structure of proteins is represented by a linear sequence of the order in which amino acids are linked together to form a polypeptide. The secondary structure of proteins consists of an a-helix in which the polypeptide backbone is arranged in a spiral (Fig. 4.1) or β-pleated sheet in which extended polypeptide backbones are side by side (Fig. 4.2), or it consists of a random coil or loop of polypeptide.

Table 4.3 *Genome size for different organisms*

Type of organism	Species	Genome size (10^6 base pairs)
Fungi	*Saccharomyces cerevisiae* (Baker's yeast)	12.1
Nematode worm	*Caenorhabditis elegans*	100
Fruit fly	*Drosophila melanogaster*	180
Mosquito	*Anopheles gambiae*	278
Rice	*Oryza sativa*	400
Chicken	*Gallus gallus*	1200
Mouse	*Mus musculus*	3454
Dog	*Canis familiaris*	2500
Human	*Homo sapiens*	3165

Source: A. M. Lesk. *Introduction to Protein Science: Architecture, Function and Genomics*, 3rd edn, p. 15. Oxford: Oxford University Press, 2016.

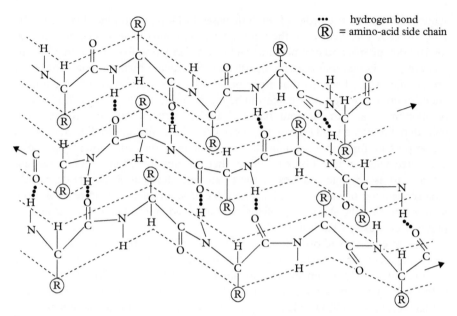

Fig. 4.1 *Structure of beta sheets.*
(R. S. Hine (ed.). Oxford Dictionary of Biology, 7th edn, p. 62. Oxford: Oxford University Press, 2015.)

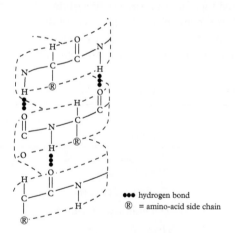

Fig. 4.2 *Structure of alpha helix.*
(R. S. Hine (ed.). Oxford Dictionary of Biology, 7th edn, p. 21. Oxford: Oxford University Press, 2015.)

A polypeptide chain can be arranged as a mixture of the different secondary structures and these separate structures can be folded and packed together to form the tertiary structure of the protein molecule. In some proteins, for example haemoglobin, different tertiary structures can be held together in a larger unit and when this happens the protein is said to have a quaternary structure.

The disruptive technology known as gene editing, or CRISPR (clustered regularly interspaced short palindromic repeats), in which a cutting enzyme can selectively remove a DNA fragment and insert a different DNA sequence into the genome, may have a significant effect on the direction taken by molecular biology in the 21st century as change to the genome can be passed from generation to generation.

4.5 Polymers in the 21st century

There has been growing interest in using renewable sources such as crops for the production of materials such as polymers in order to reduce the dependence on petrochemical sources of materials. In the case of biodegradable materials, pressures on the disposal of non-degradable plastics on land and in the ocean are reduced. Carbon dioxide, water and biomass are the products of decomposition of biodegradable polymers. Fermentation processes are known in everyday life, for example the use of yeast, a fungus, to enzymatically-catalyse the conversion of sugar to alcohol in the production of beer. This process supplies energy to the organism for growth. Genetically modified yeasts and bacteria are used in biotechnology in fermentation processes to express recombinant compounds. Lactic acid can be produced by fermentation of corn starch by bacteria or obtained from conventional resources. Lactic acid can be converted to polylactic acid (PLA), a biodegradable polymer in a conventional chemical process although polylactic acid can also be obtained directly from fermentation of food wastes. Polylactic acid can be spun into fibres for use in clothing. It is a biocompatible polymer and applications include sutures for stitching together the edges of a wound in surgical procedures and nanoparticle-based medicines for targeting diseased cells *in vivo*. Another application is in biodegradable stents that are inserted into an artery to prevent its narrowing and which slowly dissolve over a period of months. A further example of how renewable sources can be used to produce polymers relates to Sorona. Here, glucose extracted from corn is digested by genetically engineered organisms and metabolised to 1,3 propanediol. The latter is reacted with terephthalic acid to produce a polymer, Sorona, that can be spun into fibres for clothing and carpets. This example shows how biomass can be converted to higher-value products. Fermentation of glucose from corn by genetically engineered yeast or *Escherichia coli* yielded the commodity chemical succinic aid, an intermediate for the production of polyesters and polyimides. Sugars derived from the hydrolysis of cellulose can be converted chemically to hexane, thus avoiding the use of petrochemical sources for making this hydrocarbon.

Bioethanol has been produced by fermentation of sugars by enzymes and when produced in this way bioethanol is referred to as a first-generation biofuel. Bioethanol obtained from lignocellulose in agricultural waste is known as a second-generation biofuel. Development of second-generation biofuels is at an early stage. In principle lignin can be separated from cellulose and hemicellulosic feedstocks. Cellulase enzymes then break down cellulose and hemicellulose into smaller molecules with five or six carbon atoms and these C_5 and C_6 sugars are then fermented by bacteria or yeasts into ethanol for use as a chemical feedstock.

4.6   Silicon-based polymers

Silicones are polymers that contain alternating silicon and oxygen atoms in the polymer backbone, for example polydimethylsiloxanes. Silicones can be water-soluble or water-insoluble depending on the nature of the side groups on the polymer backbone. Applications for silicones include foaming agents in tooth-pastes and hair shampoos, sealants for bathroom tiles and potting agents for securing electrical wires and cables. Fluorosilicone elastomers have applications as seals, gaskets and valves in the auto and aerospace industries as they have good chemical resistance to oil and fuel as well as good thermal stability. Polycarbosi-lanes have repeating silicon–carbon linkages and can be derived from polysilanes, polymers with silicon–silicon linkages by a thermal treatment. Polycarbosilanes have an important industrial use as they can be spun into fibres that are converted into continuous fibres of silicon carbide in a heat treatment. Silicon carbide fibres are strong and are used in the fabrication of composite structures. Polysilazanes are polymers with Si–N–Si bonds, for example perhydropolysilazane. Films of poly-silazanes can be hydrolysed by steam to form layers of silica and this approach is used in the fabrication of silicon chips to produce insulating layers to prevent electrical interference between layers of metal wiring on the wafers.

4.7   Polyphosphazenes

Polyphosphazenes, or polyorganophosphazenes, are polymers with a phosphorus–nitrogen repeat unit of alternating phosphorus and nitrogen atoms. Potential applications for these polymers include hydrophobic thermoplastics, hydrophobic elastomers, fibre-forming materials and superhydrophobic coatings.

4.8   Summary

Synthetic polymers, namely plastics, are derived from petrochemical sources and are widely used in everyday life. Many plastics used nowadays were invented in the 20th century. Annual production of plastics was around 280 million tonnes in 2011, of which about 1 million tonnes were bioplastics. There is increasing interest in the 21st century in using renewable resources such as sugar from corn syrup as feedstocks for bioplastics, some of which such as polylactic acid are biodegradable. This interest arises from the requirement to move away from the use of petroleum resources as feedstocks for polymers. Deoxyribonucleic acid consists of sequences of four bases, and proteins that are expressed by DNA consist of sequences made out of twenty amino acids. Both DNA and proteins are considered to be naturally occurring polymers. Polymers based on silicon include polysilanes, polycarbosilanes and polysilazanes while polymers based on phosphorus include polyphosphazenes.

4.9 Further reading

Ashby, M. F., D. F. Balas and J. S. Coral. *Materials and Sustainable Development*. Oxford: Butterworth, 2015.

Askeland, D. K., and W. J. Wright. *The Science and Engineering of Materials*, 7th edition. Boston, MA: Cengage Learning, 2011.

Baekeland, L. H. Method of making insoluble products of phenol and formaldehyde. United States Patent 942,699, 1909.

Baldwin, G., T. Bayer, R. Dickinson, T. Ellis, P. S. Freemont, R. I. Kitney, K. Polizzi and G.-B. Stan. *Synthetic Biology: A Primer*. London: Imperial College Press, 2012.

Barber, J., and C. Rostron (eds). *Pharmaceutical Chemistry*. Oxford: Oxford University Press, 2013.

Carmichael, H. Gene genius. *Chemistry and Industry* (2015), 79, 32–35.

Carothers, W. H. Diammine-dicarboxylic acid salts and process of preparing them. United States Patent 2,130,947, 1938.

Ebnesajjad, J. (ed.). *Handbook of Biopolymers and Biodegradable Plastics*. Amsterdam: Elsevier, 2013.

Fried, J. R. *Polymer Science and Technology*, 3rd edn. Upper Saddle River, NJ: Prentice Hall, 2014.

Kwolek, S. L. Optically anisotropic aromatic polyamide dopes. United States Patent 3,671,542, 1972.

Lesk, A. M. *Introduction to Protein Science: Architecture, Function and Genomics*, 3rd edn. Oxford: Oxford University Press, 2016.

Maccioni, R. B., L. P. Navarrete and A. S. Martin. Method for preventing tau protein aggregation and treating Alzheimer's disease with a quinoline derivative compound. United States Patent 8,198,300, 2012.

Milmo, S. Celebrating ICI. *Chemistry and Innovation* (November 2016), 7–11.

Nieuwland, J. A. Vinyl derivatives of acetylene and method of preparing the same. United States Patent 1,811,959, 1931.

Papachristodoulou, D., A. Snape, W. H. Elliott and D. C. Elliott. *Biochemistry and Molecular Biology*, 5th edn. Oxford: Oxford University Press, 2014.

Plunkett, R. J. Tetrafluoroethylene polymers. United States Patent 2,230,654, 1941.

Querfurth, H. W., and F. M. Laferia. Alzheimer's disease. *New England Journal of Medicine* (2010) 362(4), 329–44.

Theyson, T. Changing face of innovation in the US chemical industry. *Chemistry and Innovation* (November 2016), 42–6.

Walton, D., and P. Lorimer. *Polymers*. Oxford: Oxford University Press, 2005.

5

Healthcare

The Benefits of Materials

5.1 Introduction

Early identification of diseases can aid the treatment of patients as the requirement for surgical procedures may be reduced and treatment with courses of drugs can start at an early stage. In recent years a range of diagnostic methods, for example those based on techniques such as magnetic resonance imaging and computerised tomography, has been increasingly used and new classes of pharmaceutical compounds have been developed. The costs of healthcare in ageing populations with chronic diseases such as diabetes and Alzheimer's disease are of concern to governments in countries with public health services. In addition, there is much debate on the causes of obesity and possible links between obesity and the types of food that people eat. Materials have an important role in the development of medical diagnostic techniques and in the treatment of patients and examples of how materials contribute to healthcare are described in this chapter.

5.2 Lasers in medicine and surgery

Albert Einstein is remembered for his Special Theory of Relativity published in 1905 and his General Theory of Relativity published in 1915. However, in 1917 he also proposed a new mechanism known as stimulated emission by which atoms and molecules could emit electromagnetic radiation. Atoms or molecules can adsorb a photon of radiation and be elevated from a ground state to an excited state with higher electronic energy. The atom can return to its ground state by spontaneous emission of light and the emitted light is incoherent; that is, there is no phase relationship between the emitted photons. In stimulated emission a photon whose energy is exactly equal to the difference between the ground and excited states can interact with the atom in its excited state, causing it to emit an identical photon and return to the ground state. Hence, the light output is amplified and the emitted light is monochromatic and coherent so that the emitted light is in the same

Materials for the 21st Century. David Segal.
© David Segal 2017. Published 2017 by Oxford University Press.
DOI: 10.1093/oso/9780198804079.001.0001

direction as the incident photon and in phase with it. Stimulated emission is a **resonance** phenomenon. The word laser is an acronym for 'light amplification by the stimulated emission of radiation' that can produce monochromatic, coherent radiation with wavelengths ranging from the infrared to ultraviolet. Two physical criteria need to be met to obtain lasing action: stimulated emission and a population inversion where there are more atoms in the excited state than in the ground state. The population inversion can be obtained by optical pumping as was used when the first practical laser was invented in 1960 by Theodore Maiman, a ruby laser based on a ruby rod whose composition corresponds to chromium-doped aluminum oxide. Lasers can be continuous or pulsed and examples include argon–fluoride (Ar–F) excimer lasers used in conjunction with photolithography in the manufacture of silicon chips while solid-state lasers based on semiconducting diodes are used to read barcodes in supermarkets and to read information on compact discs. Lasers have applications in medicine and surgery, including the reattachment of detached retinas, diabetic retinopathy, the removal of skin blemishes and tattoos.

Resonance is an oscillation of a system at its natural frequency of vibration. It is characterised by large amplitude vibrations that will result from low initial oscillations. An example refers to soldiers breaking step when marching across a suspension bridge to prevent large oscillations of the structure. Resonance can occur in atoms, molecules, mechanical systems and electrical circuits.

5.3   Magnetic resonance imaging

Magnetic resonance imaging (MRI) is a widely used diagnostic technique for producing images of soft tissue in the body. It is based on the interaction of the nuclear spins of protons in water molecules with powerful magnetic fields and how the spins are perturbed when subjected to a burst of electromagnetic radiation corresponding to wavelengths in the radio wave region. The basic principles are the same as used in nuclear magnetic resonance (NMR), an analytical technique used to determine the structure of organic chemicals and widely used in research activities in chemistry. Superconductors are materials that have no electrical resistance, and superconductivity in metals, namely in mercury, was first observed by Onnes in 1911 where there was a superconducting transition or critical temperature of 4.2 K for mercury immersed in liquid helium. A gradual increase in the superconducting transition temperature took place over the following decades, for example 22.3 K for films of niobium–germanium alloy. Magnetic fields are generated by passing an electric current through superconducting metals when immersed in liquid helium. The ability to fabricate superconducting metallic alloys into coils so that the patient lies within the coil was an important development in the construction of equipment for carrying out MRI. An explanation of superconductivity in metals known as conventional superconductivity assumes that electric current in the superconductor is carried by bound pairs of electrons known as **Cooper pairs**.

*In a **Cooper pair**, mobile electrons are coupled in pairs with opposite spins. Electrons repel each other at normal temperatures, but at lower temperatures a weak attractive force between pairs of electrons becomes significant.*

Ceramic superconductors were discovered in 1986 and become superconducting when immersed in liquid nitrogen at around 77 K. An example of a composition for these high-temperature superconductors is a mixed oxide of yttrium, barium and copper. As liquid nitrogen is much cheaper than liquid helium there are potential economic benefits in using ceramic superconductors for producing magnetic fields in MRI, although it is more difficult to fabricate ceramic powders into coils than metallic alloys.

5.4   Biopharmaceuticals

The painkiller acetylsalicylic acid, also known as aspirin, was the first drug to be prepared synthetically and has a chemical structure very similar to that of salicin, a natural product isolated from willow tree bark (Table 5.1). The initial synthesis took place at the end of the 19th century and at that time the pharmaceutical industry grew in parallel with the synthetic dye industry. Many aniline-based dyes

Table 5.1 *Examples of small molecule pharmaceuticals and biopharmaceuticals*

Type of pharmaceutical	Name
Small molecule pharmaceuticals	Acetylsalicylic acid
	Amaryl (glimepiride)
	Cozaar (losartan potassium)
	Cymbatta (duloxetine hydrochloride)
	Methotrexate
	Morphine
	Nexium (esomeprazole)
	Nicotinic acid
	Paracetamol
	Penicillins
	Simvistatin
	Tenormin (atenolol)
Biopharmaceuticals	Avastin (bevacizumab)
	Cimzia (certolizumab)
	Enbrel (etanercept)
	Humira (adalimumab)
	Remicade (infliximab)
	Symponi (golimunab)

and their derivatives were explored for their pharmaceutical properties. Throughout the 20th century potential drugs either were prepared by synthesis in a chemical laboratory or were natural products isolated from plants, soil bacteria or fungi. For example, morphine and codeine were initially obtained from the seed head of the opium poppy while the anticancer agent vincristine is found in the Madagascan periwinkle. Penicillin G was discovered by Alexander Fleming as the product of the mould *Penicillium notatum* while cephalosporins, a class of antibiotics, are produced by fungi of the *Acremonium* species. The antibacterial agent erythromycin was isolated from soil bacteria and once isolated attempts can be made to synthesise the active agents. The antibiotic vancomycin, which is effective against methicillin-resistant strains of *Staphylococcus aureus* (MRSA), was first isolated from a fermentation broth of a soil bacterium and synthesised chemically in 1998. The anticancer drug paclitaxel (Taxol) has been isolated from the Pacific yew tree but can be prepared in a semi-synthetic approach in which a chemical intermediate, 10-deacetylbaccatin, which is isolated from needles and twigs of the European yew tree, can be chemically converted to relatively large quantities of paclitaxel. Whether prepared in a chemical laboratory or isolated as natural products the molecular weights of these pharmaceutical compounds are low, on the order of hundreds at most.

In recent years, new approaches have been developed for obtaining effective drugs. In recombinant deoxyribonucleic acid (DNA) technology, specific genes are incorporated into the DNA of fast-growing bacteria or yeast to express large quantities of proteins such as human insulin for the treatment of diabetes. This approach is more commonly referred to as genetic engineering and underpins the biotechnology industry. As proteins are polymers they can be considered to be a class of material. Biopharmaceuticals which are monoclonal antibodies are attracting much attention in the 21st century as a new class of drug for the treatment of currently incurable diseases. Antibodies are a class of proteins known as immunoglobulins and are hence natural polymers based on sequences of amino acids. They are produced in B cells in the immune system and bind to antigens that enter the body in order to destroy them. A particular B cell produces one antibody in terms of binding specificity while other B cells produce other antibodies. Cesar Milstein and co-workers showed in the 1970s how to isolate the B cells that produce the desired antibody. They were fused with cancerous cells known as myeloma cells that do not themselves produce antibodies but which can multiply uncontrollably so that large quantities of antibodies could be produced. Antibodies made in this way are known as monoclonal antibodies and examples of pharmaceutical compounds based on them include Avastin and Humira. These pharmaceuticals are known as biopharmaceuticals or biologic drugs and have molecular weights in the tens of thousands, much higher than the low molecular weights of drugs such as aspirin or paracetamol synthesised in laboratories. Humira is used for the treatment of rheumatoid arthritis and targets a protein believed to be involved in the inflammation in this disease.

5.5 Drug delivery

About 80 per cent of all drug dosages are taken in oral form. Orally administered drugs must be transported to the stomach and small intestine for absorption across the gastrointestinal mucosal membrane into the blood. The efficiency of absorption can be low because of metabolism within the gastrointestinal tract. Hence, improved methods for drug delivery are being sought. For example, biodegradable nanoparticles have been considered for encapsulating and delivering nucleotide-based drugs containing DNA or ribonucleic acid (RNA) for treatment of human diseases. Gold nanoparticles have been used for the treatment of rheumatoid arthritis and if they have a particle size less than around 50 nm they can cross the brain–blood barrier. While not a drug, sliver nanoparticles have antimicrobial properties. Magnetic nanoparticles, such as iron oxide, have been used as contrast agents in MRI and, when coated with pharmaceutical agents, can target tumour sites to destroy malignant cells. Liposomes are assemblies of phospholipids consisting of a permeable lipid bilayer surrounding an aqueous core; the bilayer acts as a membrane. Liposomes can be used to encapsulate a drug in the core and transport it inside the body of a patient and they are also used to deliver cosmetic compositions to the skin. Micelles are association structures formed from surface active agents and are analogous to liposomes except that they do not have an aqueous core. Micelles are formed, for example, when washing-up liquids that contain surface active agents are added to water. The exploitation of nanoparticles in medicine is an example of the application of nanotechnology.

5.6 Nanotechnology and healthcare

The word nanotechnology is frequently mentioned in the media and is advertised as promising a future based on materials with properties not currently available. Nanotechnology is often associated with small particles that can be referred to as nanoparticles, nanomaterials and nanocrystals. There is no rigid definition of nanoparticles but particles with a diameter less than 100 nm are often referred to as nanoparticles. In order to have a clearer idea of the size of nanoparticles consider a human hair that has a diameter of around 25 μm. This diameter is equivalent to 25,000 nm, so if 1,000 nanoparticles each with a diameter of 25 nm are placed next to each other, the overall length of the assembly of nanoparticles would be approximately equal to the width of a human hair. There is much concern about the effect of carbon particulates from diesel emissions on human health. If it is assumed that potentially dangerous particles have a diameter of 2.5 nm then it can be seen that 10,000 of these particles placed next to each other would correspond to the width of a human hair. Carbon particulates are indeed very small particles. As the particle size decreases, a larger number of atoms reside in the surface and the properties of nanoparticles are affected by their particle size. Although of not direct relevance to healthcare, this size-dependency effect on properties is particularly noticeable with optical properties as observed in the change in colour with particle size for light emitted from quantum dots.

There is an overlap between nanotechnology and colloid science. Colloidal systems are those that have a dimension in the range 1 nm to 1 μm and include dispersions (sols) of particles in a liquid, emulsions where liquid droplets are suspended in a liquid, foams where the colloidal dimension is the thickness of the foam films, aerosols of liquid droplets in a gas and smokes for solid particles dispersed in a gas. Aerosols are important in the delivery of drugs to patients with asthma and breathing problems while many pharmaceutical formulations are prepared as emulsions. An important theme in colloid science is stability; hence, emulsions of pharmaceutical compounds need to be stable with respect to settling out so that consistent doses of drugs can be delivered via aerosols to patients with breathing problems.

5.7 Coronary stents

Implantable coronary stents are usually made out of an alloy of nickel and titanium known as Nitinol. This alloy undergoes a phase transition at a transformation temperature that can be varied by changing the relative amounts of nickel and titanium in the alloy. Below the transformation temperature the alloy has the martensitic phase and above the transformation temperature it has the austenitic phase. The low-temperature martensitic phase is malleable while the high-temperature austenitic phase retains the original stiff shape of the component. This alloy is an example of a shape memory alloy as the low-temperature phase 'remembers' the shape of the high-temperature phase. In use the transformation temperature is below body temperature and the stent is inserted in a compressed state with the martensitic phase. The stent then expands to its final shape with the austenitic phase as it warms up. Shape memory alloys have many applications particularly where a tight fit is required between components such as for connectors in aircraft hydraulic systems, valves, seals and even jewellery. Nitinol has good wear resistance, good erosion resistance and good corrosion resistance and is biocompatible. The austenitic phase has a body-centred cubic crystal structure while the martensitic phase has a monoclinic structure.

Table 5.2 compares the mechanical properties of the austenitic phase with the martensitic phase. The values for Young's modulus show that the austenitic phase represents a stiffer material than the martensitic phase.

> ***Young's modulus,*** *also known as the elastic modulus, is the ratio of the stress applied to a body to the strain produced, in particular the longitudinal stress and strain. Tensile stress is a measure of the force per unit cross-sectional area to break a*

Table 5.2 *Comparison of mechanical properties for the austenitic and martensitic phases of Nitinol*

Mechanical property	Magnitude of property
Young's modulus (GPa)	83 (austenitic)
	28–41 (martensitic)
Yield strength (MPa)	195–690 (austenitic)
	70–140 (martensitic)
Tensile strength (MPa)	895

Source: C. M. Agrawal, J. L. Ong, M. R. Appleford and G. Mani. *Introduction to Biomaterials: Basic Theory with Engineering Applications*, p. 127. Cambridge: Cambridge University Press, 2014.

material. If an increasing stress is applied to say a wire, the yield strength corresponds to the value of stress when the wire breaks. Stress refers to the force per unit area applied to a body that causes it to deform.

Shape memory polymers are known and have potential applications in facial reconstruction surgery although they are best known for applications as shape memory foams for mattresses.

5.8 Bioprinting

Three-dimensional, or 3D, printing has attracted much attention in recent years. This is because as an additive manufacturing technique, it allows components of intricate shapes to be fabricated without the generation of waste material. The general approach is to scan the object to be made and store its image as a computer-aided design on a computer. The image is mathematically sliced into layers and the motion of a printer is aligned to the coordinates of each layer. The component is built up layer by layer from the printer's 'ink', which is either solidified droplets of molten plastic or sintered powders of ceramic or metal, where **sintering** is usually achieved by scanning a high-powered laser over the surface of each layer before deposition of the next layer. A variant of three-dimensional printing of particular relevance to healthcare is bioprinting. Here the 'ink' is a cell culture that is printed as layers in the presence or absence of a substrate such as a hydrogel that is printed as alternate layers. The substrate can be removed as the cells continue to grow and multiply. Although at an early stage of development, bioprinting has the potential to produce organs and skin for skin grafts or for testing new drugs but this technique may grow in importance in the 21st century.

Sintering is a process in which a compacted powder is heated, sometimes under pressure. During the heat treatment diffusional processes of atoms occur that reduce the porosity, that is voids within the compact so that a dense non-porous compact is produced.

5.9   Implants

A wide variety of materials—polymer, ceramic and metal—are used in surgical implants. For example, polymers are used in interaortic balloons, coated stents for drug delivery, hydroxyapatite (a calcium phosphate) coatings on metallic hip implants and shape memory alloys for manufacture of coronary stents. In the case of hip implants, thermal spraying techniques such as plasma spraying are used to deposit porous coatings of hydroxyapatite that enable new tissue to grow into the pores. In plasma spraying, ceramic powders with a controlled particle size are passed through a high-temperature plasma and semi-molten particles deposit onto the substrate. Thermal spraying is also used to deposit thermal barrier coatings of, for example, **yttria-stabilised zirconia** onto nickel-based superalloys used for the manufacture of turbine blades.

> *If **zirconia** is heated to a high temperature, around 1000° C, it undergoes a phase change and on cooling a component will crack because of a change in dimensions associated with the reversible phase change. The addition of yttrium oxide (i.e. **yttria**) **stabilises** the zirconia in one crystallographic phase, hence eliminating the change in dimensions when heated.*

5.10   Hydrogels

Hydrogels are three-dimensional porous networks of cross-linked polymers that can be swollen by absorption and retention of water. They exhibit both liquid properties, because the major constituent is water, and solid properties, because of cross-linking during polymerisation. They can be prepared by polymerisation of acrylic acid but can also be made from natural products such as sodium alginate found in seaweed. They are often used in a dry powder form where they are known as superabsorbents and an important use is in disposable baby nappies (diapers). The ability to absorb and retain large quantities of fluid relative to their weight makes them candidates for wound dressings. Materials with potential use as wound dressings include polyurethane gels and hydroxyethyl cellulose gels, as examples.

5.11   Biodegradable materials

Polylactic acid (PLA) and polyglycolic acid (PGA) are biodegradable and are used as fixation devices for bone and soft tissue in the form of biodegradable plates, screws and anchors. PLA is a chiral compound, that is, it is not superimposable on its mirror image and exists in several so-called enantiomeric forms, designated D or L, or as a racemic mixture with equal amounts of the D and L forms. The rate of degradation of PLA *in vivo* depends on which enantiomer is used for the implant. PGA exists in only one form. Copolymers of PLA and PGA can be produced with various ratios of each polymer and the copolymers which have the trade name Vicryl are used as biodegradable sutures.

5.12   The search for antimicrobial agents

Antimicrobial resistance is a major public health threat in the 21st century. Every year 25,000 deaths in Europe are caused by drug-resistant infections, two-thirds due to **Gram-negative bacteria** such as *Escherichia coli*. There is a desperate need for new anti-bacterial agents. Some potentially promising agents are summarised here. Synthesis of analogues of spectinomycin, a naturally occurring antibiotic, has potential for treatment of tuberculosis (TB). This class of antibiotics works against TB by disrupting the function of the ribosome, the main site for protein synthesis in a cell, and binds to a site on the ribosome not shared by other TB drugs. Antimicrobial peptides (AMPs) are under development for the treatment of TB. AMPs can break through cell walls of tuberculosis cells and kill *Mycobacterium tuberculosis*. Computer modelling and simulation experiments have indicated that oxadiazoles inhibit the biosynthesis of the cell wall that enables it to resist other drugs. Oxadiazoles have shown promise in treating vancomycin- and linezolid-resistant MRSA in mouse models of infection. Another approach for the treatment of TB has been to consider the carbohydrate coating that all cells possess. In the case of TB, galactofuranose residues in the cell walls of mycobacteria cause TB. The approach has been to design small molecule inhibitors that block enzymes responsible for synthesis of galactofuranose and the growth of mycobacterial cells. Intelectin is a protein that binds to galactofuranose (not found in human cells) and has the potential to act as an antimicrobial agent. Lectins, of which intelectin is an example, are found in plants, animals, bacteria and viruses. They bind covalently to portions of sugars in cell coatings and have specificity; that is, they bind to specific regions or sequences of the sugar molecule. An example of a viral glycan-binding protein is the influenza virus lectin which binds to sialic acid-containing glycans. These are targeted by the influenza treatments Tamiflu and Relenza.

* **Gram-negative bacteria** *have thicker cell walls than Gram-positive bacteria and include* Salmonella *and* Escherichia coli *and are resistant to many antibiotics.*

5.13   Summary

Materials—polymer, ceramic and metal—have an important role in healthcare. For example, superconducting metallic alloys are used for generation of magnetic fields in magnetic resonance imaging while other alloys are used for the fabrication of coronary stents. Novel techniques based on recombinant DNA technology are under development for a new class of drugs known as biopharmaceuticals while porous hydrogels find uses as wound dressings. Lasers for surgical applications rely on suitable materials that undergo lasing action. Nanotechnology has an increasing role in drug delivery through the use of nanoparticles while bioprinting, a variant of three-dimensional printing, is under evaluation as a method for producing tissue such as layers of skin and replacement organs.

5.14 Further reading

Agrawal, C. M., J. L. Ong, M. R. Appleford and G. Mani. *Introduction to Biomaterials: Basic Theory with Engineering Applications.* Cambridge: Cambridge University Press, 2014.

Baldwin, G., T. Bayer, R. Dickinson, T. Ellis, P. S. Freemont, R. I. Kitney, K. Polizzi and G.-B. Stan. *Synthetic Biology: A Primer.* London: Imperial College Press, 2012.

Barber, J., and C. Rostron (eds). *Pharmaceutical Chemistry.* Oxford: Oxford University Press, 2013.

Davies, E. Sweet medicine. *Chemistry and Industry* (2015), 79(8), 36–39.

Dorey, E. Antibiotics revival. *Chemistry and Industry* (2014), 78(5), 24–7.

O'Driscoll, C. Bone filler for faces. *Chemistry and Industry* (2014) 78, 13.

Extance, A. Cosmetics deals push skin 3D printing. *Chemistry World* (2015) 12, 20–1.

Gavaler, J. R. Superconductivity in Nb-Ge films above 22 K. *Applied Physics Letters* (1973) 23, 480–2.

Hehenberger, M. *Nanomedicine: Science, Business and Impact.* Singapore: Pan Stanford, 2015.

Jelinkova, H. (ed.). *Lasers for Medical Applications, Diagnostics, Therapy and Surgery.* Cambridge: Woodhead, 2013.

Kresin, V., H. Morawitz and S. Wolf. *Superconducting State.* Oxford: Oxford University Press, 2013.

Langer, R. S. Biodegradable poly (β-amino esters) and uses thereof. United States Patent 8,287,849, 2012.

Melham, Z. (ed.). *High-Temperature Superconductors (HTS) for Energy Applications.* Cambridge: Woodhead, 2012.

Ngo, C., and M. van de Voorde. *Nanotechnology in a Nutshell: From Simple to Complex Systems.* Amsterdam: Atlantis Press, 2014.

Onnes, H. K. Further experiments with liquid helium. G. On the electrical resistance of pure metals. VI. On the sudden change in the rate at which the resistance of mercury disappears. *Konink Akad van Wetenschappen (Amsterdam)* (1911) 14, 818–21.

Palmer, P., T. Schreck, L. Hamel, S. Tzannis and A. Pouiatine. Small volume oral transmucal dosage forms containing sufentanil for treatment of pain. United States Patent 8,535,714, 2013.

Papachristodoulou, D., A. Snape, W.H. Elliott and D.C. Elliott. *Biochemistry and Molecular Biology,* 5th edn. Oxford: Oxford University Press, 2014.

Segal, D. *Exploring Materials through Patent Information.* Cambridge: Royal Society of Chemistry, 2014.

Wolfson, M. M. *Resonance: Applications in Physical Science.* London: Imperial College Press, 2015.

6

Let There Be Lights

6.1 Introduction

It has been estimated that over 20 per cent of the total U.S. electric energy production is consumed for lighting applications and that significant savings can be made by the introduction of high-performance illumination sources. In addition, there has been a proliferation of electronic devices that contain a screen or display. For example, smartphones, computers, whether desktop, laptop or tablet, in-vehicle navigation systems, digital cameras and game consoles. The appeal of consumer goods is made up in part by their design rather than the materials used for construction. This is evident in the appeal of certain models of smartphones and also in the availability of thin large-screen televisions compared to the bulkier television models previously available. The light sources are hidden from view and it is rapid technological changes that have given rise to novel sources of solid-state lighting that are more energy-efficient than earlier light sources, such as incandescent light bulbs and fluorescent lights, and which contribute to product designs that appeal to the consumer. This rapid technological change is underpinned by developments in materials and an overview of these changes is described in this chapter. Solid-state lighting is based on the use of light-emitting diodes, quantum dots, organic light-emitting diodes and liquid crystals. The chapter also highlights that there can be a considerable time delay between initial identification of useful materials and their commercial exploitation.

6.2 Light-emitting diodes

It is common nowadays to encounter light-emitting diode (LED) lights promoted as energy-efficient for use in the home, for consumers to purchase a LED television, to buy a LED torch or to purchase a car with LED headlamps. Toys for children can also contain LEDs. Also, it will be noticeable to purchasers of consumer goods that televisions are available in much thinner models than they were available just a few years ago. While many people are familiar with the acronym LED and the phrase 'light-emitting diode', it is not obvious what LEDs are as they do not contain incandescent wires or glass tubes characteristic of

Materials for the 21st Century. David Segal.
© David Segal 2017. Published 2017 by Oxford University Press.
DOI: 10.1093/oso/9780198804079.001.0001

conventional light bulbs or fluorescent tubes. In fact, the principle of operation of LEDs is electroluminescence, which was first observed by Henry Round in 1907 and refers to the emission of light from a material under an applied voltage. In this early work, a bright yellowish glow was noticed when a small voltage, around 10 volts, was applied to a crystal of carborundum (silicon carbide) which was impure and is a semiconductor. The glow was produced from a semiconductor **diode** although at the time it was not appreciated that this was the cause of the light emission. In fact, Henry Round referred to the light emission as a curious phenomenon. Advances in the development of LEDs were slowed until methods for fabrication of high-purity semiconductors were achieved in the 1950s and 1960s. These methods included metal-organic chemical vapour deposition (MOCVD), molecular beam epitaxy (MBE) and ion implantation.

*A **diode** is an electronic device that allows the flow of an electric current in one direction only.*

Electroluminescence is associated with electron–hole recombination at a *p–n* junction in a semiconductor diode. A *n*-type semiconductor is prepared by doping a semiconductor usually in the form of a wafer with dopant atoms, particularly those of Group V elements including arsenic, phosphorus and antimony. The doping process that can be achieved by MOVCD introduces an excess of free electrons into the semiconductor. Also, *p*-type semiconductors are prepared by introducing acceptor atoms, typically Group III elements such as indium, aluminium, gallium and boron, into the semiconductor and this process introduces an excess of holes or an electron deficit into the semiconductor. In practice one layer of semiconductor is grown and deposited onto another layer of semiconductor and a *p–n* junction is formed at the boundary between *p*- and *n*-type semiconductors. Electron–hole recombination takes place in the region of the *p–n* junction under a forward electrical bias. Examples of crystalline semiconductor alloys used for generating electroluminescence include gallium arsenide (GaAs) and indium arsenide (InAs). The recombination process involves electron transfer from conduction bands to valence bands in the semiconductor and the colour of the emitted light is determined by the bandgap energy, the energy difference between the conduction and conduction bands. It can be seen how materials expertise underpins the development of LEDs. Small changes in the preparative methods used to make the semiconductors can result in a spread of frequencies for a particular colour through changes in bandgap energy. Hence, there may be several different types of 'white' when using LEDs. Quantum dots that behave as nanophosphors and do not have an electronic band structure are more appropriate than LEDs for tuning their output for a specific colour. Colours from quantum dots can be 'cleaner' than those from LEDs. Methods used to manufacture LEDs are the same as those used for semiconductor fabrication, whereas much development is required to produce quantum dots with a controlled particle size in the nanometre range.

While a range of colours had been produced from LEDs, obtaining a blue light-emitting diode proved to be initially elusive. Examples of compositions for specific colours are indium gallium aluminium phosphide (InGaAlP) for red light (633 nm), gallium arsenic phosphide/gallium phosphide (GaAsP/GaP) for yellow light (585 nm) and gallium phosphide (GaP) on a GaP substrate for green light (565 nm). The breakthrough for blue LEDs came with incorporation of magnesium as a *p*-type dopant into a gallium nitride (GaN) substrate by Nakamura and co-workers. The availability of LEDs with emissions across the visible spectrum has expanded the use of LEDs since the 1970s and about 30×10^9 chips are made each year. LEDs are manufactured in wafer form and the latter are cut up or diced into pieces around 1 mm^2 in area when used in torches and lighting and the semiconductor chip is packaged for protection.

White light can be obtained in several ways in lighting applications. Red, green and blue emissions from LEDs can be combined to form white light. Another approach is to use ultraviolet emissions from a LED that causes fluorescence in a white-emitting diode. In the third method, a blue LED is covered with cerium-doped yttrium aluminium garnet crystals. Some of the blue light is converted to yellow light by the phosphor. As yellow is a combination of red and green wavelengths, the mix of blue and yellow produces white light. A particular application noticeable to consumers is in LED televisions. Liquid crystals are often used for displays in televisions but as liquid crystals do not emit light they require a backlight that has traditionally been a compact fluorescent tube. However, these tubes are bulky so that the television sets are also bulky. Replacement of the fluorescent tube by a LED panel for use as a backlight enables the manufacture of thin television sets. The exploitation of LEDs highlights the potential long time lag between identification of a previously unknown phenomenon and utilisation of that phenomenon in a commercial product.

6.3   Quantum dots

Phosphors are materials that emit visible light when supplied with a source of energy and this emission is known as luminescence. If the energy source is electromagnetic radiation then the emission is known as photoluminescence. If the latter process is immediate then fluorescence occurs but if the conversion of exciting radiation is slow then phosphorescence occurs. In fluorescence the emitted radiation has a longer wavelength than the exciting radiation. Many phosphors are inorganic oxides or sulphide powders and can be considered to be a class of electroceramic material. Their conventional synthesis involves mixing oxide or carbonate powders, **comminution** by milling followed by a high-temperature heat treatment to form a material containing a single crystalline phase. An example of phosphor compositions is europium-doped yttrium oxide that absorbs ultraviolet radiation and emits visible light, a process used in fluorescent tubes. Terbium-doped yttrium aluminium oxide is a cathodoluminescent phosphor that emits visible light on electron bombardment. The particle size for phosphors in use is in the range 1–2 μm and crystal quality is important for phosphors as defects such as dislocations, grain boundaries and impurities encourage non-radiative recombination of electron–hole pairs with a subsequent reduction in brightness. Features such as dislocations, grain boundaries and impurities contribute to the microstructure of phosphors.

*Note that **comminution** means a reduction in particle size, for example by grinding. In ball milling a powder is contained in a sealed jar that contains ceramic balls, for example alumina. The jar is rotated and the particles size of the powder is reduced.*

The electronic and optical properties of metals and semiconductors change sharply as the particle sizes are reduced to the nanometre range. In this size region, the electronic wave function is restricted to small regions of space in a material and this restriction is known as quantum confinement. Quantum dots, which are also referred to as quantum boxes or artificial atoms, are semiconductors where the confinement is in three dimensions. A particle with very small dimensions such as a quantum dot has discrete energy levels like an atom or molecule, in contrast to semiconductor crystals that have electronic energy bands. Quantum dots are nanoparticles and fall within the scope of nanotechnology (and colloid science) and can be called nanomaterials. Semiconductor nanocrystals such as quantum dots are referred to as artificial atoms as their structures fall between isolated atoms with discrete energy levels and semiconducting crystals that have energy bands. Properties of quantum dots, particularly optical properties, depend on both the composition and particle size.

Successful exploitation of materials depends on the ability to manufacture the materials reproducibly and in a reproducible way. Many methods have been used for the preparation of quantum dots, at least in the laboratory. For example, hydrothermal processing routes involve heating reactants such as metal salts, oxides or hydroxides as a solution or suspension in a liquid, usually water at elevated temperature. This approach has been used for synthesis of cadmium sulphide (CdS), cadmium selenide (CdSe) and cadmium telluride (CdTe). Core–shell nanoparticles have been prepared with a core of one semiconductor surrounded by a shell of a different semiconductor: for example, cadmium selenide/zinc sulphide (CdSe/ZnS), cadmium selenide/cadmium sulphide (CdSe/CdS) and indium nitride/zinc sulphide (InN/ZnS) nanocrystals. Hydrothermal processing and aerosol synthesis fall within the area of wet chemical routes to quantum dots and wet routes include sol-gel processing. In aerosol synthesis, small liquid droplets react with a gas, for example hydrogen sulphide, and this approach has been used to make cadmium sulphide particles with a diameter of 6 nm. Gas phase reactions have also been used to fabricate quantum dots, thus ion implantation, MOCVD, MBE and focused ion beams (FIB). These approaches, which are used for the fabrication of doped semiconductors for LEDs, especially MOCVD, involve deposition of a thin layer with thicknesses up to around 300 nm onto a substrate, after which regions of the layer are selectively removed or etched.

Quantum dots are semiconductor phosphors and as for all phosphors they emit wavelengths longer than that of the exciting radiation and exhibit fluorescence. Nanophosphors are fluorescent materials with diameters characteristic of quantum dots, hence between 2 and 10 nm, and particles this small contain between 10 and 5000 atoms. Quantum dots are attractive materials for solid-state lighting applications because the presence of discrete energy levels allows 'clean' colours to be produced and the wavelength (i.e. colour) of the fluorescent radiation to be tuned by control of the particle size for a particular material composition. As the size of the quantum dot increases, the colour of emitted light changes from blue to red; that is, the wavelength increases. Quantum dots with a polydisperse size distribution allow light of any wavelength to be produced. When illuminated with light from a blue diode based on a gallium nitride semiconductor with a wavelength of 450 nm, 5.5-nm cadmium selenide nanoparticles were red emitters, 4.0-nm were green emitters while 2.3-nm nanoparticles were blue emitters. It is difficult to obtain light with a pure single frequency from LEDs because tight control of semiconductor chemistry is required. Quantum dots have the potential to overcome this drawback of LEDs. White light can be obtained by primary colours of red, green and blue light emitted from quantum dots. At the time of writing quantum dots have not reached the level of commercial exploitation that has been achieved by LEDs although they have the potential, when used in displays, to produce cleaner colours than LEDS and as they can be used as a coating contribute to the design of thin portable devices. As semiconductors, quantum dots have potential for use in solar or photovoltaic cells.

6.4   Liquid crystals

Molecules in solution can form association structures. For example, solutions of surface active agents (surfactants) such as sodium dodecyl sulphate form spherical micelles for concentrations above the critical micelle concentration. The micelles have diameters of several nanometres and have a structure in which hydrocarbon chains of the surfactants point inward to the core and the head groups of the surfactants point outwards and interact with the aqueous phase. Surface active agents are components of washing-up liquids and the role of micelles is to solubilise dirt and fats from surfaces. Molecules in solution can also form association structures known as liquid crystalline phases that exist between crystalline solid phases and the fully disordered liquid state. Crystals have a long-range positional order while liquids have neither positional nor orientational order. Molecules in liquid crystalline phases exhibit liquid-like properties such as low viscosity together with crystal-like properties such as anisotropy. Qualitatively the formation of liquid crystals can be described according to thermodynamic considerations. Hence, the entropy of a system decreases with the formation of liquid crystals, hence raising the free energy, but hydrogen bonding, or the attractive van der Waals force between molecules, is maximised when molecules are aligned and formation of a preferred close-packed array of aligned molecules contributes to lowering the free energy of the system, maintaining thermodynamic stability. In the smectic phase, molecules are arranged side by side in layers and have an elongated or cigar-like structure. The smectic phase has orientational order and one-dimensional positional order. The nematic phase has orientational order but no positional order and ordering is along the long molecular axis and these liquid crystals have a thread-like structure. So, in a nematic liquid crystal, molecules or polymers are aligned parallel to each other in the same direction but in a random way. In the cholesteric phase, molecules are ordered in nematic layers, each layer rotated relative to the layers above and below. Polymer molecules can form liquid crystals and Kevlar fibres contain nematic phases.

Liquid crystals, when used in liquid crystal displays, do not emit light and require a backlight. The latter has traditionally been a compact fluorescent tube but increasingly LED panels are being used as backlights in electronic devices, enabling thinner devices to be manufactured, for example thinner television sets. The key property of liquid crystals that makes them attractive for use in displays is their birefringence and ability to cause a rotation of the angle of polarisation in the incident light of 90°. Cyanobiphenyls that were developed by George Gray and co-workers in the 1970s form stable nematic phases at or near room temperature and these phases allowed liquid crystal displays to operate at these temperatures. When nematic liquid crystals are used in a display, switching an electric field on and off changes the orientation of the liquid crystals in a thin layer and this effect changes a rotation of the angle of polarisation of incident light. The overall effect is that the display screen will appear either bright or dark to an observer. Twisted nematic liquid crystals such as *n*-4-ethoxybenzylisene-4-aminobenzonitrile (PEBAB) has been used in displays where its optical activity is controlled by an applied voltage that causes the helical structure to untwist and control light transmission that determines the image seen by an observer.

6.5   Organic light-emitting diodes

Materials underpin the development of solid-state lighting technology for use in the home, in business, in the environment and in electronic displays such as smartphones and televisions (Table 6.1). Any material that gives sharper images and a wider range of colours can give a competitive advantage to manufacturers, improving product design that will appeal to potential purchasers. The acronym OLED refers to organic light-emitting diode and emits light by the process of electroluminescence. OLEDs are based on organic molecules, both small molecules and polymers. OLEDs are particularly attractive for use on flexible electronic displays or on curved monitors that have a wider field of view. Work on small molecule OLEDs, known as SMOLEDs, dates to the early 1950s and these species include anthracene and naphthalene. Flexible polycrystalline films of poly-acetylene could be doped by exposure to vapours of electron–acceptor dopants

Table 6.1 *Materials with use or potential use in solid-state lighting*

Application	Material
Light-emitting diodes	Doped gallium nitride
	Doped gallium arsenide
	Indium gallium aluminium phosphide
Quantum dots	Cadmium sulphide
	Cadmium selenide
	Cadmium selenide/zinc sulphide
Liquid crystals	Cyanobiphenyls
SMOLED	Anthracene
	Naphthalene
PLED	Poly (p-phenylenevinylene)

such as arsenic pentafluoride (AsF_5) so that electrical conductivities could be adjusted to lie between the conductivities or semiconductors and metals. In general, SMOLEDs had to be deposited onto substrates from the vapour phase.

The polymer poly (p-phenylenevinylene) (PPV), which was identified in the early 1990s, laid the foundations for a new generation of flat panel displays based on electroluminescence from conjugated polymers. PPV is known as a polymer light-emitting diode (PLED) and unlike SMOLEDs it can be applied to a substrate by spin-coating from solution. The potential flexibility of organic polymers makes them ideal candidates for coating flexible substrates. While electroluminescence in inorganic semiconductors such as gallium nitride can be explained in terms of valence and conduction bands this approach does not apply to conjugated polymers. Injected electrons and holes recombine in the organic layer and emit light whose wavelength is dependent on the properties of the organic material when the OLED is subject to forward bias. Electroluminescence in semiconductor polymers is described in terms of the highest occupied molecular orbital (HOMO) and lowest unoccupied molecular orbital (LUMO) rather than valence and conduction bands whereby electron–hole pairs form **excitons** which exhibit electroluminescence on decay.

*An **exciton** is an electron–hole pair in a solid that is bound in a way analogous to the electron and proton in a hydrogen atom.*

6.6 Summary

Developments in materials underpin the growth of solid-state lighting in the home, in offices and in the environment as well as the use of solid-state lighting sources in electronic devices such as smartphones, tablet computers and televisions. It has taken approximately one hundred years since the discovery of electroluminescence for the widespread use of energy-efficient light-emitting diodes, illustrating the potential large time lag before commercial exploitation of discoveries. Liquid crystals have been used widely in displays such as in televisions, and backlights of light-emitting diodes are increasingly replacing compact fluorescent tubes, enabling thinner models to be produced. It is anticipated that quantum dots and organic light-emitting diodes will be increasingly exploited as solid-state lighting in future years.

6.7   Further reading

Dunmur, D., and T. Sluckin. *Soap, Science and Flat-Screen TVs: A History of Liquid Crystals.* Oxford: Oxford University Press, 2014.

Hack, M., M-Hao M-Lu and M. S. Weaver. Organic light-emitting devices for illumination. United States Patent 8,100,734, 2012.

Helfrich, W., and M. Schadt. Optical device. United Kingdom Patent Application 1 372 868, 1974.

Jones, R. A. L. *Soft Condensed Matter.* Oxford: Oxford University Press, 2013.

Nakamura, S., N. Iwasa and M. Senoh. Method of manufacturing p-type compound semiconductor. United States Patent 5,468,678, 1995.

Ngo, C., and M. Van de Boorde. *Nanotechnology in a Nutshell: From Simple to Complex Systems.* Amsterdam: Atlantic Press, 2014.

Segal, D. *Chemical Synthesis of Advanced Ceramic Materials.* Cambridge: Cambridge University Press, 1989.

Segal, D. *Exploring Materials Through Patent Information.* Cambridge: Royal Society of Chemistry, 2014.

Segal, D. L., and A. Atkinson. Wet chemical synthesis of cathodoluminescent powders. *British Ceramic Transactions* (1996) 95(3), 103–6.

Solymar, L., D. Walsh and R. A. Syms. *Electrical Properties of Materials,* 9th edn. Oxford: Oxford University Press, 2014.

Tilley, R. J. D. *Understanding Solids: The Science of Materials.* London: Wiley, 2013.

7

Energy Supplies for the 21st Century

7.1 Introduction

Greenhouse gases, in particular carbon dioxide, methane and nitrous oxide, absorb infrared radiation (heat) that is reflected from the Earth's surface, atmosphere and clouds so that the radiation does not escape into space. Global warming results from an increase in the temperature of the Earth's atmosphere and contributes to climate change. There is a worldwide quest to identify technologies that can reduce carbon dioxide emissions into the environment for limiting the effects of global warming and an increasing desire among governments to utilise renewable energy source such as solar, hydroelectric, wind, tidal and geothermal, thus the use of clean energy sources. In addition, nuclear power is under active consideration as it is not a source of carbon dioxide, a contributing gas for global warming. This approach involves decarbonisation, thus reducing the dependency on fossil fuels such as coal and oil which are considered to be major polluters with respect to the amounts of carbon dioxide they produce. The use of clean energy sources is carried out in parallel with the increased interest in using renewable sources as feedstocks for materials, thus reducing dependency on oil in petrochemical sources that have been used as feedstocks for synthetic polymers, namely plastics throughout the 20th century and for power generation. An aspiration that was agreed upon by participants at the Government's Committee on Climate Change in Paris in December 2015 was to limit a rise in the Earth's temperature to $1.5°C$ by the end of the 21st century, a reduction from a $2°C$ rise that had previously been considered necessary to limit global warming. Indeed, the United Kingdom has already made a commitment to phase out coal-fired power stations by 2025. Rigorous adherence to these targets may result in phasing out gas-fired power stations unless a way of capturing their CO_2 emissions in what is known as carbon capture and storage. In this chapter aspects of selected energy sources are described with particular reference to the role of materials in their operation.

Materials for the 21st Century. David Segal.
© David Segal 2017. Published 2017 by Oxford University Press.
DOI: 10.1093/oso/9780198804079.001.0001

7.2   Global electricity consumption

Global electricity consumption in 2014 was 21×10^{12} kWh (21,000 TWh), equivalent to an average continuous power consumption of 2.4 TW. Over 80 per cent of electricity is generated by burning fossil fuels releasing about 10 billion tonnes of carbon emissions per year. In 2011, 75 per cent of electricity generation in the United Kingdom was achieved by burning fossil fuels, 19 per cent from nuclear power and 6 per cent from renewable sources. In Norway 1 per cent of electricity was generated from burning fossil fuels and the remaining 99 per cent from renewables. In the USA 71 per cent of electricity was generated from burning fossil fuels, 19 per cent from nuclear and 10 per cent from renewables. The United Kingdom is legally committed to obtaining 15 per cent of its energy demand from renewable sources by 2023.

7.3   Nuclear power

The first civilian nuclear power station was opened at Calder Hall in the United Kingdom in the early 1950s and since then a variety of reactors have been built and operated and all use fuel based on enriched uranium oxide. Uranium ores contain about 99.3 weight per cent of non-fissionable ^{238}U and about 0.7 weight per cent of fissionable ^{235}U.

*The superscripts **238** and **235** refer to the atomic weight.*

The fuel contains uranium dioxide pellets enriched with about 3 weight per cent of ^{235}U. Enrichment can be achieved by first converting the ore into uranium hexafluoride, UF_6, and then separating out the lighter $^{235}UF_6$ by high-speed centrifugation. The pellets are fabricated in a shielded facility to protect operators from radiation. It is interesting to note that the first application of the newly invented polymer polytetrafluoroethylene was for the separation of uranium isotopes by gaseous diffusion in the Second World War in the Manhattan Project as this polymer was chemically resistant to corrosive fluorine-containing gases. Neutron bombardment of fissile material produces a range of highly radioactive fission products such as ^{92}Kr and ^{141}Ba and the difference in masses between the fission products and ^{235}U is converted to energy. The release of energy is used to convert water to steam and drive turbines in order to generate electricity. The following nuclear reaction also occurs in thermal nuclear reactors in which fissionable ^{239}Pu is produced by neutron bombardment of non-fissionable ^{238}U along with β-decay:

$$^{238}\text{U} \text{-----n}\beta\text{-----} \rightarrow \ ^{239}\text{U} \text{-----} \rightarrow \ ^{239}\text{Np} \text{-----} \rightarrow \ ^{239}\text{Pu} \qquad (7.1)$$

Another nuclear fuel cycle is based on thorium which is more abundant than uranium in the Earth's crust. Thorium is found in the mineral monazite. In this fuel cycle non-fissile ^{232}Th is converted to fissile ^{233}U through the reaction

$$^{232}\text{Th}\text{-----n}\gamma\text{---}\rightarrow \text{ }^{233}\text{Th----}\beta\text{-----}\rightarrow^{233}\text{Pa---}\beta\text{---}\rightarrow^{233}\text{Pu} \qquad (7.2)$$

The **half-life** for decay of ^{233}Th to ^{233}Pa is 23.5 minutes and the half-life for decay of ^{233}Pa to ^{233}U is 274 days. The decay of ^{232}U that is associated with ^{233}U is associated with daughter products, for example ^{208}Tl and ^{212}Bi, that emit high-energy γ-rays. In a thermal nuclear reactor neutrons are slowed down by a moderator which, depending on the type of reactor, can be water, graphite or **heavy water**. However, commercial nuclear reactors are not based on the use of this thorium fuel cycle.

The time for half of an initial number of radioactive atoms to decay is known as the **half-life** *for that particular atom or nuclide. A radioactive atom or nuclide, after radioactive decay, can result in formation of a daughter atom. The daughter atom is formed from the decay of a parent atom. In beta decay, often written as β-decay, an unstable atomic nucleus decays into a nucleus of the same atomic weight but different atomic number with the release of beta particles, namely electrons.*

Note that **heavy water** *is deuterium oxide, D_2O, in which deuterium and hydrogen are isotopes. They have the same atomic number but different atomic weights.*

Materials are of crucial importance in the operation of nuclear reactors. Thus, zirconium alloys (with iron, chromium and tin) known as Zircaloy are used as fuel rods to contain fuel pellets. Zircaloy has a low neutron-capture cross-section, which means it does not absorb neutrons that are intended to take part in the fission process. In pressurised water reactors the pressure vessel must be constructed of high-quality steel that clearly must not fail over a working reactor lifetime of many years and the outer concrete shell of the reactor building needs structural integrity. While spent fuel rods are stored under water to cool down before possible reprocessing, existing stockpiles of highly radioactive waste will need to be immobilised for long-term storage. It should be noted that exposure of materials to radiation such as neutrons in a nuclear reactor can cause radiation damage; that is, the crystalline structure of structural materials can change on irradiation. This irradiation can weaken the materials such as steels. A family of materials known as oxide-dispersion-strengthened steels in which oxide particles are embedded in alloys such as high-chromium iron-based alloys has potential as radiation-tolerant structural materials for nuclear reactors. The oxide particles are based on compositions of yttrium and titanium.

The use of nuclear-generated electricity produces highly radioactive fission products as well as fissionable ^{239}Pu. There will be continuing concerns among governments that so-called 'dirty bombs' might be constructed from radioactive fission products and that weapons-grade ^{239}Pu (approximately 93 per cent of ^{239}Pu) will fall into the hands of those with criminal intent.

7.4 Solar cells

The demand for solutions to worldwide concerns over global warming and energy supplies has generated much research and development on solar or photovoltaic cells. In a solar cell, photons emitted from the sun are absorbed by a semiconductor and excite electrons from the semiconductor's valency band across the bandgap and into the conduction band. Electron–hole pairs are separated in the solar cell to generate a photocurrent and photovoltaic effect. Photons with energies equal to or greater than the bandgap energy are absorbed but absorption does not take place if the photon energy is less than the bandgap energy. Single crystal silicon has traditionally been used as a rigid substrate for photovoltaic panels. Because silicon is an indirect bandgap semiconductor and hence not an efficient absorber of light, a thick layer of silicon is required for its use in solar cells. Commercial solar panels have an efficiency of around 20 per cent and a lifetime of 25 years. Single crystals of silicon are made by the Czochralski technique that involves growing the crystal from a melt onto a seed crystal and there is a search for materials that can be produced economically and have high efficiencies for conversion of solar energy to electricity when used in a photovoltaic cell. Copper indium gallium diselenide, a semiconductor, is one such material that can be deposited onto cheap flexible substrates. Examples of cheap substrates include glass, stainless steel and plastics and this area is known as thin film photovoltaics. Quantum dots are semiconductors and also have potential for use in solar cells and examples of these quantum dots are lead sulphide (PbS), lead selenide (PbSe) and lead selenide sulphide (PbSSe). In dye-sensitised photovoltaic cells, a dye, for example a ruthenium complex, is adsorbed onto nanoparticles of a semiconducting oxide such as anatase titanium dioxide. The dye absorbs light which releases an electron into the conductor band of the semiconductor and then into an external circuit. Dye-sensitised materials have the potential to increase the efficiency of solar cells. Another class of material that has the potential to increase the efficiency of solar cells are **perovskites**, for example the semiconductor methylammonium lead triiodide. The perovskite is adsorbed onto anatase particles that form the anode in the photovoltaic cell and absorbs light that releases electrons into an external circuit. The perovskite can be deposited by spin-coating a solution, a relatively easy process.

*A **perovskite** is a specific type of crystal structure represented by the formula ABO_3.*
Examples of minerals with this structure are calcium titanate and barium titanate,
$CaTiO_3$ and $BaTiO_3$, respectively.

7.5   Wind energy

Many conventional materials are used in wind turbines, for example carbon steel, stainless steel, concrete, copper aluminium and polymer matrix composites. However, the generators in wind turbines use neodymium-boron magnets and it has been stated that a 2-MW wing turbine contains 25 kg of neodymium. Neodymium is considered to be a critical material with respect to its supply so annual construction of 50,000 new turbines per year would require a supply of 1,250 tonnes of neodymium per year.

7.6   Geothermal energy

Geothermal energy arises from heat generated within the Earth from volcanic activity and decay of radioactive isotopes over long periods of time and it is abundant and sustainable. Electricity generated from geothermal sources does not directly produce emissions as no combustion is used, unlike the combustion of fuels in power plants. Geothermal energy is associated with columns of hot water and steam emerging from fissures in the ground. There are several ways in which electricity is generated. In the first method, superheated hydrothermal water flashes to steam that drives turbine. Another approach uses hot water to vaporise a secondary liquid which drives a turbine while combinations of these two methods can also be used to drive turbines. A fourth approach uses dry steam obtained directly from the ground to drive turbines. The oldest geothermal plant in the USA is at the Geysers in California and was built in 1960 and uses dry steam to drive turbines for electricity generation. From a materials perspective, corrosion is a potential problem for geothermal energy power plants due to species in the superheated water. Carbon steel, copper and its alloys and stainless steels are used as well as non-metallics including polymers. Dissolved chloride salts, carbon dioxide, hydrogen sulphide species, microbes and sulphate ions can contribute to corrosion of a geothermal plant. Corrosion can be controlled by the use of material selection, protective coatings, chemical treatment to inhibit corrosion and cathodic protection. Conventional hydrothermal geothermal energy sources are, however, limited to the presence of particular geological conditions that trap steam or pressurised hot water in permeable geological formations.

7.7   Tidal energy

Oceans cover 70 per cent of the Earth's surface and a number of technologies have been proposed for extracting energy from the oceans:

(a) Wave energy—Wave energy is the energy carried by movements of the ocean surface.

(b) Tidal energy—Tidal energy is driven by the gravitational force between the Earth and the Moon. The available potential energy is based on the difference between low tide and high tide.

(c) Current energy—Current energy is generated due to ocean water flow. Submerged turbines can in principle capture kinetic energy from the water flows.

(d) Ocean thermal energy conversion (OTEC)—Ocean thermal energy conversion uses differences in ocean temperatures from the surface to the depths to extract energy through a heat engine.

(e) Osmotic energy—This approach uses the difference in salinity between seawater and freshwater and generates power by use of an ion-specific membrane.

While extraction of energy from the oceans is not at an advanced stage, materials that can withstand the harsh conditions in the oceans will be required to exploit this energy source.

7.8   Hydroelectric power

Pumped hydroelectric technology is a mature technology and its application is constrained to regions where there is a significant difference in elevation between the upper and lower reservoirs. No special materials of construction are required and standard materials such as steel, copper and concrete are used.

7.9   Carbon capture

The quest for clean renewable energy sources runs in parallel with decarbonisation. The latter process refers to reducing or eliminating carbon dioxide emissions from burning fossil fuels particularly in power plants (i.e. coal, gas and oil) as these emissions contribute to global warming. The aim of this reduction is to move from a high-carbon to low-carbon economy. The phrases carbon capture and sequestration are often used in conjunction with carbon capture.

Sequestration refers to trapping carbon dioxide emissions in underground geological formations. Sedimentary basins are porous and consist of layers of sand, silt, clay and carbonates such as chalk and have been built up over a period of millions of years and are often filled with salt water. Sedimentary basins are found near reserves of oil and gas. There is much worldwide experience in pumping carbon dioxide into oil reservoirs during enhanced oil recovery. In the future it is possible that geological sequestration of carbon dioxide could take place in deep sedimentary basins.

There are three current approaches for capturing carbon dioxide emissions from power plants that burn fossil fuels:

(a) Post-combustion capture—Chemical solvents or organic amines remove carbon dioxide from flue gases in a power plant, after which the solvents can be regenerated. Note that ionic liquids also have the potential for selectively absorbing gases. Post-combustion capture units can be retrofitted to existing power plants. Post-combustion capture is a mature technology.

(b) Pre-combustion capture—Fossil fuel is reacted with steam, air or oxygen to produce synthesis gas, a mixture of carbon monoxide and hydrogen. Carbon monoxide then reacts with steam to produce carbon dioxide and hydrogen and the latter is separated from the mixture and used for electricity generation while carbon dioxide undergoes sequestration.

(c) Oxygen combustion—The fossil fuel is burned in oxygen rather than in air to generate electricity. In this way flue gases consist mainly of carbon dioxide and water so that carbon dioxide can be separated relatively easy. However, an electricity supply is required to separate oxygen from air.

7.10 Lithium-ion batteries

Lithium-ion batteries are secondary batteries; that is, they are rechargeable. They are lightweight with a high specific energy, have a small self-discharge and can be manufactured in different shapes. They are widely used in mobile phones, laptop and tablet computers, power tools and electric cars. They are a source of 'clean' electricity although the electricity used to recharge them may come from conventional fossil-fuel power plants. In a lithium-ion cell, the negative electrode (anode) is a porous carbon whereas the cathode (positive cathode) is a metal oxide, frequently lithium cobalt oxide ($LiCoO_2$). The electrolyte is a lithium salt such as $LiPF_6$ dissolved in a polar solvent, for example a mixture of ethylene carbonate and dimethylcarbonate. Lithium cobalt oxide is an example of an **intercalation compound.**

*A **intercalation compound** is a compound that can trap atoms, ions or molecules between layers in a crystal lattice.*

During discharge, lithium ions (Li^+) are extracted from the cathode, move through the electrolyte and are inserted or intercalated into the carbon anode.

*Here, **during discharge** means 'while in use'.*

The reverse process takes place during charging. These processes can be represented by the equations

$$LiCoO_2 \text{----} \rightarrow xLi + +Li_{X-1}CoO_2 + xe^- \tag{7.3}$$

$$xLi^+ + xe^- + 6C \text{----} \rightarrow Li_X C_6 \tag{7.4}$$

Pioneering work on the intercalation of lithium ions was carried out by John Goodenough at the University of Oxford in the 1970s. There have been some safety concerns over lithium-ion batteries relating to fire or explosion caused by overheating in thermal runaway and overcharging but protective electronic circuits are included in the design to prevent overcharging. Thus, in a lithium-ion battery the lithium ions move in both directions between the anode and cathode.

7.11 Fuel cells

Fuel cells are, in the context of energy supplies for the 21st century, a source of clean electricity. All fuel cells require a continuous supply of oxygen and hydrogen to a cathode and anode, respectively, separated by an electrolyte. While oxygen can be extracted from the air, different sources of hydrogen have been considered, for example storing hydrogen as liquid or a pressurised gas, in a metal hydride or as a source that can be catalytically converted to hydrogen, thus methanol, methane or petrol as examples. Fuel cells rely on the overall reaction of hydrogen and oxygen to give water and a cell voltage and the different types of fuel cell are described according to the electrolyte used:

(a) Alkaline fuel cell (AFC)—The electrolyte is aqueous potassium hydroxide that transports hydroxyl ions (OH^-) between electrodes. The anode reaction is

$$H_2(g) + 2OH^-(aq) \text{----} \rightarrow 2H_2O \ (l) + 2e^- \tag{7.5}$$

And the cathode reaction is

$$O_2(g) + 2H_2O \ (l) + 4e^- \text{----} \rightarrow 4OH^-(aq) \tag{7.6}$$

The overall reaction is represented by

$$2H_2(g) + O_2(g) \text{----} \rightarrow 2H_2O \ (l) \tag{7.7}$$

Catalysts are deposited onto the electrodes to speed up the reactions, for example dispersed platinum particles.

(b) Proton exchange membrane fuel cell (PEMFC)—This uses a solid polymer electrolyte that conducts hydrogen ions (H^+).

(c) Phosphoric acid fuel cell (PAFC)—This uses phosphoric acid as the electrolyte that is distributed within a silicon carbide matrix to transport H^+.

(d) Molten carbonate fuel cell (MCFC)—This uses molten lithium carbonate as electrolyte in a lithium aluminate ($LiAlO_2$) matrix transporting carbonate ions ($CO_3{}^{2-}$).

(e) Solid oxide fuel cell (SOFC)—This uses a solid yttria-stabilised zirconia electrolyte that allows transport of oxygen ions ($O_2{}^-$). The operating temperature is around $650°C$ as diffusion of oxygen ions at room temperature is slow.

7.12 Summary

There is a worldwide quest to identify technologies that can reduce carbon dioxide emissions into the environment in order to limit the effects of global warming and an increasing desire among governments to utilise renewable energy sources such as solar, hydroelectric, wind, tidal and geothermal. While nuclear power does not contribute to carbon dioxide levels in the atmosphere, highly radioactive fission products as well as the production of fissile plutonium in thermal nuclear reactors need to be stored in secure environments and eventually encapsulated for long-term disposal. Solar or photovoltaic solar cells represent an established technology, based initially on silicon panels. However, developments in materials have allowed the production of solar cells using cheap or flexible substrates. Wind turbines are increasingly used for power generation but a critical material, neodymium, is a component of magnets used in the generator and is not available in unlimited quantities. Hydroelectric and geothermal power generation are established technologies and conventional materials of construction are used, whereas tidal energy is at an early stage of development. Carbon capture and carbon sequestration are under consideration for treatment of emissions from power plants or long-term storage of carbon dioxide in underground formations of sedimentary rocks. Carbon capture is an established technology and can be retrofitted to existing power plants and chemical absorbents such as amines are important in the removal of carbon dioxide from flue gases.

7.13 Further reading

Arnold, E. D. Radiation hazards for recycled ^{233}U-thorium fuels. In *Proceedings of the Thorium Fuel Cycle Symposium, Gatlinburg, Tennessee, 5–7 December*, TID 7650, book 1, 1962.

Ashby, M. F., D. F. Balas and J. S. Coral. *Materials and Sustainable Development*. Oxford: Butterworth-Heinemann, 2015.

Boyle, G. (ed.). *Renewable Energy: Power for a Sustainable Future*, 3rd edn. Oxford: Oxford University Press, 2012.

Burke, M. Solar future looks bright. *Chemistry and Industry* (Sept. 2011) 19–21.

Christie, R. M. *Colour Chemistry*, 2nd edn. Cambridge: Royal Society of Chemistry.

Eisberg, N. Solar bananza. *Chemistry and Industry* (Nov. 2014) 68, 38–41.

Extance, A. The power of perovskites. *Chemistry World* (2014) 11, 46–9.

Ginley, D. S., and D. Cahen (eds). *Fundamentals of Materials for Energy and Environmental Sustainability.* Cambridge: Cambridge University Press, 2012.

Hu, Y. H., U. Burghaus and S. Qiao (eds). *Nanotechnology for Sustainable Energy*, ACS Symposium Series 1140. Washington, DC: American Chemical Society, 2013.

Hummel, R. E. *Electronic Properties of Materials*, 4th edition. New York: Springer 2012.

McLarty, L., and M. J. Reed. The US geothermal industry: three decades of growth. *Energy Sources* (1992) 4, 11–19.

Nozik, A. J., M. Beard, M. G. Law and J. M. Luther. Solar cells based on quantum dots or colloidal nanocrystals films. United States Patent Application, 2011/0146766, 2011.

Segal, D. *Chemical Synthesis of Advanced Ceramic Materials.* Cambridge: Cambridge University Press, 1989.

Tilley, R. J. D. *Understanding Solids.* Chichester: Wiley, 2013.

8

The Preparation of Materials

8.1 Introduction

A recurring theme throughout this book is the importance in the 21st century of materials in the lives of people and national economies. But very little has been said on the preparation of materials and the time lag from isolation of new materials to commercial exploitation. Hence while solid-state lighting, in particular the use of light-emitting diodes, is used in everyday life for home use and in offices and also for light sources in electronic displays, the underlying physical process, electroluminescence, was discovered in 1907. Certain materials have been described as 'wonder materials', for example high-temperature ceramic superconductors that were discovered in 1986 and graphene that was isolated in 2004; the promise of successful exploitation of 'wonder materials' have not been realised at the time of writing. Some materials, particularly certain metals, are designated as critical materials and their long-term supply cannot be guaranteed. There is a drive to use environmentally friendly methods to produce materials by the use of 'green chemistry'. Without the ability to produce materials with exceptional chemical purity the age of digital electronics in the 21st century would not have been possible. In this chapter some aspects of material preparation are discussed for selected entries in the Glossary.

8.2 Critical materials

Critical materials are those that are essential for national security or important economically and Table 8.1 lists materials that are considered to be critical by the U.S. government. Table 8.2 lists the amounts of some critical elements used in mobile phones.

Indium is a critical material and it has an important role as a transparent electrically conducting electrode especially for portable devices where it is coated onto the back surface of the glass display. Worldwide production of indium was around 640 tonnes per year (in 2011). Neodymium, a component of magnets used in wind turbines, is also a critical material. It may be necessary to consider carefully

Materials for the 21st Century. David Segal.
© David Segal 2017. Published 2017 by Oxford University Press.
DOI: 10.1093/oso/9780198804079.001.0001

Table 8.1 *The U.S. list of critical materials*

Antimony	Beryllium	Bismuth
Cerium	Chromium	Cobalt
Dysprosium	Erbium	Europium
Gallium	Germanium	Indium
Iridium	Lanthanum	Lithium
Manganese	Neodymium	Osmium
Palladium	Platinum	Praseodymium
Rhodium	Ruthenium	Samarium
Scandium	Tantalum	Tellurium
Terbium	Thulium	Tin
Tungsten	Yttrium	

Source: Ashby, M. F., D. F. Balas and J. S. Coral. *Materials and Sustainable Development*. Oxford: Butterworth-Heinemann, 2015.

Table 8.2 *Concentrations of five critical elements in mobile phones*

Critical elements	Grams per tonne in mobile phones
Platinum	70
Gold	140
Silver	1,300
Cobalt	19,000
Copper	70,000

Source: Ashby, M. F., D. F. Balas and J. S. Coral. *Materials and Sustainable Development*. Oxford: Butterworth-Heinemann, 2015.

the availability of critical materials as the 21st century progresses, especially as the number of electronic devices increases.

8.3   Pure materials

Silicon of exceptional chemical purity is required for fabrication of silicon chips; otherwise, impurities will have a deleterious effect on the electronic properties of the semiconductor and semiconductor grade silicon is made in the Czochralski process. Here a single crystal is pulled from a melt using a rotating seed crystal and the crystal is then cut into slices or wafers for use in semiconductor fabrication. In recent years, ionic liquids have been considered as solvents for purification of elemental materials such as sulphur, phosphorus, selenium and tellurium.

8.4   Fine powders

Interest in the use of quantum dots for solid-state lighting has highlighted the requirement to produce powders with sizes around 1 μm or less, in particular with sizes in the nanometre range. Two methods with the potential to produce small particle sizes are the following:

(a) Hydrothermal synthesis—**Hydrothermal synthesis** involves heating reactants often metal salts or oxides or hydroxides as a solution or suspension in a liquid, usually but not necessarily water at elevated temperature and pressure up to around 573 K and 100 MPa. For non-oxide materials, chemicals such as thiourea or thioacetamide that decompose to give sulphur-containing gases can be used as reactants. Nucleation and growth processes takes place under hydrothermal conditions.

   *A **hydrothermal reactor** is similar to a pressure cooker.*

(b) Aerosol synthesis—Aerosols are colloidal systems and consist of liquid or solid droplets in a gaseous dispersion medium. Aerosol routes to powders fall into two categories. In the first, a supersaturated vapour of reactant is produced followed by homogeneous nucleation. The second method involves generation of liquid droplets by, for example, ultrasonic atomisation or electrostatic atomisation. The liquid droplets are then subjected to a heat treatment. The diameter of droplets generated by ultrasonic cavitation of a solution can be written in the form

$$d = 0.34(8\pi\gamma/\rho f^2)^{1/3}, \tag{8.1}$$

where d is the calculated droplet size, γ is the surface tension, ρ is the density of the solution and f is the ultrasonic frequency used for cavitation. A general problem associated with all methods for preparing powders is to obtain powders with a controlled particle size in an economic way.

8.5   Thin films in the semiconductor industry

The fabrication of integrated circuits frequently depends on the deposition of thin films onto a silicon wafer or injection of ions into the surface region of the wafer (Table 8.3). For example, sputtering, which is an example of physical vapour deposition (PVD), is carried out in a vacuum chamber. A gas such as argon at low pressure in the chamber is ionised and accelerated towards the material to be deposited and atoms from this material are dislodged and deposit onto the substrate. In this way, metallic coatings can be produced and sputtering is used to deposit interconnects onto silicon wafers. In physical vapour deposition a low-pressure vapour of the material to be deposited (e.g. a vapour of metal atoms) is

Table 8.3 *A range of techniques used for deposition of coatings*

Technique	Description
Chemical vapour deposition (CVD)	Gaseous reactants in a vacuum chamber adsorb onto a hot substrate and react to form a coating. The gaseous reactants can also undergo chemical reactions in the gas phase before deposition onto the substrate or react on the substrate after absorption. CVD is used in the semiconductor industry for introducing dopants into a substrate.
Metal-organic chemical vapour deposition (MOCVD)	Gas molecules such as trimethyl gallium are used to carry in metal elements at low temperature onto a surface.
Evaporative deposition	The target is melted in a vacuum and vapour then deposits onto the substrate.
Ion implantation	Energetic ions are injected into the surface of a solid target to modify or change the chemical composition near the surface of the substrate. Ion implantation is used in the semiconductor industry to introduce dopant atoms into a semiconductor and is also used to modify the hardness of metals.
Physical vapour deposition (PVD)	A range of techniques in which the material to be deposited (the target) is vaporised and the vapour is then deposited onto the substrate.
Plasma spraying	A thermal spraying technique where a direct current gas plasma is generated between a water-cooled tungsten cathode and a water-cooled copper anode. The temperature in the plasma can be around 5000°C. Powder particles in the carrier gas are fed through the plasma, partially melt and deposit onto the cold substrate. Plasma spraying is used to deposit thermal barrier coatings, for example yttria-stabilised zirconia or alumina onto turbine blades and coatings of hydroxyapatite onto orthopaedic implants. Coatings made by plasma spraying tend to be porous. Plasma spraying is carried out at atmospheric pressures although a variant is known as vacuum plasma spraying where a reduced atmosphere is present which affects the microstructure of the deposited coatings, such as their density.
Plasma-enhanced chemical vapour deposition (PECVD)	A variation of the CVD process, chemical reaction rates are enhanced in the presence of a plasma that is generated by application of a radiofrequency field to a low-pressure gas. Reactants collide with electrons in the gas phase, producing energetic species through ionisation and these species adsorb onto the substrate.

PECVD is used in the semiconductor industry and is a lower-temperature process than CVD. It is suitable for substrates that do not have chemical stability at high temperature. PECVD has been used for producing diamond-like coatings (DLC), which are hard coatings with high electrical resistivity and a low coefficient of friction.

Pulsed laser deposition	A form of physical vapour deposition in which the target material is evaporated by a focused pulsed laser beam.
Radiofrequency (RF) sputter deposition	This technique uses an alternating current source compared to a direct current source in the sputtering process. It is useful for deposition of non-conducting target materials such as ceramics.
Radiofrequency (RF) magnetron sputter deposition	The efficiency of RF sputter deposition is enhanced in the presence of a moving magnetic field.
Sol-gel processing	A wet chemical technique for preparation of ceramic materials in the form of coatings, powders, fibres and monoliths. Starting reactants are either aqueous colloidal dispersions of oxides or solutions of metal alkoxides. For both types of reactants, a gelation process is promoted followed by a calcination step to yield the oxide. Sol-gel processes, particularly when based on aqueous sols, have been used to fabricate porous coatings of catalyst support materials, for example alumina supports. Anti-reflectance coatings for architectural glass have been manufactured using alkoxide solutions. Coatings can be deposited by dipping, spinning and spraying.
Sputtering	A type of PVD process carried out in a vacuum chamber. Atoms are dislodged from the target by an ionised gas such as argon in a vacuum chamber and deposit onto the substrate.
Thermal spraying	Powders are carried in a gas stream through a plasma and deposit onto a substrate. Thermal spraying includes the techniques of plasma spraying.

produced in a vacuum chamber and the vapour condenses onto the substrate. Ion implantation is a method for injecting energetic ions into the surface region of a target to modify or change the surface composition near the surface. Modifications can occur in mechanical properties such as hardness, stiffness and corrosion behaviour and in the case of integrated circuits electronic properties can be modified as implanted elemental atoms can be concentrated in a specific region of the surface. In chemical vapour deposition (CVD) the material to be deposited is present as a gas and heated in a vacuum chamber in the presence or absence of other reactants. Chemical reactions can take place between gas-phase species and the product of the reaction then deposits onto the substrate. Metal-organic vapour chemical deposition (MOCVD) is a variant of chemical vapour deposition. In electrodeposition, coatings are electrochemically deposited from an electrolyte.

8.6   Synthetic polymers

Detailed methods for the preparation of polymers are given in reference books but comments on several general methods are given here. Aliphatic polyamides are known as nylons that are condensation polymers. In the latter, two monomers A and B react to form a product with the repeat unit A-B in a **step-growth** polymerisation.

   Step-growth *means monomers added to a growing polymer chain.*

There are many types of nylon and one type, nylon-6, 6, is made by the condensation of adipic acid with hexamethylene diammine in an interfacial reaction. (The nomenclature for nylon 6, 6 means that there are six carbon atoms between the amine groups for both monomers.) The initial product is a crystalline salt, hexamethylene diammonium adipate, that undergoes polymerisation with the repeat unit

$$-(-CO - (CH_2)_4 - CO - NH - (CH_2)_6 - NH-)- \qquad (8.2)$$

Aromatic polyamides are known as aramids and an example is Kevlar (poly (p-phenylene terephthalamide)) prepared by a condensation reaction between terephthaloyl chloride and p-phenylene diamine to give the repeat unit

$$-(-NH - C_6H_4 - NH - CO - C_6H_4-CO-)- \qquad (8.3)$$

In contrast with condensation reactions, in emulsion polymerisation monomer droplets are dispersed in an aqueous medium containing a surface active agent that forms micelles and also contains a solution of an initiator such as potassium persulphate. Monomer molecules from the droplets and initiator diffuse through the aqueous environment into the micelles where polymerisation takes place eventually, leading to a latex dispersion.

8.7   Polymerase chain reaction

The polymerase chain reaction, which was invented by Kary Mullis, is used to amplify, that is copy, a strand of deoxyribonucleic acid (DNA) *in vitro* so that millions of copies can be produced. Single strands of DNA are separated by heating duplex DNA, after which chemically synthesised DNA primers are added. The primers are oligonucleotides containing about twenty base pairs and attach themselves to priming sites at each end of the separated DNA strands so that they flank the section to be amplified. The process of attachment is known as hybridisation. A heat-sensitive enzyme, DNA polymerase, for example *Thermus aquaticus* (known as *Taq* polymerise), enzymatically replicates the selected section using four added deoxynucleotide triphosphates (dNTPs) namely dATP, dGTP, dCTP and dTTP. The primer is required as DNA **polymerase** cannot by itself initiate new chains.

*Note that names for enzymes end in **-ase**.*

In practice the **amplification** process requires cycles of heating and cooling, the first to separate the duplex DNA and the second to carry out the amplification step.

*In this context, **amplification** means copying.*

The process is automated and if there are N cycles of heating and cooling the number of copies equates to 2^N so that millions of copies can be produced. The polymerase chain reaction can be viewed as a method for polymer synthesis and has various applications. It is particularly associated with forensic science in the techniques known as DNA fingerprinting.

8.8 Fibres

Many materials are in the form of fibres and examples are shown in Table 8.4. Oxide fibres have been prepared by 'blowing' from a melt and these so-called blown fibres (e.g. aluminosilicates) have applications as thermal insulation but the ability to blow fibres from a melt is limited by the viscosity of the molten oxide. Continuous oxide fibres such as those based on alumina and silica have been

Table 8.4 *Examples of materials in fibrous form*

Subject area	Examples of fibre
Foods	Candy floss (cotton candy) made by spinning
Medicine	Amyloid fibrils in Alzheimer's disease
Oxide fibres	Glass fibres from a melt by spinning Oxide fibres such as Nextel from solution precursors
Non-oxide fibres	Continuous silicon carbide fibres from polycarbosilanes Carbon fibre from carburisation of polyacrylonitrile fibres (PAN)
Polymers	Continuous Kevlar fibres made by spinning Polyester fibres by spinning

prepared from solution precursors in sol-gel processes, for example Nextel. Oxide fibres have been used as thermal insulating tiles on the Space Shuttle. Non-oxide fibres, for example silicon carbide, have been made by spinning solutions of **polycarbosilanes** that are then pyrolysed in an inert atmosphere while carbon fibres are obtained on carbonisation of continuous polyacrylonitrile fibres (PAN) in a controlled temperature range in an inert atmosphere.

> **Polycarbosilanes** *contain carbon–silicon bonds in the polymer backbone.*

Applications for non-oxide fibres include carbon–carbon composites for brake pads in aircraft and racing cars and silicon carbide–silicon carbide composites for protection of hot components of spacecraft. Textile fibres can be spun from solutions of synthetic polymers, for example polyester and polyacrylonitrile, the latter known as **acrylic fibres**, while Kevlar is produced from spinning continuous fibres from solutions of polymer in concentrated sulphuric acid. When fibres of a polymer are produced, they are often placed under a tensile stress to orientate molecules in the damp fibre in the spinning process in order to enhance the strength of the fibres.

> **Acrylic fibre** *are fibres that are derived from acrylonitrile monomers.*

In a completely different area, protein fibres known as amyloid fibres are associated with Alzheimer's disease. Proteins have a secondary and tertiary structure in which the protein chains can form helical structures (α-helix) and folded β-sheets. While folding of proteins into compact structures is essential for their normal functioning, in Alzheimer's disease mis-folding of proteins occurs, resulting in aggregation of protein fibrils to form plaque in the brain with devastating consequences for the patient's health. The search for solutions to Alzheimer's disease has become a pressing medical quest in the 21st century.

8.9   Summary

While materials are important in the lives of people and nations, some materials known as critical materials are essential for national security or important economically. Indium is an example of one such material. The ability to prepare materials with high chemical purity is crucial in the semiconductor industry that underpins the manufacture of electronic devices. In the biological sciences, the polymerase chain reaction, which is a way of amplification or copying DNA sequences, is particularly associated with forensic science, in particular DNA fingerprinting. Many materials are often used in the form of powders. In the case of powders, the ability to reproduce them with a controlled size is important particularly for quantum dots.

8.10 Further reading

Agrawal, C. M., J. L. Ong, M. R. Appleford and G. Mani. *Introduction to Biomaterials: Basic Theory with Engineering Applications*. Cambridge: Cambridge University Press, 2014.

Ashby, M. F., D. F. Balas and J. S. Coral. *Materials and Sustainable Development*. Oxford: Butterworth-Heinemann, 2015.

Askeland, D. H., and W. J. Wright. *The Science and Engineering of Materials*, 7th edn. Boston, MA: Cengage Learning, 2016.

Earle, M. J., E. Boros, K. R. Seddon, M. A. Gilea and J. S. Vyle. Elemental solvents, United States Patent Application 2010/0178229, 2010.

Fried, J. R. *Polymer Science and Technology*, 3rd edn. Upper Saddle River, NJ: Prentice Hall, 2014.

Lang, R. J. Ultrasonic atomization of liquids. *Journal of the Acoustic Society of America* (1962) 34, 6.

Page, T. J. Ion implantation. In *Concise Encyclopaedia of Advanced Ceramic Materials*, edited by R. J. Brook, pp.252–7. Oxford: Pergamon Press, 1991.

Papachristodoulou, D., A. Snape, W. H. Elliott, and D. C. Elliott. *Biochemistry and Molecular Biology*, 5th edn. Oxford: Oxford University Press, 2014.

Segal, D. *Chemical Synthesis of Advanced Ceramic Materials*. Cambridge: Cambridge University Press, 1989.

9

Disruptive Technologies

9.1 Introduction

The expansion of the Internet in the past twenty years has seen many businesses, both established and new, migrate to the Internet in order to develop digital formats. Some of these businesses have been disruptive in the sense that they have impacted on existing businesses and business models. For example, electronic mail (email) has resulted in a decrease in the numbers of letters delivered manually, electronic books have affected sales of paper books, online banking has removed the necessity for having physical branches while online shopping allows peoples to shop from home without visiting shops. Online services have also created job opportunities and while the number of letters that are delivered may be decreasing, goods that are purchased online must be delivered by delivery drivers or by existing or new postal services. The ability to purchase and download music files or hear music streamed on specific websites has over the years significantly reduced the sales of conventional records. One area where competition has been particularly acrimonious relates to the use of taxis. In a city such as London there has always been competition between conventional taxi drivers and minicab drivers but in recent years a new way of requesting a vehicle for transport has increased in popularity in London and in cities across the world. The mobile phone app from the company Uber has the potential to affect the livelihoods of existing drivers. Hence disruptive technologies are proliferating across business areas fuelled by the increasing use of the Internet and ways of carrying out transactions that were unavailable just a few years ago. Disruptive technologies also occur in the area of materials and receive much attention in the media although it can be difficult for the layman and also the specialist to ascertain whether the publicity is 'hype' or genuine. When assessing the claims made for disruptive technologies it is appropriate to recall the words spoken by the White Queen to Alice in the book *Through the Looking Glass* by Lewis Carroll: The rule is, jam tomorrow and jam yesterday—but never jam today. Examples of potentially disruptive material technologies are described in this chapter.

Materials for the 21st Century. David Segal.
© David Segal 2017. Published 2017 by Oxford University Press.
DOI: 10.1093/oso/9780198804079.001.0001

9.2   Disruptive technologies

Gene editing, also known as genome editing, allows the introduction of deoxyribo-nucleic acid (DNA) sequences into a gene to correct a genetic defect. Its formal name is clustered regularly interspaced short palindromic repeats, or CRISPR, in which a cutting enzyme known as CAS9 can target a region of the genome with precision, remove a DNA fragment and insert a different DNA sequence into the genome. Thus, the technique, more formally CRISPR/CAS9, is an editing tool that involves cut and paste processes for selected DNA targets. While a potentially disruptive technology there are ethical questions associated with it, because while editing DNA in humans could provide cures for diseases, changes to the human genome could be introduced that would be passed from generation to generation. CRISPR is not the only gene-editing technique and other methods include zinc finger nucleases (ZFNs) and transcription activator-like effector nucleases (TALENs).

It may not be obvious why CRISPR is mentioned in a monograph on materials. Its presence arises because DNA can be considered to be a polymer, in particular a heteropolymer, and polymers generally, such as nylon, polyester and polytetra-fluoroethylene, are considered to be materials although these are synthetic materials.

Three-dimensional (3D) printing has attracted much publicity in recent years and has been described as a disruptive technology. It allows manufacture of an object to be made layer by layer out of plastics, ceramics or metal powders and also by use of living cells. When the latter are used, there is the long-term potential for producing artificial organs and this approach is sometimes referred to as bioprinting. 3D printing is an additive manufacturing method or rapid prototype technique and a feature of the process is that there is minimal wastage of material compared to conventional subtractive manufacturing techniques such as milling and drilling. It is particularly suited for high-value, low-volume specialist products such as turbine blades made out of nickel-based superalloys that have intricate assemblies of cooling channels and also for metallic implants for patients undergoing reconstruction surgery. However, 3D printing is not a new technique and its origins reside in stereolithography, a process in which a photopolymerisable liquid is cured layer by layer by the controlled movement of a laser across the liquid surface to build up the object.

9.3   Wonder materials

The discovery of high-temperature ceramic superconductors in 1986 resulted in unprecedented worldwide research interest and the materials were frequently described in national newspapers. The almost frenzied interest arose from potential applications and the temperature at which they develop superconductivity, that is zero electrical resistance, which was around liquid nitrogen temperatures. Up until their discovery superconductivity in metals was observed around liquid helium temperatures and in fact the large magnetic fields generated in magnetic resonance imaging (MRI) that is widely used nowadays involve coils of metallic alloys immersed in liquid helium. The latter is more expensive than liquid nitrogen. Potential applications for ceramic superconductors included magnetically levitated trains, power transmission without energy losses and compact, powerful and efficient electric motors. The new materials were truly 'wonder materials'. However, since their discovery it is fair to say that these future applications have not yet been realised.

Graphene, a two-dimensional sheet of carbon that is one atomic layer thick, has attracted much publicity in the public domain and was isolated in 2004 by Sir Andre Geim and Sir Konstantin Novoselov at the University of Manchester in the United Kingdom. It also has been described as a wonder material. Graphene is the thinnest known material and has a hexagonal array of carbon atoms. The excitement surrounding graphene is related to its potential applications based on its electronic and mechanical properties. Hence, some applications include as components of electrodes in fuel cells, lithium-ion batteries and supercapacitors; conductive transparent coatings to replace indium tin oxide coatings in electronic displays; integrated circuits; field effect transistors; solar cells; flexible display devices; protective coatings and fillers or reinforcements for composites. A sheet of graphene that is rolled up into a cylinder forms a carbon nanotube. Whether graphene lives up to its description as a wonder material remains to be seen.

Another class of material that is mentioned here relates to auxetic materials although these have never been referred to as wonder materials. The inclusion relates to their negative Poisson ratio, which means that the materials become thicker in the direction perpendicular to the direction of stretch, which is a counter-intuitive concept. Certain alloys of copper and brass exhibit auxetic behaviour as do some polymer foams.

The interaction of electromagnetic radiation with materials occurs in a number of ways. Hence, MRI involves radio waves, radar utilises microwaves and the optical properties of stained glass windows relies on the interaction of visible light with small metallic particles such as gold. Metamaterials are a class of materials that have not been referred to as wonder materials but have intriguing properties that would probably qualify for their description as wonder materials. The refractive index, n, of a material at a particular frequency can be expressed by the equation

$$n = (\varepsilon\mu)^{1/2}, \tag{9.1}$$

where ε is the permittivity and μ is the permeability. The refractive index is often assumed to be positive. A theoretical study by Viktor Veselago that was published initially in Russian in 1967 discussed the properties of a material that would arise if both the permittivity and permeability were negative. He concluded that the material would have a negative refractive index and such a material is known as a metamaterial. However, this study remained largely unknown for many years until the early years of the 21st century and interest in metamaterials has grown since then. Metamaterials are artificial materials in the sense that they do not occur in nature and have the potential ability to render objects invisible by deflecting radiation around them and generating an invisibility cloak as associated with the fictional character Harry Potter in the novels by J. K. Rowling. Metamaterials have been constructed by using elements smaller than the wavelength of the electro-magnetic wave in which a combination of metallic rods and **split-ring resonators** are used in combination to modify the permittivity and permeability, respectively. Clearly, as the wavelength decreases, the dimensions of the rods and resonators need to decrease also. The split-ring resonators are analogous to the resonators used for the magnetron cavity that had a central role in the development of radar. Metamaterials have the potential to be used as flat lenses with improved optical properties compared to those of existing lenses.

> *A **split-ring resonator** consists of two concentric rings with gaps on opposite sides and has an inductance and capacitance. They were invented in 1981 and a detailed description is given in W. H. Hardy and L. A. Whitehead, Split-ring resonator for use in magnetic resonance from 200–2000 MHz,* Review of Scientific Instruments *(1981) 52, 213–16.*

It is envisaged that other wonder materials will be prepared throughout the 21st century that will generate excitement and promises of widespread application. A quotation from the eminent physicist Ernest Rutherford made to the British Advancement of Science in 1933 is relevant when considering the exploitation of wonder materials: 'We cannot control atomic energy to an extent which would be of any value commercially, and I believe we are not likely ever be able to do so.'

9.4 Summary

Developments on materials do not usually receive much publicity in the public domain. However, the technique of three-dimensional printing and very recently the analytical method known as CRISPR have attracted much attention. Both of these techniques represent disruptive technologies; the first relates to the manu-facture of components while CRISPR has profound implications on the sequen-cing of the human genome. High-temperature ceramic superconductors and graphene have received much publicity and been described as wonder materials because of their potential applications, but their promises have yet to be realised. Auxetic materials and metamaterials have not been described as wonder materials but have intriguing properties that justify their description as wonder materials.

9.5 Further reading

Anonymous. CRISPR. In *Oxford Dictionary of Biology*, ed. by R. S. Hine, 7th edn, p.148. Oxford: Oxford University Press, 2015.

Boot, H. A. H., and J. T. Randall. Historical notes on the cavity magnetron. *IEEE Transactions on Electronic Devices* (1976) 23(7), 724–9.

Carmichael, H. Gene genius. *Chemistry and Industry* (Oct. 2015) 79, 32–5.

King, A. Gene editing fears. *Chemistry and Industry* (Nov. 2015), 79, 6.

Megget, K. The cutting edge of gene editing. *Chemistry World* (Feb. 2016) 13, 62–6.

Narlikar, A. V. *Superconductors*. Oxford: Oxford University Press, 2014.

Papachristodoulou, D., A. Snape, W. H. Elliott and D. C. Elliott. *Biochemistry and Molecular Biology*, 5th edn. Oxford: Oxford University Press, 2014.

Parrington, J. *Redesigning Life: How Genome Editing Will Transform the World*. Oxford: Oxford University Press, 2016.

Qiu, X. G. (ed.). *High-Temperature Superconductors*. Cambridge: Woodhead, 2011.

Sachs, E. M., A Curodeau, T. Fan, J. F. Bredt, M. Cima and D. Brancazio. Three dimensional printing system. United States Patent 5,807,437, 1998.

Solymar, L., and E. Shamonina. *Waves in Metamaterials*. Oxford: Oxford University Press, 2014.

Veselago, V. G. The electrodynamics of substances with simultaneously negative values of ϵ and μ. *Soviet Physics-Uspekhi* (1968) 10, 509–14.

Warner, J. H., F. Schaffel, A. Bachmatiuk and M. H. Rummeli. *Graphene: Fundamentals and Emerging Applications*. Amsterdam: Elsevier, 2013.

10

The Importance of Microstructure on Material Properties

10.1 Introduction

The colours observed in the wings of butterflies can be an enchanting sight but it may come as a surprise to those observers that the wings of butterflies are in fact colourless and do not contain pigments. A vase when dropped onto the floor can break into many pieces while a plastic, Kevlar, can stop a bullet when used in protective body vests and molluscs have shells that behave as ceramic armour. The behaviour of materials is affected by a property known as microstructure and just as superhydrophobicity is a property of surfaces rather than material composition, two identical material compositions can have different microstructures and hence different properties. Plastics, ceramics and metals are characterised by their microstructures and many living organisms have complex microstructures in their bodies; in many materials microstructures are often associated with the concept of toughness. Microstructural control can, in principle, allow the properties of materials to be tailored. In this chapter a brief overview of the concept of microstructures is given with particular reference to entries mentioned in the Glossary.

10.2 Microstructure: a definition

The properties of a material are determined by its microstructure, that is, the nature, quantity and distribution of structural elements or phases that make up the material. These properties include optical properties, strength, toughness, stiffness (Young's modulus), hardness, elasticity and electronic properties such as conductivity.

10.3 Toughness

Consider ceramic systems. They are strong under compression, a property used in steel reinforcement of concrete structures, but weak in tension and are brittle

Materials for the 21st Century. David Segal.
© David Segal 2017. Published 2017 by Oxford University Press.
DOI: 10.1093/oso/9780198804079.001.0001

materials. A. A. Griffith explained the low tensile strengths of brittle solids compared to theoretical values who considered growth of a crack with an elliptical cross-section through such a material. He showed that when the surface energy for formation of new fracture faces was balanced by release of elastic strain energy then the tensile fracture strength, S, could be written as

$$S = (2E\gamma/\pi C)^{1/2}, \tag{10.1}$$

where E is Young's modulus for the material, γ the surface energy and $2C$ the crack length corresponding with the major elliptical axis. The fracture toughness or stress intensity factor, K_{IC}, for a solid was defined as

$$K_{IC} = (2E\gamma)^{1/2} \tag{10.2}$$

when the energy balance used for deriving Eq. (10.1) is applicable. Fracture in ceramics occurs after crack initiation and growth until a critical size is reached, several tens of micrometres. Applied forces are concentrated up to 1000 times at the crack tip, sufficient for rupturing atomic bonds in the solid. The significance of toughness to materials is that while a ceramic cup or saucer may break when dropped onto the floor, the toughness of ceramic and metallic matrices can be increased through reinforcement with fibres through changes in microstructures. Ductile metals can absorb around 100,000 times more energy when a crack grows through the material compared to brittle ceramics, and approaches to increase the ability of ceramics to absorb energy of an advancing crack involve the development of fibre-reinforced matrices or particle-reinforced materials.

10.4   Composite materials

Composite materials or ceramic matrix composites are widely used. Some examples include silicon carbide fibre/silicon carbide composites (SiC/SiC) as potential replacements for superalloys in gas turbines due to their higher operating temperatures, alumina fibres in an alumina matrix, glass fibres in a a polyester resin known as Fibreglass, glass-fibre felts and mats impregnated with epoxy resins for fairings and trailing edge panels on the Boeing 747 and carbon fibre/carbon composites for brake discs.

*In a **composite**, one component, for example fibres, is dispersed into a medium known as the matrix.*

A feature of all of these systems is that fibres deflect cracks and increase the fracture toughness, thus limiting crack propagation and fracture. Many living organisms contain intricate structures of minerals obtained by a mineralisation process from soluble species interspersed with proteins. These structures are often natural composites whose toughness and strength help to protect the organism. For example, egg shells contain ordered crystals of calcium carbonate in a matrix of natural polymers and the skeletons of sponges comprise silica rods. A composite structure of alternating layers of protein and calcite (calcium carbonate) in the abalone shell has a high fracture toughness compared to calcite alone and helps to protect the mollusc from predatory attack. In terms of nomenclature, biomaterials refer to materials found in living or dead organisms while biomimetic materials are materials made in the laboratory with the aim of utilising the structures of naturally occurring materials in organisms in order to develop improved material properties. Some fibres contain liquid crystals. For example, the aramid fibre Kevlar contains polymer molecules that are closely aligned in the fibre with hydrogen bonding between fibres. This combination of closely packed fibres with attractive forces between fibres contributes to the strength of the material.

Transformation toughening is another way of controlling the microstructure to tailor the properties of materials and is associated with the dispersion of fine zirconia particles in an alumina matrix. Toughening is associated with the phase change from the tetragonal to monoclinic phase of zirconia particles dispersed in a brittle ceramic. The volume and shape changes associated with the martensitic transformation of discrete zirconia particles reduce the stress intensity at the crack tip. Although the examples shown here relate to ceramics, microstructural control is just as important in metals for tailoring material properties.

10.5   Opals, butterflies and photonic crystals

The gemstone opal has an opalescent milky-white appearance and a range of colours due to impurities. Its microstructure consists of a three-dimensional array of close-packed spheres of silica in a periodic structure. Wings in a butterfly contain structures of small protein struts interspaced with small air gaps in periodic structures. The structures in opal and butterflies act as diffraction gratings for incident light that gives rise to optical interference and the colours seen by observers. These periodic structures are characteristic of photonic materials in which optical diffraction occurs with visible light within the materials in which the periodicity is on the order of the wavelength of light. Photonic materials are also known as photonic crystals and the latter have a photonic gap that is analogous to the bandgap in semiconductors. Photons that have energies within the photonic gap cannot propagate through the crystal and are confined to defect regions. Hence, photonic crystals act as waveguides for transporting light around corners as light is repelled from the bulk crystal. Interference takes place between diffracted components of visible light. Photonic crystals can be made in the laboratory, for example by forming a close-packed structure of silica spheres around 1000 nm in diameter. Mica is an aluminosilicate material and mica flakes with a thickness around 300–600 nm, when coated for example with titanium dioxide, are used as an iridescent pigment, for example as an 'effect' pigment in car paints. In these pigments which are not photonic crystals, interference between components of incident light gives rise to the colours.

10.6   Traditional ceramics

Chemical reactions occur between reactants during heating in a kiln in the manufacture of traditional ceramics such as tableware. The microstructure of cups and saucers as examples consist of crystalline material such as mullite, pores and glassy phases that act as a glue to hold components together. The presence of pores lowers the strength of the products, which helps to explain why tableware breaks easily when dropped. In contrast with the porosity in tableware, porcelain such as Meissen porcelain is dense. A consequence of the absence or presence of porosity is that glazes are used on tableware before applying decoration and to prevent ingress of water into the body of the ceramic, whereas glazes are not necessarily required for porcelain.

Macro-**defect**-free cement, unlike Portland cement, has a lower fraction of voids (i.e. pores) and the resulting microstructure enables coiled springs to be produced. Whereas conventional cements contain pores with sizes around 1 mm, the pore size in macro-defect-free cements is around a few micrometres. The effect of this drastic reduction in pore size is reflected in a comparison in the values of strength for conventional cements that are around 5 MPa compared to 150 MPa for macro-defect cements. Hence it can be noted that the method of processing material can affect their properties.

Defect *refers to pores or impurities.*

10.7   Metamaterials

Metamaterials have an engineered microstructure or nanostructure where the structure does not occur in nature and the structure has a periodicity. They can be referred to as artificial materials. Metamaterials have a negative refractive index and have the ability to guide electromagnetic radiation, visible or other wavelengths, for example microwaves around an object rendering it invisible. While an 'invisibility cloak' can be produced in theory and has been much publicised by the fictional character Harry Potter, such cloaks have not been made in practice.

10.8   Shape memory alloys

Coronary stents made out of a nickel–titanium alloy known as Nitinol are used for implantable coronary stents. The key microstructural feature of this alloy that makes them stable for stents is a reversible phase transition from the martensitic to the austenitic phase at around body temperatures.

10.9 DNA

There is much interest in the 21st century on personalised medicine that involves sequencing the genome of an individual, that is, in determining the sequence of four bases, adenine, cytosine, guanine and thymine in the polymer chain of the nucleic acid. In the double helical structure of deoxyribonucleic acid (DNA), hydrogen bonds link adenine to thymine and cytosine to guanine. The linkages can be considered to be essential in forming the microstructure of DNA, that is, in promoting the helical structure.

10.10 Optical properties

Optical properties of materials such as metals and quantum dots depend not just on their composition but on their particle size. The optical properties of small particles such as gold are utilised in stained-glass windows. Quantum dots behave as nanophosphors where the wavelength of emitted light depends on the particle size, moving to longer wavelengths as the particle size increases as a result of quantum effects known as **quantum confinement**.

Quantum confinement *means the electronic wave function is confined to small regions of space.*

10.11 Summary

The properties of a material are determined by its microstructure, that is, the nature, quantity and distribution of structural elements or phases that make up the material (Table 10.1). The microstructure affects material properties such as strength and toughness. The toughness of composites is increased when fibres are incorporated into the matrix. Shape memory alloys, DNA, metamaterials, traditional ceramics and photonic crystals are all characterised by microstructures.

Table 10.1 *Properties of materials that can be affected by microstructure*

Material property
Creep
Electrical conductivity
Fracture strength
Fracture toughness
Hardness
Stiffness
Thermal conductivity
Thermal expansion

A common example of how microstructure affects material properties relates to the colour of the wings of butterflies that do not contain pigments and the opalescent nature of opal gemstones that contain periodic arrays of silica spheres.

10.12 Further reading

Bonsal, N. P., and J. Lamon (eds). *Ceramic Matrix Composites: Materials, Modeling and Technology*. Chichester: Wiley, 2015.

Brook, R. J. (ed.). *Concise Encyclopaedia of Advanced Ceramic Materials*. Oxford: Pergamon Press, 1991.

Davidge, R. W. *Mechanical Behaviour of Ceramics*. Cambridge: Cambridge University Press, 1979.

Griffith, A. A. The phenomena of rupture and flow in solids. *Philosophical Transactions of the Royal Society of London* (1920) A221, 163–98.

Lee, M. (ed.). *Remarkable Natural Material Surfaces and Their Engineering Potential*. New York: Springer, 2014.

Novotny, L., and B. Hecht. *Principle of Nano-Optics*, 2nd edn. Cambridge: Cambridge University Press, 2006.

Ozin, G. A., A. C. Arsenault and L. Cademartiri. *Nanochemistry: A Chemical Approach to Nanomaterials*. Cambridge: Royal Society of Chemistry, 2009.

Segal, D. *Chemical Synthesis of Advanced Ceramic Materials*. Cambridge: Cambridge University Press, 1989.

Solymar, L., D. Walsh and R. A. Syms. *Electrical Properties of Materials*, 9th edn. Oxford: Oxford University Press, 2014.

11

Patents, Patent Trolls and Intellectual Property

'When I use a word, it means just what I choose it to mean, neither more nor less.'

—*Humpty Dumpty, in* Through the Looking Glass, *by Lewis Carroll.*

11.1 Introduction

When companies, government organisations, universities and individuals spend time and money on developing inventions they will want to protect their inventions from being copied and will hope to obtain a financial return on their investment. This can be achieved by building up a portfolio of intellectual property and patents; a form of intellectual property is particularly relevant to activities on materials. This chapter gives guidelines to readers who may not be familiar with patents but it should be stressed that anyone contemplating taking out patents to protect their inventions should seek professional advice from qualified specialists such as patent attorneys as I am not qualified to offer professional advice on patents.

11.2 Patents

A patent is a legal document and conveys a right to its owner, that is, the owner of the invention described in the patent. This right allows the owner to prevent others from practicing the invention, including manufacturing or importing goods, for sale into a country where the patent is enforced for a limited period of time, usually twenty years from the filing date for the patent. The patent owner does not have an automatic right to exploit the invention. The patent is a contract between the patent owner and a national government and in exchange for the right to exclude others from practicing the invention, the patent owner has to disclose the details of the invention in the public domain. Thus, the applicant is given a monopoly for a limited period of time. There are no world patents and a patent is validated in

Materials for the 21st Century. David Segal.
© David Segal 2017. Published 2017 by Oxford University Press.
DOI: 10.1093/oso/9780198804079.001.0001

individual countries. Patents are also a vast source of technical information; for example, there are approximately 9,000,000 granted United States patents and the first United States patent was granted on 31 July 1790. There is no restriction on the length of a patent and the language used in patent documents can appear to be arcane and sentences can be long. One patent describes one invention and, in general, patents describe a product, a method for making the product or a system. As a form of property, patents can be sold, licensed or allowed to lapse so that anyone can practice the invention. The quotation from *Through the Looking Glass* at the beginning of this chapter can seem relevant to the interpretation of patent documents because words and phrases in a specification do not have to have the same meaning they have in everyday life. Their meaning is determined by the way they are defined in the specification.

All patent documents have the same structure:

(a) Front page—The title, abstract, name(s) of inventor(s), applicant (the owner), patent classification, priority date, filing date and application date.

(b) Background to the invention—Description of the prior art, that is, earlier published documents of relevance to the invention. This section is a good source of references when finding out about a new field of technology.

(c) Summary of the invention—Brief description of the invention.

(d) Embodiments—Several ways of producing the invention, along with the preferred way.

(e) Drawings—Labelled drawings are supplied at the front or back of the document.

(f) Claims—The legal part of the patent document that describes the scope of the rights given to the applicant (owner). Independent claims are claims that do not refer to other claims. The first claim, namely claim 1, is always an independent claim. Other claims are known as dependent claims because they can refer to an independent claim or to other dependent claims. For example, the second claim, claim 2 in the list of claims, can be a dependent claim. Note that a patent document can contain more than one independent claim.

The description, drawings and claims for a patent application are the patent specification and only features described in the specification can be claimed.

Patent infringement can occur when someone has a product, method or system that is described in part in at least one claim of another patent. Two patent documents that have similar claims are said to potentially interfere with each other but patent documents do not infringe other patent documents.

The general approach to filing a patent document is outlined in the following stages:

(a) The specification is prepared and then filed or deposited at the United Kingdom Intellectual Property Office (UKIPO) if prepared in the United

Kingdom, or at the United States Patent and Trademark Office (USPTO) if prepared in the USA. A filing date is then issued by UKIPO. Additions, that is, new material to the specification, can be made within 12 months of filing and additions are given a new filing date. The applicant can request a search report from UKIPO and this report indicates prior art documents that may affect patentability.

(b) Twelve months after initial filing of the original specification, filings are combined and a decision is taken by the applicant as to which route to take in order to obtain granted patents. For example, the applicant may decide just to let the filing process proceed in the United Kingdom or file for an international application or a regional filing at, for example, the European Patent Office. Filings in individual countries such as the USA, China or Japan are made at this 12-month stage. Translations of patent documents in some countries are made at this stage.

(c) The patent application is published 18 months after the initial filing.

(d) Examination of the specification takes place at national patent offices over a period of several years, for example four years. In order to obtain a granted patent, the invention must be novel, which is not disclosed to the public at the time of filing the patent application. Also the invention must be inventive or non-obvious, so it could not be predicted how the invention works by consideration of published documents in the public domain before the filing date. In addition to being novel and inventive, the invention must be capable of industrial application.

(e) Grant and renewal fees are paid, usually each year, to keep the patent in force. The lifetime of a granted patent is usually twenty years from the filing date.

Patent infringement relates to manufacturing in or importing an invention into a country where a patent protects the invention. Lawsuits are held before a jury in the USA and before a judge in the United Kingdom. While traditionally lawsuits are brought about between two manufacturers, lawsuits brought about by non-practicing entities (NPEs) have been a feature of cases in recent years. NPEs are individuals or firms who own patents but do not directly use their patented technology to produce goods or services. Instead NPEs assert their patents against companies that do produce goods and services. NPEs are sometimes referred to as 'patent trolls', but when used in this way the phrase has derogatory overtones.

11.3 Patents and intellectual property

Patents are a form of intellectual property originating from creations of the mind. Other forms of intellectual property include copyright, trademarks, designs, databases and trade secrets. When a patent has lapsed and the limited monopoly collapses, a third party can manufacture the product; this approach happens in the pharmaceutical industry for generic drugs. When a patent lapses, a manufacturer can continue to use a **trademark** for brand recognition. As an example, although patent protection for Lycra has lapsed, the manufacturer can still use the trademark and although the identical product can be produced by third parties, the latter cannot use the trademark without permission.

A registered **trademark** *is represented by a superscript ® while an unregistered trademark is represented by a superscript* ™.

11.4   Patents as a source of technical information

Patents are a major source of technical information and often the results of scientific research are only published in the patent literature. Table 11.1 lists some patent numbers and the subject of the inventions in order to give an idea of the scope of technologies covered by the patent literature.

Table 11.1 *Patents as a source of technical information*

Patent number	Subject area	Year of grant
US 3633	Vulcanisation of rubber	1844
US 644,077	Acetylsalicylic acid	1900
US 942,699	Bakelite	1909
US 1,266,766	Cellophane	1918
CA 234336	Insulin	1923
US 1,811,959	Neoprene	1931
US 1,929,453	Polyvinyl chloride	1933
US 2,130,947	Nylon	1938
US 2,133,235	Glass fibre	1938
US 2,171,765	Polymethyl methacrylate	1939
US 2,277,013	Xerography	1939
US 2,230,654	Polytetrafluoroethylene	1941
US 2,752,339	Cortisone	1956
US 3,353,115	Ruby laser	1960
GB 854211	Hovercraft	1960
US 3,083,737	Velcro	1963
US 3,174,851	Nitinol	1965
US 3,541,541	Computer mouse	1970
US 3,691,140	Post-it notes	1972
US 3,922,464	Post-it notes	1975
US 3,671,542	Kevlar	1972

US 3,953,566	Gore-Tex	1976
US 4,253,132	Taser stun gun	1981
US 4,683,202	Polymerase chain reaction	1987
US 7,345,671	iPod	2008

11.5   Summary

Patents are a form of intellectual property and describe inventions. In order to obtain a granted patent the invention must be novel, inventive and capable of industrial application. A patent grants the patent owner a limited monopoly for exploiting the invention for a limited period of time in exchange for disclosing details of the invention in the public domain. Patents are a primary source of technical information; for example, there are around 9,000,000 granted United States patents. In recent years, a feature of lawsuits on patent infringement is the participation of non-practicing entities (NPEs).

11.6   Further reading

Currano, J. N., and D. L. Roth (eds). *Chemical Information for Chemists: A Primer.* Cambridge: Royal Society of Chemistry, 2014.

van Dulken, S. *Introduction to Patent Information*, 3rd edn. London: British Library, 1998.

Hopkins, S. Improvement in the making of Pot Ash by a new apparatus and process, United States Patent, X000001, 1790.

Jackson Knight, H. *Patent Strategy for Researchers and Research Managers.* Chichester: Wiley, 2013.

McManus, J. *Intellectual Property: From Creation to Commercialisation.* Oxford: Oak Tree Press, 2012.

Segal, D. *Exploring Materials Through Patent Information.* Cambridge: Royal Society of Chemistry, 2015.

Tyrell, J. A. *Fundamentals of Industrial Chemistry: Pharmaceuticals, Polymers and Business.* Chichester: Wiley, 2014.

12

Everyday Products

The Role of Materials

12.1 Introduction

The previous chapters have shown how materials are essential for the functioning of everyday life with applications in consumer goods, healthcare and foods as examples. Important materials may only form a small part of a product, for example a transparent indium tin oxide coating as an electrode in a smartphone. Many materials are associated with complex technology, for example superalloys and their use in gas turbine blades. However, in everyday life people routinely purchase products and these products often contain materials that have analogous properties to other materials that are used in what may be described as more advanced applications. In this chapter, some everyday products are described with respect to certain of their ingredients. However, the chapter does not include data sheets for all ingredients and the comments should not be interpreted as endorsement for any particular product.

12.2 Sunscreens

The function of a sunscreen is to prevent ultraviolet (UV) light from the sun from penetrating the skin and causing damage such as sunburn and skin cancer. There are two types of ultraviolet wavelengths: UV-A, which causes the skin to darken which is desirable, and UV-B, which produces red and raw skin which is undesirable. Ingredients of a sunscreen can absorb radiation that can cause a small amount of local heating but are associated with the presence of small particles of titanium dioxide or zinc oxide. These particles are nanoparticles, around 50 nm in diameter, and they can reflect and, more importantly, scatter components of sunlight such as UV wavelengths away from the skin. It is important not to block all sunlight from the skin as it promotes the formation of vitamin D, a deficiency of which can cause rickets.

Nanoparticles and their interaction with electromagnetic radiation such as visible light occur in many applications involving materials, for example the

Materials for the 21st Century. David Segal.
© David Segal 2017. Published 2017 by Oxford University Press.
DOI: 10.1093/oso/9780198804079.001.0001

optical properties of quantum dots or the colours in the wings of butterflies, which arise from scattering in photonic crystals. However, whereas photonic crystals have periodic structures, titanium dioxide powders in sunscreens do not form photonic crystals on the skin and are randomly packed. As the particles are smaller than the wavelength of visible light, they appear invisible on the skin; that is, the sunscreen is 'non-streaky' in appearance. A number of methods have been used to prepare nanoparticles. For example, homogeneous nucleation from the gas phase or from solution can result in distributions of monodisperse nanoparticles. Hydrothermal synthesis, which involves nucleation from solutions at elevated temperatures and pressures, can yield distributions of nanoparticles. Titanium dioxide can be present in one of two crystalline phases, anatase and rutile. Anatase is photoactive and in the presence of sunlight can produce free radicals. It has applications in dye-sensitised solar cells. However, for sunscreen formulations, rutile tends to be the preferred phase as it limits the production of free radicals.

12.3   Washing-up liquids and surfactants

Anyone who has purchased a bottle of washing-up liquid or hair shampoo or toothpaste may have noticed that the list of ingredients often refers to the presence of a surfactant, an abbreviation for surface active agent. Surfactants are amphiphilic molecules consisting of a hydrophobic part, which is commonly referred to as a hydrocarbon 'tail', and a 'headgroup'. The tail part of the molecule is soluble in hydrocarbon and non-polar solvents while the hydrophilic headgroup is soluble in aqueous media. Examples of surface active agents are sodium dodecyl sulphate, an anionic surfactant; dodecyl trimethyl ammonium bromide, a cationic surfactant; and dodecyl pentaethylene glycol ether, a non-ionic surfactant. Zwitterionic surfactants contain both positively charged and negatively charged head groups. Two properties of surfactants are important when dissolved in water. Firstly, they adsorb at the air–liquid interface, a process that lowers the surface tension and, secondly, they form association structures in solution known as micelles. Thus, the molecules undergo self-assembly. The surfactant concentration at which micelles form is known as the critical micelle concentration. In dilute solutions, these micelles can be spherical in shape with a diameter around twice the molecular length, thus several nanometres in diameter. In an aqueous environment the hydrocarbon part of the molecule points inwards and the headgroup is in contact with the aqueous phase. Solutions of surfactants foam when shaken, as is well-known from washing-up liquids and dirt and grease are removed by solubilisation, in which particles are absorbed into the micelles.

However, association structures occur in other materials, for example in the formation of liquid crystals. As an example, the aramid fibre Kevlar contains liquid crystals of aligned polymer molecules that contribute to its strength.

12.4 Cosmetics

Cosmetics cover a wide range of products: for example, hand creams, shampoos, shaving foams, lipsticks, eye shadows, nail varnishes and removers, hair shampoos and conditioners, antiperspirants, shower gels, moisturizers and creams to produce a younger-looking skin. In section 12.3 micelles and liquid crystals were described as association structures. Liposomes are also association structures consisting of an aqueous core surrounded by a hydrophobic lipid bilayer. Products that are soluble in water can be trapped in the core because they cannot pass through the bilayer. Liposomes can have sizes ranging up to several micrometres and for sizes less than 100 nm they are known as nanoliposomes; they can be used to encapsulate cosmetic ingredients in order to penetrate outer layers of the skin known as the stratum corneum and release active ingredients. Liposomes can also be used for encapsulation of pharmaceutical compounds for drug delivery. The use of liposomes for delivery of cosmetics or pharmaceuticals is called nanoencapsulation.

Colloid science and nanotechnology underpin the development of cosmetic formulations, for example as emulsions. There is much interest in the development of superhydrophobic surfaces, for example in self-cleaning textiles and buildings. Hydrophobic surfaces are important in cosmetics particularly for lipsticks and mascara to prevent 'running' of the cosmetic while adding particles to cosmetic products can modify the appearance of the skin. This modification depends on optical interference for light that is scattered and reflected by the particles. Hydrogels are used in moisturisers to retain water.

12.5 Disposable nappies (diapers): the role of hydrogels

Hydrogels are porous materials and consist of three-dimensional networks of hydrophilic homopolymers or copolymers that can be swollen by absorption and retention of water. They exhibit both liquid properties, because the major constituent is water, and solid properties, because of cross-linking during polymerisation. Although the polymers are soluble in water, the hydrogel is insoluble because of cross-linking during polymerisation. Hydrogels have also been described as superabsorbent and superporous and their pores can have diameters in the micrometre to millimetre range. Hydrogels can absorb 99.9 weight per cent as water without losing structural integrity. Examples of hydrogels are those based on polyacrylic acid and they are often used in powder form after they have been dried, crushed and classified. Hydrogels can be prepared from natural products, for example from sodium alginate from seaweed.

An important application for hydrogels is in disposable nappies (diapers) while other applications include wound dressings, moisturizers for use as cosmetics, sanitary towels and perhaps surprisingly disposable contact lenses.

12.6 Hard candy (boiled sweets) and fudge: the role of microstructure

Examples have been given throughout the text on the importance of microstructure on the properties of materials. For examples, the colours of the wings of butterflies arise not from the presence of pigments but from a periodic structure of components in the wings that form photonic crystals. Also, the double helical structure of deoxyribonucleic acid (DNA) and the toughening of materials by incorporation of fibres are examples of the role of microstructure in material properties. Glucose has an important role in the chemical structure of carbohydrates such as cellulose and starches and an important role in the preparation of sweets. In fact, sweets represent materials where microstructure is also important in determining physical properties.

Many sweets are derived from heating solutions of granulated sugar (i.e. sucrose) and water. Hard candy is a term used in the United States for a variety of sweets including drops, lollipops and rock candy and is analogous to the phrase 'boiled sweets' used in the United Kingdom. When a syrup of granulated sugar and water is cooked at the upper end of the temperature range of 135–70°C in what is referred to as the hard crack stage so that the sugar syrup contains less than 2 weight per cent water, then the syrup forms a clear hard mass when cooled. It is important to prevent crystallisation of sugar in the syrup when producing hard candy as the crystals or 'grain' can affect the transparency of the product. Sucrose can break down or invert into its components, glucose and fructose, during the heating process and the inversion promotes crystallisation. Inverting agents can be added to control the inversion; thus, the microstructure, for example upon adding corn syrup, does not crystallise. Acids such as tartaric acid and citric acid in lemons also prevent crystallisation of sugar. The concentrated syrup can be poured into molds and allowed to solidify. Adding butter to the syrup results in butterscotch.

When a mixture of granulated sugar and milk is boiled to the softball stage at a temperature between 113 and 115°C, after which butter is added and the mixture allowed to cool and then beaten to encourage the growth of sugar crystals, then the resultant mixture is known as fudge, formally a crystalline candy. The formation of sugar crystals in fudge is analogous to the promotion of crystallisation in glass ceramics. Caramel is produced by heating sugar syrup, milk and butter to temperatures of 160°C or above, by which it browns and develops a burnt flavour; in contrast to hard candy, caramel is opaque. Caramels have a texture ranging from soft to medium hard and chewy. The heat treatment to produce caramel causes a complex decomposition process in the sugar solution and decomposition products, such as acrolein (CH_2=CHCHO), are incorporated into the caramel.

It can be seen how the control of sugar crystallisation affects the properties of the final product, whether hard candy, fudge or caramel.

12.7   Liquid crystal thermometers

During the winter months, governments such as that in the United Kingdom advise people, especially elderly people, to keep the temperature in their homes at specific levels to maintain good health. For example, 18°C in a bedroom and 21°C in a living room are recommended. It is possible to obtain a thermometer that resembles a black plastic strip about 10 cm in length and 1 cm wide and divided into segments, each with a length around 1 cm. A number is present in each segment representing a temperature but only one number can be seen at any one time. Cholesteric liquid crystals have a twisted structure. Their rod-like molecules do not align in a parallel way as in nematic and smectic liquid crystals but are twisted with respect to each other. The pitch of the twist represents the distance over which the twist completes a full turn. In the thermometer each segment contains liquid crystals with a slight difference in the twist. When the pitch corresponds to the wavelength of incident light, reflection occurs and the relevant number can be read as the temperature. A variation in this approach involves using mixtures of cholesteric liquid crystals that change colours over different ranges of temperature.

12.8   Breathable garments

Membranes are porous media that selectively allow species to pass through them. For example, reverse osmosis membranes are used for desalination of salty and brackish waters and these membranes have been derived from polysulphones. Ultrafiltration membranes can be used for removal of bacteria from aqueous environments and can be polymer or ceramic. Polytetrafluoroethylene (PTFE) was first prepared by Roy Plunkett in 1938 and its applications reflect its properties. Thus, chemical inertness favoured its use in the Second World War in the Manhattan Project on development of the atomic bomb as it was inert to corrosive gaseous uranium hexafluoride. Its low coefficient of friction encouraged use as bearings. However, it was found that expanded PTFE tape, abbreviated to ePTFE, contained pores of a controlled size and this material has been used in garments that allow the fabric to 'breathe'. This is because the pores do not allow water droplets to pass through them but are permeable to water vapour. An example of porous PTFE is known as Gore-Tex.

12.9   Acrylic textiles and carbon fibre

The production of textile fibres from synthetic polymers is carried out on an industrial scale. A general method of preparation involves forcing the polymer solution (or melt) through a spinerette and then stretching the fibre to orientate molecules, which enhances the fibre strength. Removal of the solvent in the fibre is carried out either by evaporation, a process known as dry spinning, or by precipitation into a solvent bath, a process known as wet spinning. The latter is used in the production of Lyocell (Tencel). Acrylic fibres based on polyacryloni-trile (PAN) and derived from acrylonitrile monomer ($CH_2=CHCN$) are widely used as a textile fibre in garments such as pullovers and are often blended with other fibres, synthetic and natural. However, anyone who purchases clothing containing acrylic fibres may be unaware of an important application of PAN fibres, that is in the production of carbon fibres. PAN fibres are first heated in the range 200–300°C to produce a thermally stable structure of carbon hexagons known as a cyclised ladder structure. This process results in ring-closing in the polymer chain. Carbonisation of the fibre is carried out under an inert atmosphere around 1000–1500°C and carbon fibres are produced with a **turbostratic struc-ture** with a preferred orientation of carbon crystallites with respect to the fibre axis. A further heat treatment removes residual nitrogen from the structure. Carbon fibres have also been produced from viscose rayon and mesophase pitch pre-cursors. Carbon fibres are used for lightweight composites, in particular for carbon–carbon composites, for braking pads in aircraft and racing cars. They are also used in carbon-reinforced silicon carbide composites for protection against oxidation, for example in spacecraft.

Turbostratic structure *means the crystals in the fibre are oriented.*

There is much interest nowadays in encouraging young people to pursue a career in so-called STEM (science, technology, engineering and mathematics) subjects. Role models are highlighted; for example, Rosalind Franklin is frequently men-tioned in connection with photograph 51 from an X-ray crystallographic study on DNA fibres obtained by her and Raymond Gosling. It is useful to mention here that Franklin also carried out crystallographic studies on the characterisation of coals for their fuel efficiency in combustion.

12.10 Decaffeinated coffee

Drying is an important industrial process when materials are produced. For example, synthetic fibres that are made by spinning from solution can be dried by evaporation of the solvent from the fibres or by treatment of the damp fibres with a solvent. Coffee can be decaffeinated by treatment with solvents but a 'greener' process involves treatment with a supercritical fluid. The latter is a substance above its critical temperature where it remains as a single fluid phase. An increase in pressure affects fluid density but does not produce a separate fluid phase. Supercritical fluids are good solvents. Carbon dioxide has values of critical temperature and pressure of 304.1 K and 7.38 M Pa, respectively, much lower than those for water. Materials dried using supercritical carbon dioxide tend to have a 'fluffy' texture. Aerogels, for example silica aerogels, are made by super-critical drying and they are extremely lightweight with high porosity.

12.11 Spider silk

The preparation of synthetic fibres, for example textile fibres, is a large industry but a natural fibre that will be familiar to many people is silk that is spun into polymeric fibres by silkworms, scorpions and most commonly spiders. Silkworms produce silk fibres to weave their cocoons. Silk fibres consist mainly of the fibrous protein fibroin surrounded by the protein sericin, which acts as binder to maintain structural integrity. Fibroin is a semi-crystalline polymer, whereas sericin is amorphous. Spider silks are lightweight but strong and elastic with a fracture strength of around 1100 MPa and contain predominantly alternating sequences of glycine and alanine, whic self-assemble into β-sheets. The microstructure of the fibre consists of crystalline regions associated with β-sheets and amorphous regions associated with segments of amino acids with bulky side-groups. Silk is a biomedical material and has been used as sutures. Anyone who has seen a spider's web on a rainy day may have noticed droplets of water equally spaced along the strands. This distribution of droplets arises due to an instability (Rayleigh instability) of a cylindrical film of water on the strands.

12.12 Stainless steel

Stainless steel is extremely common in everyday life with wide-ranging applications including trays in refrigerators, panels in cars and structures in buildings. By definition stainless steels are iron-based alloys containing at least 10 weight per cent chromium. They were developed at the beginning of the 20th century and at that time considerable research and development in Germany and in Sheffield in England led to these materials. Stainless steels are alloys and can be characterised by their microstructure. Martensitic stainless steels have a body-centred tetragonal structure and after hardening by heat treatments the alloys have good mechanical properties and moderate corrosion resistance. The chromium content ranges from the minimum up to 18 weight per cent and the carbon content is no greater than 1–2 weight per cent. Additions of other elements such as tungsten increase the toughness of the steel. Applications of martensitic stainless steels include surgical and dental instruments such as forceps and scalpels.

Ferritic stainless steels have a body-centred cubic crystal structure and the chromium content can be as high as 30 weight per cent. Sulphur or selenium can be added to aid machinability but unlike martensitic steels, ferritic stainless steels cannot be strengthened by heat treatment.

Austenitic stainless steels have a face-centred cubic crystal structure. The chromium, nickel and manganese contents can vary from 15 to 20 weight per cent, 3 to 14 weight per cent and 1 to 7.5 weight per cent, respectively. These alloys cannot be hardened by heat treatment but can be hardened by cold working.

316L stainless steel consists mainly of iron (60–5 weight per cent), chromium (17–20 weight per cent), nickel (12–14 weight per cent) and smaller amounts of molybdenum, manganese, copper and other elements. The low carbon content (less than 0.03 weight per cent), designated by the letter 'L', provides good corrosion resistance.

Duplex stainless steels contain equal amounts of ferrite and austenitic phases in their microstructure. They have higher strengths than austenitic steels and improved toughness and ductility compared to ferritic steels.

12.13 Summary

Many materials are associated with complex technology, for example nickel-based superalloys for gas turbine blades, nickel–titanium alloys for implantable coronary stents and quantum dots for solid-state lighting in displays. However, products used in everyday life may not represent advanced technologies, but considerable underlying development is required to produce them. Examples of these everyday products include sunscreens, washing-up liquids, cosmetics, disposable nappies, boiled sweets and liquid crystal thermometers.

12.14 Further reading

Agrawal, C. M., J. L. Ong, M. R. Appleford and G. Mani. *Introduction to Biomaterials: Basic Theory with Engineering Applications*. Cambridge: Cambridge University Press, 2014.

Atkins, P. *Atkins' Molecules*, 2nd edn. Cambridge: Cambridge University Press, 2014.

Bansal, N. P., and J. Lamon (eds). *Ceramic Matrix Composites: Materials, Modelling and Technology*. Washington, DC: American Ceramic Society, 2015.

Buzak, S., and D. Rende. *Colloid and Surface Chemistry: A Laboratory Guide for Exploration of the Nano World*. Boca Raton, FL: CRC Press, 2014.

Dunmur, D., and T. Sluckin. *Soap, Science, and Flat-Screen TVs: A History of Liquid Crystals*. Oxford: Oxford University Press, 2014.

Emsley, J. *Chemistry at Home*. Cambridge: Royal Society of Chemistry, 2015.

Franklin, R. E. The structure of graphitic carbons. *Acta Crystallographica* (1951) 4(3), 253–61.

Goldstein, D. (ed.). *The Oxford Companion to Sugar and Sweets*. Oxford: Oxford University Press, 2015.

Misra, A., T. Vats and J. H. Clark. *Microwave-Assisted Polymerisation*. Cambridge: Royal Society of Chemistry, 2016.

Ngo, C., and M. Van de Voorde. *Nanotechnology in a Nutshell: From Simple to Complex Systems*. Amsterdam: Atlantis Press, 2014.

Segal, D. *Exploring Materials through Patent Information*. Cambridge: Royal Society of Chemistry, 2014.

Stewart, I. C., and J. Lamont. *The Handy Chemistry Answer Book*. Canton, MI: Visible Ink Press, 2014.

13

Conclusions

13.1 Introduction

Periods of human history are often described in terms of 'ages'. In the Stone Age, a period of prehistory around 300,000 years ago, stone (a complex mixture of silicate, aluminates and calcium carbonate minerals), bones, flint, wood and animal skins were common materials and flint and quartz could be shaped into tools. In the Bronze Age, a period of history which corresponds to around 3500 BC, alloying of tin with copper was achieved and produced bronze. The latter is a harder material than copper and could be used for tools and weapons. At around this time, glass, cement and pottery were invented. It was the Iron Age, and in particular the development of blast furnaces around 1500 AD, that enabled the construction of complex structures of cast iron, for example bridges, buildings and railway stations in the early part of the 19th century. The use of cast iron in construction heralded the start of the Industrial Revolution aided by the development of the Bessemer process in 1856 for making steel on an industrial scale. Cast iron contains at least 3 weight per cent carbon that modifies the viscosity of the molten iron, making it easy to cast, and around 1 weight per cent silicon. In this book emphasis has been placed on the development of selective materials from the beginning of the 19th century up to the present time. This is not to ignore the great impact that materials used in previous ages have had and continue to have to this day. In this chapter summaries of different classes of material are given with particular reference as to whether they can be described as 'ages' in the 21st century.

13.2 The commercial exploitation of materials

There can be considerable delay from when a new phenomenon is first observed or when a material is invented to their exploitation. For example, electroluminescence was first observed in 1907 by Henry Round, who observed it in an impure sample of the semiconductor silicon carbide, but the application of light-emitting diodes (LEDs) whose operation is based on electroluminescence has only reached widespread usage in, for example, solid-state lighting in recent years. In fact, many

Materials for the 21st Century. David Segal.
© David Segal 2017. Published 2017 by Oxford University Press.
DOI: 10.1093/oso/9780198804079.001.0001

members of the general public will be familiar with the acronym 'LED' if not the technical details of how they work. The delay arose partly due to the lack of chemical techniques for producing very pure materials and for depositing dopants into this material such as gallium nitride. The arrival of these techniques such as metal-oxide chemical vapour deposition (MOCVD), ion implantation and physical vapour deposition facilitated the exploitation of light-emitting diodes. Similarly, superconductivity in metals was first observed by Kamerlingh Onnes in 1911, who measured a transition temperature of 4.2 K for mercury in liquid helium. The transition temperature for metals increased slowly over the years and nowadays the application most associated with superconducting metals is their use for generation of strong magnetic fields in the medical diagnostic technique of magnetic resonance imaging (MRI). The ability to exploit superconducting metals in this application depends in part on the ability to form suitable superconducting alloys, their fabrication into coils and the application of mathematical methods to reassemble images obtained in the scanning process. As for LEDs, members of the general public routinely refer to MRI scans although the details of how they are obtained are probably unknown to many people.

The discovery by Georg Bednorz and Karl Muller of high-temperature ceramic superconductors in 1986 based on complex oxides of lanthanum, barium and copper with a transition temperature around 30 K caused great excitement in scientific circles and also in the public domain. The superconducting transition temperatures for ceramic superconductors are now around 100 K, meaning that liquid nitrogen (at 77 K), which is cheaper than liquid helium, can be used to generate the effect. However, the exploitation of ceramic oxide superconductors has been slow, perhaps partly due to the inherent difficulty in fabricating ceramic powders into shapes such as coils by sintering; this example illustrates that caution should be applied to comments made in the media relating to the exploitation of materials.

The isolation of graphene in 2004 has led to much excitement about potential applications, but at the time of writing they are mainly still possible applications. Caution is required in interpreting possible applications in an age when the expansion of the Internet and social media have allowed for the dissemination of information among people across the world and given a 'voice' to individuals or organisations that previously may have only attracted a limited audience. It is important to consider carefully the claims made for future exploitation of newly discovered materials and processes, for example three-dimensional printing, that attract much attention as a disruptive technology.

13.3   The Silicon Age?

The invention of the point-contact transistor in 1947 by John Bardeen, Walter Brattain and William Shockley in which a voltage was applied across two metal wires in contact with a doped semiconductor, namely germanium, had an immediate impact for potential applications as it acted as a rectifier. Transistors could replace valves and led to the development of integrated circuits that were pioneered by Gordon Moore and co-workers in the 1960s. Integrated circuits underpin all aspects of modern life in the second decade of the 21st century, whether in communications such as in smartphones; in business (ecommerce, online banking, online shopping); in manufacturing, for example in automation and machine and process control (e.g. chemical plant, nuclear reactors); in information storage; in imaging (e.g. digital cameras, medical imaging); and in the emerging field of autonomous vehicles and engine control systems (e.g. aircraft and road vehicles).

Perhaps the 21st century can be described as the Silicon Age but surprisingly of all the materials mentioned in this book, silicon chips and integrated circuits are completely hidden in equipment away from the view of the general public.

13.4　The age of specialist alloys

In the Bronze Age, alloying tin with copper was found to produce a hard material, bronze, that could be used as tools and weapons and which was harder than copper alone. Specialist alloys have a crucial role in the 21st century but are largely out of sight of the general public. In the area of transport, nickel-based superalloys are used to produce blades in gas turbines, allowing higher operating temperatures and more fuel-efficient aircraft. Nickel–titanium alloys have been developed as shape memory alloys and were first invented in 1961. They are used as implantable coronary stents. Zircaloy, an alloy of zirconium, is used as fuel rods to contain uranium oxide pellets, the fuel in thermal nuclear reactors, while alloys are also critical for LEDs, for example gallium arsenide. It is envisaged that specialist alloys will continue to be used in niche markets throughout the 21st century.

13.5　The Genomic Age?

X-ray crystallography on deoxyribonucleic acid (DNA) fibres under a humid atmosphere was carried out by Rosalind Franklin and Raymond Gosling; in particular, their 1952 photograph, known as photograph 51 which showed a pattern of spots from the diffraction of X-rays, in the shape of an 'X', was a crucial piece of information that enabled Francis Crick and James Watson to propose in 1953 that DNA had a double helical structure. This elucidation of the DNA structure gave birth to the subject of molecular biology which is advancing at a rapid pace. Recombinant DNA technology invented in the early 1970s is associated with the technology known as genetic engineering and is used to produce pharmaceuticals such as human insulin and bioethanol by fermentation processes while the ability to produce specific monoclonal antibodies on a large-scale has led to a new class of biopharmaceuticals based on antibodies. The very recently developed technology known as gene editing, or CRISPR, has the potential to help cure currently incurable diseases but it also raises ethical issues as changes to the human genome can be transmitted down through future generations. There is much interest in the development of personalised medicine in which an individual's genome can be targeted by specific medicines and treatments. DNA and the proteins that they express can be viewed as polymers and hence, there is an overlap between molecular biology and polymer science.

Whether the 21st century is referred to as the Genomic Age remains to be seen.

13.6　The Polymer Age

In the 19th century a range of useful products was obtained by the chemical treatment of cellulose. Examples of these materials, which are known as cellulosics, are cellulose nitrate (gun cotton), cellulose acetate, collodion, Parkesine, celluloid, viscose rayon and acetate rayon. Cellulose is a natural polymer and cellulosics are also polymers. Another type of polymer developed in the 19th century is vulcanised rubber latex obtained from trees; vulcanised rubber was used in the developing transport industry as tyres in motor vehicles. Another cellulosic developed in the second half of the 20th century is Tencel, an artificial fibre with a silk-like texture used in textiles. However, the availability of oil as a source of chemicals from the start of the 20th century led to the development of a range of synthetic polymers over a period of around 50 years and these are widely used nowadays. These polymers, sometimes made in the form of fibres, form the basics of the plastics industry and include nylon, polyethylene and polyvinyl chloride as examples. Many of these polymers are commodity items, are produced on the industrial scale and are widely used in the 21st century.

From the perspective of materials, the 20th century could be referred to as the Polymer Age. However, in the 21st century there is a trend in reducing the use of petrochemical sources and to develop materials from renewable sources. In addition, many synthetic fibres are not biodegradable and there is increasing emphasis on preparing biodegradable polymers to limit pollution of the environment. To some extent the activities on polymers have come full circle since the invention of cellulosics from cellulose in the 19th century as there is increasing interest in using cellulose as a source of biofuels as well as useful materials.

Whether the 21st century becomes known as the New Polymer Age remains to be seen.

13.7　The role of climate change in materials development

Ways to mitigate the effects of global warming are likely to increase in importance throughout the 21st century. The attempts to reduce global warming run parallel with attempts to reduce carbon dioxide emissions and this approach may eventually phase out the use of coal, natural gas and oil as sources of fuel in power stations and may affect the use of oil for the synthesis of plastics. Whether this approach encourages the use of renewable materials remains to be seen.

13.8   The role of international conflicts and wars

Developments on materials have also been driven by international conflicts and wars. For example, the Cold War and Space Race of the 1950s and 1960s between the USA and Soviet Union were an impetus for the development of the microelectronics industry and integrated circuits in the USA. During the Second World War, the Manhattan Project, which developed the atomic bomb, led to enhanced activity on the transuranic elements such as plutonium and the separation of isotopes by gas diffusion. Polytetrafluoroethylene had a crucial role in isotope separation due to its chemical inertness. Polyethylene, which was invented in 1933 by Reginald Gibson and Eric Fawcett, had an important application as an electrical insulator for cables in fighter aircraft of the Royal Air Force that contained the newly developed compact radar systems based on the magnetron cavity. Also during the Second World War, work carried out at Bletchley Park in the United Kingdom by Alan Turing and others greatly contributed to the emerging field of computer science and the development of computers based on the use of electronic valves as switches. In the First World War, acetone was produced on an industrial scale by fermentation processes and used as an intermediate for the explosive cordite by the British Army.

Nowadays governments are concerned with international terrorism and there are pressing concerns with the decryption of encoded messages, whether on smartphones or in so-called chatrooms on the Internet. Monitoring and decoding of these messages require considerable computing power and ultimately depends not only on the skill and expertise of government employees but on the ability to pack logic circuits based on transistors onto silicon chips that may have the ability to break encrypted messages in a realistic timescale.

13.9   Summary

Periods in human history have been described in terms of 'ages': for example, the Stone Age, the Bronze Age and the Iron Age. The conversion of cellulose to useful products such as celluloid and cellulose acetate that could be spun into the textile fibre rayon was achieved in the 19th century. The availability of sources of oil in the early part of the 20th century led to the production of a wide range of synthetic polymers, commonly called plastics over a period of about 50 years. Examples of these polymers are nylon and polyethylene. The 20th century may, from the viewpoint of materials development, be referred to as the Polymer Age. However, the invention of the transistor in 1947 led to the fabrication of silicon chips and integrated circuits that underpin all aspects of societies in the early part of the 21st century, whether in commerce (e.g. online banking, online shopping), communications (e.g. smartphones, tablet computers, social media), medical diagnostics (e.g. magnetic resonance imaging) or consumer goods (e.g. televisions). Although integrated circuits are incorporated into devices and hidden from the view of the user the 21st century may turn out to be the Silicon Age. However, the elucidation of the double helical structure of DNA in 1953 led to the development of molecular biology and genetic engineering, enabling the production of pharmaceuticals such as human insulin for treatment of Type 2 diabetes and a new class of drugs known as biopharmaceuticals. It is possible that the 21st century will be the Genomic Age.

Irrespective of how the 21st century is eventually described, this text has shown how important materials are to 21st-century economies and their importance is likely to continue.

13.10 Further reading

Ashby, M. F. *Materials and the Environment: Eco-Informed Material Choice*, 2nd edn. Oxford: Butterworth-Heinemann, 2013.

Reed, R. C. *The Superalloys: Fundamentals and Applications*. Cambridge: Cambridge University Press, 2008.

Glossary

Listing of 500 materials in alphabetical order

abrasives

Hard materials are used for grinding and cutting wheels for metals.[1,2] Representative media for wheels include cubic boron nitride (BN), silicon carbide (SiC) and alumina that are bonded together in a glass or polymer matrix. Industrial diamonds, natural or synthetic, are used for grinding and polishing and as coatings for cutting tools. New cutting surfaces are exposed as the hard materials are worn down. Advances in manufacturing methods such as three-dimensional printing avoid the use of grinding media during the fabrication of components. Three-dimensional printing (3D printing) is a disruptive technology (see Chapter 9) in which components are fabricated layer by layer. Ceramic powders, metal powders, plastics in the form of strips that are melted and deposited as molten droplets and biological cells can be used as 'inks' in the printing process. 3D printing not only avoids the use of grinding media but produces minimal wastage of materials.

See: *three-dimensional printing*

1. Askeland, D. K., and W. J. Wright. *The Science and Engineering of Materials*, 7th edn, p. 622. Boston, MA: Cengage Learning, 2011.
2. Somiya, S. (ed.). *Handbook of Advanced Ceramics*. San Diego, CA: Academic Press, 2013.

acetylsalicylic acid

Salicylic acid was obtained in crystalline form from the bark of the willow tree by Raffaele Piria around 1830[1] and while it had an antipyretic action, side-effects included an unpleasant taste and irritation to the stomach. At the end of the 19th century, Felix Hoffmann reacted salicylic acid (HOC_6H_4COOH) with acetic anhydride [$(CH_3CO)_2O$] to produce acetylsalicylic acid that retained the beneficial effects of salicylic acid without the side effects.[2] Acetylsalicylic acid was sold under the name of aspirin and continues to be a very beneficial pharmaceutical in the 21st century not only for pain relief and reduction of fever but as an aid to prevent heart attacks. Derivatives of salicylic acid, for example 5-amino salicylic acid, have applications for the treatment of ulcerative colitis.

See: *ZMapp*

1. Balzani, V., and M. Venturi. *Chemistry: Reading and Writing the Book of Nature*. London: Royal Society of Chemistry, 2014.
2. F. Hoffmann. Acetyl salicylic acid. United States Patent 644,077, granted 27 February 1900.

acrylic fibres

Acrylic fibres are widely used in clothing and textiles, especially when blended with natural or synthetic fibres such as cotton.[1] Polyacrylonitrile derived from acrylonitrile (CH_2CHCN) can be spun into fibres from solutions in dimethylformamide but not melt-spun as the nitrile group in the polymer undergoes cyclisation on heating.

See: *carbon fibre*

> 1. Fried, J. R. *Polymer Science and Technology*, 3rd edn. Upper Saddle River, NJ: Prentice Hall, 2014.

acrylic paints

Acrylic paints are used by both professional and amateur artists. They use an emulsion or resin based on polymerised acrylic acid ($CH_2=CHCOOH$) combined with pigment powders to give a wide colour spectrum. For example, greens can be produced by use of chlorinated copper phthalocyanine, black by use of synthetic iron oxide, whites by use of titanium dioxide and magenta by use of quinacridone violet.[1] Gold and silver colours use titanium dioxide-coated mica.

> 1. Winsor & Newton (Harrow, England). Make your mark, catalogue number 7543452, available at http://www.winsornewton.com.

additive colour mixing

Colours are an integral part of everyday life whether in newspapers and magazines, in textiles or in electronic displays such as in mobile phones or in television and computer screens. Visible light contains a narrow range of wavelengths within the electromagnetic spectrum from around 360 to 780 nm. Violet has a wavelength in the region of 400 nm while red has a longer wavelength around 750 nm. Red, green and blue are known as primary additive colours as they cannot be obtained by mixing of light of other colours (i.e. wavelength). Additive mixing of red and blue produces magenta, blue and green gives cyan while mixing of red and green gives yellow.[1] Magenta, cyan and yellow are complementary colours that arise when one component of visible light is absorbed and the other components are transmitted or reflected. Thus, if blue light is absorbed, the reflected light contains red and green components that appear as yellow. Additive mixing is used in displays with light-emitting diodes or quantum dots in the presence or absence of phosphors when electroluminescence produces emitted light in the red, green and blue regions. Additive mixing of these colours produces white light.

See: *light-emitting diodes; quantum dots; subtractive colour mixing*

> 1. Christie, R. M. *Colour Chemistry*, 2nd edn, p. 23. London: Royal Society of Chemistry, 2015.

additive manufacturing

Additive manufacturing is a technique in which three-dimensional objects are built up layer by layer without producing waste material as occurs in subtractive machining.[1] Three-dimensional printing is an additive manufacturing technique that is also referred to as a rapid prototyping method.

See: *subtractive machining; three-dimensional printing*

1. Andersson, C. Innovation in rapid prototyping and additive manufacturing. *European Medical Device Technology* (2012) (January/February).

advanced ceramic materials

Advanced ceramic materials are derived from chemical synthetic routes or from naturally occurring materials that have been highly refined.[1] For example, aluminium oxide is obtained from the mineral bauxite in the Bayer process by dissolution of the mineral in sodium hydroxide followed by precipitation of aluminium hydroxide and separation of the precipitate from impurities in the bauxite. Examples of advanced ceramic materials include silicon carbide powders, silicon carbide fibres, silicon nitride, sialons, stabilised zirconias and boron nitride. These materials are noted for their strength and hardness, a characteristic of structural ceramics. Electroceramics also belong to the class of advanced ceramic materials, examples of which are piezoelectric materials such as lead zirconate titanate and lithium niobate for electro-optic devices.

See: *ceramic materials*

1. Somiya, S. (ed.). *Handbook of Advanced Ceramics* San Diego, CA: Academic Press, 2013.

aerogels

Aerogels are lightweight, highly porous materials with nanometre-sized pores and can be described as solid foams.[1–3] They were discovered in 1932 by Samuel Kistler who exchanged water for ethyl alcohol in a silica hydrogel and then removed the alcohol at elevated temperature and pressure under supercritical conditions in an autoclave; drying the gel in this way maintained the pore structure in the resulting aerogel.[1] The latter has around 95% open interconnected porosity, high surface area, low thermal conductivity useful for thermal insulation, optical transparency and thermal stability. Examples of aerogels include inorganic aerogels such as silica and alumina. Applications include catalyst supports, anti-reflectance coatings, composite materials and moulds for metal casting. Aerogels have also been used in space exploration by NASA as a trap for capturing micrometeoroids during space flights and in the Mars Pathfinder probe in the 1980s.

See: *organic aerogels*

1. Kistler, S. S. Method of making aerogels. United States Patent 2,249,767, granted on 22 July 1942.

2. Reade, L. Full of air. *Chemistry and Industry* (2013) 77, 32–5.
3. Felice, M. Silica aerogel. *Materials World* (2013) 21, 54–5.

aerosol-derived powders

Aerosols are dispersions of particles in a gas and the particles can have a colloidal dimension.[1,2] When the particles are liquid, the aerosol is described as a fog or mist but when particles are solid the aerosol can be referred to as a smoke. Aerosols are particularly important for pharmaceutical compounds when used in inhalers for the treatment of asthma. The active pharmaceutical ingredient requires a specific particle size and size distribution to be carried effectively into the lungs. In general, fine particles on the order of colloidal dimensions can be prepared either by comminution of larger particles or by growth from solution species.

See: *colloidal systems*

1. Segal, D., and A. Atkinson. Wet chemical synthesis of cathodoluminescent powders. *British Ceramic Transactions* (1996) 95, 103–6.
2. Booker, D. R., B. R. Bowsher and D. A. V. Morton. The synthesis of advanced ceramics in bult quantities using aerosol routes. *Journal of Aerosol Science* (1992) 23, S819.

after-glow pigments

After-glow pigments are used in safety signs and dials and rely on the property of phosphorescence rather than a source of electrical power to generate light.[1] Phosphorescence is a form of photoluminescence in which conversion of the exciting radiation is slow. A representative phosphorescent material is a doped strontium aluminate, $SrAl_2O_4$, doped with europium (Eu) and dysprosium (Dy), which has a blue-green glow (485 nm). The phosphorescence can last for around 10 hours. A qualitative explanation involves daylight becoming absorbed, promoting electrons to undergo electronic transitions and to become temporarily trapped in the crystal lattice, after which the electrons are released through thermal energy. The visible glow occurs by electrons undergoing transitions between energy levels. Another phosphorescent composition is cobalt (II)-doped ZnS:Cu with a green emission at 530 nm.

See: *phosphors*

1. Lucas, J., P. Lucas, T. Le Mercier, A. Rollat and W. Davenport. *Rare Earths: Science, Technology, Production and Use*, p. 312. London: Elsevier, 2015.

algal biomass

Algae are a source of cellulose and can be hydrolysed to sugars using fungal enzymes, for example from the phylum Neocallimastigomycota.[1] Examples of algae that are a potential source of biofuels include *Spirulina*, *Chlorella*, *Arthrospira* and *Dunaliella*, as well as the blue-green algae known as cyanobacteria.[2] The algae strains are susceptible to genetic modification to sustain growth under a range of aquatic conditions and improve yields of lipids and other hydrocarbons. The resulting sugars can be fermented to bioethanol. Ionic liquids have been used to produce cell lysis in algae cells and harvest lipids, hydrocarbons and carbohydrates contained within the cells for conversion to biofuels.

See: *biomass; ionic liquids*

1. Menetrez, M. Y. An overview of algae biofuel production and potential environmental impact. *Environmental Science and Technology* (2012) 46, 7073.
2. King. Plant power. *Chemistry and Industry* (2013) 77, 28–31.

aliphatic polyamides

Aliphatic polyamides, for example those formed by reaction of a diamine with a dicarboxylic acid, are referred to as nylons.[1] Aromatic polyamides, or aramids, include Kevlar and Nomex as examples.

See: *nylon*

1. Walton, D., and P. Lorimer. *Polymers*, p. 97. Oxford: Oxford University Press, 2005.

amino acids

Amino acids are water-soluble organic compounds with the general formula $R-CH(NH_2)$ COOH, where R is hydrogen or an organic group that determines the properties of a particular amino acid.[1,2] Amino acids join together through formation of peptide bonds to form short chains, namely peptides or longer chains known as polypeptides. The peptide bond is formed by the reaction between adjacent carboxyl $(-COOH)$ and amino $(-NH_2)$ groups. The repeating unit in peptides and polypeptide polymers can be represented by $(-N-CHR-C=O-)$. Proteins consist of polypeptide polymers with varying amounts of 20 amino acids and the characteristic sequence of these acids determines the shape, properties and biological function of the proteins. The latter can have as few as 124 amino acids (the enzyme ribonuclease) to many thousands. Enzymes and hormones are also proteins that are expressed by DNA. These acids are alanine, arginine, asparagine, aspartic acid, cysteine, glutamic acid, glutamine, glycine, histidine, isoleucine, leucine, lysine, methionine, phenylalanine, proline, serine, threonine, tryptophan, tyrosine and valine.

See: *DNA; peptides; polymers; polypeptide; proteins*

1. Daintith, J. (ed.). *A Dictionary of Chemistry*, 6th edn. Oxford: Oxford University Press, 2008.
2. Fromm, H., and M. Hargrove. *Essentials of Biochemistry*. New York: Springer, 2012.

AMOLED

AMOLED refers to an active matrix organic light-emitting diode (OLED). In an OLED, semiconducting organic species, small molecules or polymer is used in displays. The layer of semiconductor is divided into pixels and each pixel acts as a light-emitting diode.[1] In an AMOLED a backing layer of conductors is in contact with the OLEDs for addressing the pixels. The conductors form an active matrix and each pixel is backed with a thin-film transistor (TFT) to store its state during transitions in the energising voltage. An AMOLED is sometimes referred to as a TFT display.

See: *organic light-emitting diodes; PMOLED; thin-film transistors*

1. Platt, C., and F. Jansson. *Encyclopaedia of Electronic Components*, vol. 2. San Francisco, CA: Maker Media, 2014.

amorphous metal foams

Amorphous metal foams also known as foamed metallic glasses and have potential as structural materials because of their strength and hardness, but their lack of ductility limits their use. They have been prepared by casting molten metal into a porous bed of sintered leachable particles of barium fluoride (BaF_2) or a matrix of carbon beads that can be removed on calcination. Compositions of foamed metallic glasses include $Zr_{57}Nb_5Cu_{15.4}N_{12.6}Al_{10}$ and $Pd_{43}Cu_{27}Ni_{10}P_{20}$.[1]

See: *amorphous metals*

1. Liu, P. S., and G. F. Chen. *Porous Materials*. London: Elsevier, 2014.

amorphous metals

Amorphous metals are usually alloys that have a disordered non-crystalline glass-like structure and have also been referred to as metallic glasses.[1] They have been produced by squirting a jet of molten metal against a rotating disc of copper cooled very quickly with liquid nitrogen. The individual metallic components have differing atomic sizes and crystal structures that favour the formation of glasses. Examples of metallic glasses are $Au_{75}Si_{25}$ and $Fe_{72}Al_5Ga_2P_{11}C_6B_4$ and amorphous metals have been used as magnetic materials in, for example, transformer cores. Vitreloy is an amorphous metal with the approximate composition $Zr_{46.75}$ $Ti_{8.25}$ $Cu_{7.5}$ $Ni_{10}B_{27.5}$, has high strength arising from a lack of dislocations and has applications in sports equipment such as golf-club heads.

1. Tilley, R. J. D. *Understanding Solids: The Science of Materials*. Chichester: Wiley, 2013.

amorphous silicon

Single crystal silicon is used for the manufacture of silicon chips. Amorphous silicon (a-Si) is a disordered form of the semiconductor silicon consisting of randomly oriented small crystallites. This structure contains 'dangling' bonds where unpaired electrons can trap electrons and holes.[1] Amorphous silicon can be deposited onto a substrate from a gas discharge. The number of dangling bonds can be reduced by incorporating hydrogen

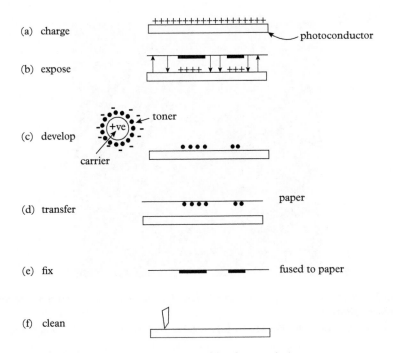

Fig. G.1 *Schematic representation of the xerographic (photocopying) process.*
(R. M. Christie. Colour Chemistry, *2nd edn, p. 292. Cambridge: Royal Society of Chemistry, 2015.)*

into the gas discharge which results in a hydrogenated amorphous silicon (a-Si:H) where hydrogen atoms neutralise the dangling bonds. Amorphous silicon is cheaper to produce than single crystal silicon and is used for large-area solar cells. Amorphous semiconductors can be photoconductors; that is, they become more electrically conducting on exposure to light and are used in the xerographic process. The latter is associated with photocopiers. The xerographic process was invented by Chester Carlson in 1938, which was further developed by the Haloid Company (later to become the Xerox Corporation) in which dry powder would stick to charged areas of the semiconductor that were exposed to light before being transferred to paper (Fig. G.1).

See: *silicon chips; silicon-based solar cells*

1. Solymar, L., D. Walsh and R. R. A. Syms. *Electrical Properties of Materials*, 9th edn. Oxford: Oxford University Press, 2014.

amyloid fibrils

There is much concern in the 21st century about the potential increase in Alzheimer's disease among an ageing population, particularly in countries with advanced healthcare services such as the United Kingdom and the associated cost to national economies.[1] There is currently no cure for Alzheimer's disease. Proteins are polymers in which the amino acids

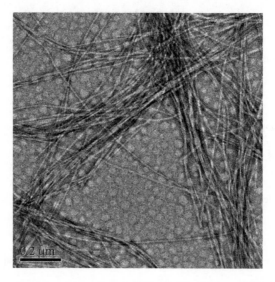

Fig. G.2 *Electron microscope picture of amyloid fibrils of the kind seen in the brains of victims of Alzheimer's disease.*
(W. Gratzer. Giant Molecules: From Nylon to Nanotubes, *p. 73. Oxford: Oxford University Press, 2013.)*

in the polymer backbone are the monomer units and the proteins undergo folding to form compact structures in the body (Fig. G.2). Alzheimer's disease is associated with protein fibrils (fibres or filaments) that for unknown reasons do not undergo folding and can aggregate or clump together with other fibrils to form a plaque that interferes with the functioning of nerve cells in the human brain. These fibrils are known as amyloid fibrils although it is not understood whether amyloid fibrils are the cause of Alzheimer's disease or are just associated with it.

See: *polymers*

1. Gratzer, W. *Giant Molecules: From Nylon to Nanotubes,* p.73. Oxford: Oxford University Press, 2013.

ancient works of art

Ancient works of art can be highly valued in the 21st century and the materials used to make them as well as the material properties can add to their aesthetic appeal. An example is the Lycurgus cup that was made by ancient Roman artists.[1-3] When illuminated by white light from behind, the cup shows a rich shade of colours ranging from deep green to bright red. The transmitted light is red while observing scattered light perpendicular to the direction of incident light indicates a greenish colour in the cup. These colours often arise from incorporation of nanoparticles such as gold into the glass.

See: *nanoparticle-based art*

1. Novotny, L., and B. Hecht. *Principles of Nano-Optics*, 2nd edn, p.401. Cambridge: Cambridge University Press, 2006.
2. Louis, C., and O. Pluchery. *Gold Nanoparticles for Physics, Chemistry and Biology*. London: Imperial College Press, 2013.
3. Extance, A. Plasmons with a purpose. *Chemistry World* (Sept. 2012) 56–9.

anti-ageing skin treatments

Proteins, for example collagen, are polymers and as such can be described as materials. The cosmetics industry is a global enterprise and products that are frequently a mixture of materials are widely advertised. The US market for cosmetics and beauty products in 2014 was approximately $38 billion.[1] One group of products purports to help restore the suppleness and bounce of skin or helps to modify the appearance of the user's skin. These products are usually applied from a cream to, for example, the face but a recent development involves hydrolysed collagen (e.g. bovine or fish collagen) that is taken as a drink, sometimes in conjunction with other compounds such as hyaluronic acid.[2] Aspects of skincare, in particular how the appearance of the skin may be modified, are discussed in reference (3).

See: *collagen*

1. Eisberg, N. Oil be back. *Chemistry and Industry* (Feb. 2015) 79, 42–5.
2. Stacey, S. *You Magazine*, 77, 10 May 2015, available at http://You.co.uk.
3. Howes, L. Here comes the science bit. *Chemistry World* (Oct. 2012) 9, 54–7.

antibodies

Antibodies are generated *in vivo* in response to foreign pathogens or molecules (antigens) and bind to the antigen in order to neutralise and destroy it.[1] Antibodies are Y-shaped glycoproteins with molecular weight around 150,000 (150 kDa). Pioneering work carried out by Cesar Milstein and co-workers in the early 1970s showed that it was possible by using the new technique of hybridoma technology to clone single mouse cells (known as B-cells) and the single antibody made by that cell.[2] The monoclonal antibody recognised only a single antigen and could be produced in large quantities. The sequences in the protein that are similar but not identical to those in human monoclonal antibodies corresponded to antibodies that are referred to as chimeric. Monoclonal antibodies form the basis of biological drugs or biopharmaceuticals (e.g. Avastin) rather than small molecules, as is the case for many existing pharmaceutical compounds.

See: *biopharmaceuticals; glycoproteins; polymers; proteins.*

1. Marks, J. D. Molecular engineering of antibodies. In R. Rapley and D. Whitehouse (eds), *Molecular Biology and Biotechnology*, 6th edn, pp. 167–9. Cambridge: Royal Society of Chemistry, 2015.

2. Kohler, G., and C. Milstein. Continuous cultures of fused cells secreting antibody of predefined specificity. *Nature* (1975) 256, 495–7.

anti-reflection coatings

Purchasers of spectacle lenses are sometimes offered anti-reflection coatings either as a free add-on or for an extra cost. These coatings are not restricted to spectacle lenses and have many applications including architectural glass and specialist lenses. Coatings have been produced by physical vapour deposition such as sputtering and by wet chemical techniques, for example sol-gel processing.[1] The refractive index of the coating should be less than the substrate and for glass lenses the refractive index is about 1.5. Magnesium fluoride with a refractive index of 1.28 has been used as an anti-reflection coating and deposited by physical vapour deposition. A lower index is required to reduce the reflectance below 1% and silica aerogels are attractive candidates to fulfil this criterion.

See: *aerogels; sol-gel processes*

1. Suzuki, M., T. Shiokawa, K. Yamada, H. Nakayama, H. Yamaguchi and A. Maryta. Production methods of silica aerogel film, antireflection coating and optical element. United States Patent 7,931,940, granted on 26 April 2011.

aramids

Aromatic polyamides such as Kevlar (poly (p-phenylene terephthalamide)) are referred to as aramids.[1] Aliphatic polyamides are known as nylons and can be prepared by a condensation reaction between a diammine with a dicarboxylic acid.

See: *Kevlar*

1. Atkins, P. *Atkins' Molecules*, 2nd edn, p. 94. Cambridge: Cambridge University Press, 2003.

artemisinin

Artemisinin and its derivatives are used for the treatment of Falciparum malaria and it has been obtained from *Artemisia annua*, a slow-growing plant that produces a low yield of the compounds.[1,2] Recombinant DNA technology has been used in fermentation processes with microbes to produce a high yield of the important artemisinin precursor, amorpha-4,11-diene by an economic route, thus lowering the cost per dose for a patient. The yield was increased by using *Escherichia coli* in which synthetic biology is applied to tailor the function of the microbe's genomics.

See: *recombinant DNA technology; synthetic biology*

1. Tsuruta, H., C. J. Paddon and D. Eng. High-level production of amorpha-4,11-diene, a precursor of the antimalarial agent artemisinin, in *Escherichia coli*. *PLos One* (2009) 4(2), e4489.

2. May, P., and S. Cotton. *Molecules That Amaze Us*, pp. 30–6. Boca Raton, FL: CRC Press, 2015.

artificial fertiliser

Sources of fertiliser for crops are likely to be a high priority for a growing world population in the 21st century. The production of ammonia in the Haber–Bosch process was described in 1910 and in initial work hydrogen and nitrogen were reacted over an osmium catalyst.[1] Nowadays, about 100 million tons of ammonia are produced per year in this process using an iron catalyst and reaction conditions around 500°C and 200 atm.[2,3]

1. Haber, F., and R. Le Rossignol. Production of ammonia. United States Patent 971,501, granted on 27 September 1910.
2. Gross, M. Fertile ground. *Chemistry and Industry* (Jan. 2014) 28, 24–7.
3. Busca, G. Heterogeneous Catalytic Materials: Solid State Behaviour, Surface Chemistry and Catalytic Behaviour. London: Elsevier, 2014.

artificial nails

Materials occur widely in consumer products, sometime visible to the purchaser and sometimes hidden from view. Cosmetics and associated products are advertised widely and their benefits widely promoted. Artificial nails are a popular product available in kit form for an individual to apply or they can be applied in a nail bar. One way the artificial nail is attached to the natural nail is to apply a liquid between the attachment and the natural nail and to cure the liquid by exposure to light from, for example, a light-emitting diode. Curing causes polymerisation of the liquid to, for example, an acrylic resin. The liquid acts as a photopolymer.[1]

See: *photopolymers*

1. Tiwari, A., and A. Polykarpov (eds). *Photocured Materials*. Cambridge: Royal Society of Chemistry, 2014.

artificial sweeteners

Artificial sweeteners are used in sugar-free products including carbonated soft (fizzy) diet drinks, chewing gum, puddings and fillings, desserts, yoghurt and cough mixtures.[1] Examples of artificial sweeteners are saccharin, sucralose (a chlorinated sucrose) sold under the trade name Splenda and aspartame that has been marketed under the name Nutrasweet. On a weight per weight basis, aspartame is about 200 times sweeter than table sugar (sucrose). Unlike aspartame, sucralose is not heat sensitive and can be used in cooking, frying and baking. Natural products, for example steviol glycosides from the stevia plant, are increasingly used as sugar substitutes.

1. Brazil, R. The sweet and the lowdown. *Chemistry World* (June 2015) 11, 50–63.

associated gas

Associated gas is natural gas (methane) found in combination with oil deposits.[1] It has been flared-off in the past but there is increasing interest in converting the gas to useful products such as synthetic diesel. Gas deposits in remote regions are referred to as stranded gas. Synthetic diesel can be prepared by first converting natural gas to a mixture of carbon monoxide and hydrogen known as synthesis gas by, for example, steam reforming. The reaction with steam at around 1000°C in the presence of a catalyst, often platinum. Carbon monoxide and hydrogen are then reacted over a nickel-based catalyst in the Fischer–Tropsch reaction to synthetic diesel.

See: *synthetic diesel*

1. Chianelli, R. R., X. C. Kretschmer and S. A. Holditch. Petroleum and natural gas. In D. S. Ginley and D. Cahen (eds), *Fundamentals of Materials for Energy and Environmental Sustainability*, pp. 106–16. Cambridge: Cambridge University Press, 2012.

auxetic materials

Poisson's ratio is the ratio of the lateral strain to the longitudinal strain in a stretched rod.[1,2] A positive Poisson ratio corresponds to a contraction perpendicular to the direction of stretch or an expansion perpendicular to the direction of compression when compressed. An auxetic material has a negative Poisson ratio. Put simply, it tends to become thicker in the direction perpendicular to the direction of stretch. Examples of auxetic materials include polymer foams[3] and a composite structure of carbon nanotube films embedded in a polymer matrix (e.g. polydimethylsiloxane, polyurethane, epoxy resin and polymethylmethacrylate) in which carbon nanotubes in one film are oriented at 90° to nanotubes in a neighbouring film. Some metals exhibit auxetic behaviour, particularly those with a cubic crystalline phase including the alloy β-brass, an alloy of copper and zinc.[3] Qualitatively, auxetic behaviour may be explained by assuming the bonds between elements of the structure, such as crystallites in a polymer, are of constant length but the links between the elements are flexible. On stretching the materials the regions of crystallinity move apart and the flexible links adjust their orientation so that the material becomes thicker.

1. Daintith, J. (ed.). *A Dictionary of Physics*, 6th edn. Oxford: Oxford University Press, 2009.
2. Chen, L.-Z., C.-H. Liu, J.-P. Wang and S.-S. Fan. Method for using a Poisson ratio material. United States Patent 8,545,745, granted on 1 October 2013.
3. Tilley, R. J. D. *Understanding Solids: The Science of Materials*. Chichester: Wiley, 2013.

backlights

Liquid crystals do not emit light. In a liquid crystal display a white backlight is used and combined with red, green and blue filters to display full colours.[1] Conventional backlights can be replaced by red, green and blue light-emitting diode (LED) pixels that eliminate the requirement for colour filters. This approach has been used in LED LCD televisions. A compact fluorescent light (CFL) source has been used conventionally as a backlight in LCD televisions. However, CFL sources are bulky and their replacement by LEDs enables thinner television sets to be manufactured when using a liquid crystal display.

1. Lucas, J., P. Lucas, T. Le Mercier, A. Rollat and W. Davenport. *Rare Earths: Science, Technology, Production and Use*, pp. 302–3. London: Elsevier, 2015.

bamboo

Bamboo is a fast-growing grass that can be harvested for construction purposes when around four years old.[1] Major growers and producers of bamboo are China, India, Myanmar and Nigeria while world production of bamboo is about 1.4 billion tonnes per year. Bamboo board is made from compressed bamboo strips cut from bamboo poles and arranged in stacks with the grain of neighbouring strips at right angles to each other. The stack is impregnated with a resin and then pressed and cured. Bamboo board is used for flooring while bamboo poles are used for scaffolding.

1. Ashby, M. F., D. F. Balas and J. S. *Coral. Materials and Sustainable Development*, p. 198. Oxford: Butterworth, 2016.

band gaps

Energy levels in a crystalline solid such as a semiconductor are arranged in energy bands. The band gap is the energy difference between the so-called valence band and conduction band and the band gap in a semiconductor decreases as the size of the atoms in the semiconductor increases.[1] Outer electron orbitals of larger atoms overlap and can produce wider bands. For example, the band gaps for silicon and germanium are 1.12 and 0.66 eV, respectively. The band gaps for gallium nitride (GaN) and gallium arsenide (GaAs) are 3.34 and 1.35 eV, respectively. Electrons are elevated from the valence to conduction bands when subjected to an electric field or to external radiation or by thermal energies. Electrons in the conduction band have a delocalised arrangement and have mobility throughout the crystal structure. The band gap is an important parameter in determining the wavelength of emitted light from semiconductor quantum dots and light-emitting diodes and for emission to occur the band gap must correspond to visible wavelengths. A larger band gap is a reflection of stronger bonding between the different atoms in the semiconductor. Shorter wavelengths are emitted as the band gap energy increases.

See: *quantum dots; extrinsic semiconductors; intrinsic semiconductors; semiconductors*

1. Tilley, R. J. D. *Understanding Solids: The Science of Materials*. Chichester: Wiley, 2013.

bioadhesives

Bioadhesives are natural adhesives.[1] An example of a bioadhesive concerns the ability of mussels, barnacles and limpets to remain attached to surfaces under water.[2] The blue mussel attaches itself to a rock through a foot and extrudes a protein gel and a hardening solution. The cross-linking agent is L-DOPA (dihydroxyphenylalanine), a derivative of the amino acid tyrosine; L-DOPA has uses in the treatment of Parkinson's disease. Adhesive proteins undergo cross-linking in the presence of dihydroxyphenylalanine.

See: *biodegradable hydrogels; epoxy adhesives; hairy adhesives; Post-it notes; superglue*

1. Burke, M. Stick insects. *Chemistry and Industry* (Feb. 2015) 79, 24–7.
2. Gratzer, W. Giant Molecules: From Nylon to Nanotubes. Oxford: Oxford University Press, 2013.

bio-based chemicals

The nomenclature used for processing renewable materials can cause confusion. Bio-based chemicals refer to materials derived from renewable resources.[1] Thus bio-based chemicals include biodiesel, biofuels (produced in bio-refineries), fermentation processes on sugars and use of genetically engineered organisms. Examples of bio-based chemicals include polyhydroxyalkanoates, 3-hydroxypropionic acid for conversion to acrylic acid, conversion of carbohydrate feedstocks into dicarboxylic acids, diammines and amino acids for polyamides including nylon as well as succinic acid for polyesters.

See: *biodiesel; bio-1,3-propanediol; carbohydrates; nylon; bio-succinic acid; recombinant DNA technology*

1. Trager, R. Bio-based chemicals on the rise in the US. *Chemistry World* (Dec. 2014) 11, 14.

bio-based resins

Synthetic resins such as epoxy resins have applications in composite materials. Bio-based resins have potential for production of composites:[1] for example, flax fibre–bio resin composites for vehicle interiors such as parcel shelves, car floors and door rims where the resin is derived from sugar, corncob and oat bagasse. The use of these composites makes vehicles lighter and saves fuel.

See: *composite materials; epoxy adhesives*

1. Redahan, E. Needing a resin for change. *Materials World* (Nov. 2013) 21, 10.

biobutanol

Compared to ethanol, butanol (C_4H_9OH) is less hygroscopic, less soluble in water and less flammable and has a low vapour pressure and a higher energy density. It can also be blended with petrol. Biobutanol has been produced by bacterial fermentation using natural microbial strains including *Clostridium acetobutylicum* although butanol is a toxic material for microorganisms. Butanol was produced industrially for use as a biofuel in 1916 by a process pioneered by Chaim Weizmann[1] and molasses, the residue left over from sugar refining, could be used as a feedstock for the process. Acetone was also produced on the industrial scale by fermentation processes[2] and it was an important chemical intermediate for the production of the explosive cordite that was used by the British Army.

See: *bioethanol*

1. Northen, T. R. Biofuels and biomaterials from microbes. In D. S. Ginley and D. Cahen (eds), *Fundamentals of Materials for Energy and Environmental Sustainability*, p. 316. Cambridge: Cambridge University Press, 2012.
2. Freemantle, M. The Chemists' War. *Chemistry World* (Sept. 2015) 12, 55–9.

biocomposites

Biocomposites are composites that include a natural material.[1] Examples of the latter are flax, hemp, coir (fibre from coconut shells), willow, cotton, sugar and sheep wool.

See: *composite matrials; flax fibres*

1. Thakur, V., and A. Singha (eds). *Biomass-Based Biocomposites*. Shrewsbury, UK: Smithers Rapra, 2013.

biodegradable hydrogels

Biodegradable hydrogels have applications in medicine as supports for slow-release of drugs and as adhesives for tissues during surgical procedures and will dissolve over time, the rate of dissolution dependent on the pH and composition of the hydrogel.[1] For example, polyethylene glycol (molecular weight 10,000) was reacted with glycolide oligomers (derived from glycolic acid $CH_2OHCOOH$) after which the copolymer was end-capped with acryloyl chloride ($CH_2CHCOOCl$). Polymerisation and crosslinking by exposure of the end-capped acrylate to ultraviolet or visible light in the presence of a free-radical initiator such as ethyl eosin produced the hydrogel.

See: *bioadhesives*

1. Hubbell, J. A., C. P. Pathak, A. S. Sawhney, N. P. Desai and J. L. Hill-West. Photopolymerizable biodegradable hydrogels as tissue contacting materials and controlled-release carriers. United States Patent 5,567,435, granted on 22 October 1996.

biodegradable hygiene products

There is much interest in limiting the amount of waste products that are taken to landfill sites. In the area of hygiene products, disposable nappies are made out of hydrogels and are not routinely biodegradable or renewable. In addition 'wet wipes', which are used for general cleaning purposes and for applying and removing cosmetics from the skin, can be but are not always biodegradable as some are made out of polyester fibres. Flushing wet wipes down lavatories can cause blockages in sewers. Biopolymers have been used to prepare disposable absorbent articles for potential use in sanitary towels and incontinence products.[1,2] For example, a mixture of a biodegradable aromatic-aliphatic copolyester mixture (Ecoflex resin) with starch and wheat gluten could be converted into a biodegradable film.

See: *engineering fabrics*

1. Shi, B. Biodegradable and renewable films. United States Patent 8,329,601, granted on 11 December 2012.
2. DeRosa, T. F. *Engineering Green Chemical Processes: Renewable and Sustainable Design.* New York: McGraw-Hill, 2015.

biodegradable nanoparticles

Biodegradable nanoparticles of poly (β-amino esters) have been prepared by condensing piperazine with selected bis(acrylate esters).[1] The poly (β-amino esters) had a range of molecular weights from around 4,000 to 31,000 Da. The biodegradable particles could be used to encapsulate and deliver nucleotide-based drugs containing DNA or RNA for treatment of human diseases.

See: *nanoparticles*

1. Langer, R. S., et al. Biodegradable poly (β-amino esters) and uses thereof. United States Patent 8,287,849, granted on 16 October 2012.

biodegradable plastics

Biodegradable plastics are materials broken down in soil by the action of microbes or by heat from a composting process or *in vitro*.[1,2] These plastics include polymers derived from natural resources such as corn starch or sugar, for example polylactic acid. Products from degradation are water, carbon dioxide and biomass. Potential applications for biodegradable plastics include nanoparticle-based medicines for targeting diseased cells *in vivo*.

See: *nanoparticle-based medicines; polylactic acid*

1. Reade, L. Plastic war. *Chemistry and Industry* (July 2014) 78, 20–3.
2. Felice, M. Biodegradable plastics. *Materials World* (Oct. 2013) 21, 50–1.

biodegradable stents

Materials are having an increasing impact on surgical procedures. For example, stents based on shape memory alloys, in particular Nitinol, are widely used. In a materials development, a stent made out of polylactic acid, a biodegradable plastic, can be inserted into an artery to prevent narrowing of the artery and possible heart attack. The stent (known as the Abbott absorb stent) contains a coating of slow-release drugs to help prevent inflammation of tissue, has the shape of a coil around 3 mm wide and about 3 cm in length and is inserted through the radial artery in the wrist.[1] The stent expands when it reaches the coronary artery and dissolves over a period of about 12 months.

See: *polylactic acid; shape memory alloys*

1. Davis, C. It's magic! Tiny tube unblocks your arteries—then vanishes! *Daily Mail*, 28 Apr. 2015.

biodegradable sutures

Surgical procedures often involve sutures in which stitching together the edges of a wound is carried out. In addition, meshes are used in some operations, for example those involving hernias. Biodegradable materials have been used for stitching a wound and for meshes and these materials dissolve in the body over a period of time.[1] An example of a biodegradable material for these applications is known by the trade name Vicryl.

1. Ethicon. Wound closure: Coated Vicryl (polyglactin 910) suture, available at http:// www.ethicon.com/healthcare-professionals/products/wound-closure/absorbable-sutures/coated-vicryl-polyglactin-910-suture)

biodiesel

Biodiesel refers to the mono-alkyl esters of long-chain fatty acids derived from vegetable oils or animal fat, for example the methyl ester of rapeseed oil.[1–3] Transesterification of animal fat followed by separation of glycerine (also called glycerol, a trihydric alcohol $HOCH_2CH(OH)$ CH_2OH)) from the fatty acid methyl ester or esterification of vegetable oils followed by separation of the water by-product are used for its preparation. Biodiesel is free of sulphur and aromatics and can be added to petroleum-based diesel to produce a biodiesel blend. Biodiesel and bioethanol are biofuels. Rudolf Diesel, who invented the diesel compressor engine in 1897, used peanut oil for fuel but this early use of biodiesel was surpassed by the use of petroleum as the fuel in the 1920s. An early patent for a self-propelled motor vehicle that used a mixture of hydrocarbons known as ligroin as the fuel is given in reference (3).

See: *bio-based chemicals; biofuels; green chemistry; polylactic acid*

1. Earle, M. J., K. R. Seddon and N. V. Plechkova. Production of bio-diesel. United States Patent Application 2009/0235574, published on 24 September 2009.
2. Sarin, A. *Biodiesel: Production and Properties*. Cambridge: Royal Society of Chemistry, 2012.
3. Benz & Company. Fahrzeug mit Gasmotorenbetrieh. German Patent 37435, granted on 2 November 1886.

bioethanol

Bioethanol is a biofuel and is produced by fermentation of crops such as sugar cane by using microorganisms, for example yeast.[1] Cellulose is broken down by hydrolysis using enzymes from micro-organisms into simple sugars that can be fermented into ethanol for use as biofuels. Bioethanol is produced in commercial quantities and is blended with petrol for automotive use; a 6% limit on the use of crop-based biofuels including bioethanol applies throughout the European Union for ground transport since September 2013.[2] Bioethanol obtained from crops is known as a first-generation biofuel.

See: *biobutanol; biodiesel; cellulosic ethanol*

1. Harfouche, A., R. Meilan and A. Altman. Tree genetic engineering and applications to sustainable forestry and biomass production. *Trends in Biotechnology* (2011) 29, 9–17.
2. Bellof, M., and R. Maas. Enzymes on display. *Chemistry and Industry* (Mar. 2014) 78, 36–9.

biofilms

There is worldwide concern in the 21st century on the increase in antimicrobial resistance. The bacteria *Staphylococcus aureus*, known as a 'Superbug' is a methicillin antibiotic resistant strain (MRSA). Over 2 million associated *Staphylococcus* infections occur annually in the USA with 40–50% associated with MRSA, the antibiotic resistant strain.[1] It has been estimated that by 2050 antimicrobial resistance may lead to 300 million premature deaths with up to £64 trillion lost to the global economy.[2] Biofilms are coatings found on a wide variety of surfaces including pipes and other parts of industrial equipment and can led to the build-up of mussels, corrosion and blockages. The films result in colonies of dangerous pathogens such as *S. aureus* and can form in implants such as heart valves, catheters and orthopaedic implants as well as surfaces in hospitals including doors, surgical equipment and beds. While there is much research on the identification on new antibiotics, another approach to tackling antimicrobial resistance is from consideration of the surface texture of shark skin.

See: *shark skin*

1. O'Driscoll, C. Antibody attack on *Staphylococcus*. *Chemistry and Industry* (Feb. 2015) 79, 10.
2. King, A. Antimicrobial resistance will kill 300 million by 2050 without action. *Chemistry World* (Dec. 2014) 12, 6.

biofuels

Biofuels are fuels derived from plant biomass and include bioethanol and biodiesel.[1] Bioethanol produced by fermentation of sugar from sugar cane by micro-organisms involving enzymatic hydrolysis is known as a first-generation biofuel. There is interest in second-generation biofuels obtained from sources of cellulose such as wood chips, although the technology is not as advanced as for first-generation biofuels.

See: *bioethanol; biodiesel; biomass*

1. Northen, T. R. Biofuels and biomaterials from microbes. In D. S. Ginley and D. Cahen (eds), *Fundamentals of Materials for Energy and Environmental Sustainability*, pp. 315–35. Cambridge: Cambridge University Press, 2012.

biomass

Biomass refers to natural organic materials such as plants, sawdust, agricultural waste, namely wheat straw, bamboo, corn stover, rice straw, rice hulls, hemp, flax, jute, cotton, sugar cane bagasse and wood chips that are sources of renewable energy (biofuels such as bioethanol) or renewable materials as alternatives to materials derived from petrochemical sources or which can be burnt to produce heat and power.[1]

See: *lignocellulose; cellulose; hemicellulose*

1. Thakur, V., and Singha, A. (eds). *Biomass-Based Biocomposites*. Shrewsbury, UK: Smithers Rapra, 2013.

biomaterials

Biomaterials refer to materials found in living (or dead) organisms.[1,2] Examples include a composite structure of alternating layers of protein and calcite (calcium carbonate) in the abalone shell, in which the high fracture toughness compared to calcite alone protects this mollusc from predatory attack.[1] The colours of butterfly wings and peacock wings arise not from the presence of pigments but from interference of diffracted light from periodic structures of biomaterials in the wings with a periodicity around the wavelength of light.

See: *iridescent organisms; iridescent pigments; nacre; photonic materials*

1. Ozin, G. A., A. C. Arsenault and L. Cademartiri (eds). *Nanochemistry: A Chemical Approach to Nanomaterials*. Cambridge: Royal Society of Chemistry, 2009.
2. Kulshrestha, A., A. Mahapatro and L. Henderson (eds). *Biomaterials*, American Chemical Society Symposium Series. Oxford: Oxford University Press, 2011.

biomimetic materials

Many living organisms contain intricate structures of minerals obtained by a mineralisation process from soluble species interspersed with proteins.[1] These structures are often natural composites whose toughness and strength help to protect the organism. Examples of these structures include diatoms, single-celled animals that have an outer shell of silica with a well-defined pore size. Egg shells contain ordered crystals of calcium carbonate in a matrix of natural polymers. The skeletons of sponges comprise silica rods while the shells of molluscs contain aggregates of aragonite crystal. Bone is a composite of the protein collagen and hydroxyapatite. The teeth of limpets contain fibres of goethite (α-$Fe_2O_3.H_2O$) in a strong composite structure.[2] Biomimetic materials are materials made in a laboratory with the aim of utilising the structures of naturally occurring materials in organisms in order to develop improved material properties such as greater toughness.

See: *composite materials; biomaterials, hydroxyapatite*

1. Calvert, P. Biomimetic processing. In R. J. Brook, *Materials Science and Technology*, vol. 17B, *Processing of Ceramics*, Part II, pp. 51–82. Weinheim, Germany: VCH Publishers, 1996.

2. Spencer, B. Nothing's as tough as the teeth of a limpet. *Daily Mail,* p.3, 18 Feb. 2015.

biopharmaceuticals

Throughout the 20th century many drugs for therapeutic purposes were based on small molecules prepared by a chemical synthesis, for example acetylsalicylic acid (molecular weight 180). There has been increasing interest since the last two decades of the 20th century on the development of biopharmaceuticals or biologic drugs and these are based on monoclonal antibodies and prepared by techniques involving yeasts or bacteria in recombinant DNA technology.[1] The biopharmaceuticals have sequences of amino acids in their structure and are monoclonal antibodies. These techniques have been called protein engineering. Examples of biopharmaceuticals are Herceptin (trastuzumab) for the treatment of breast cancer, Avastin (bevacizumab) for the treatment of colon cancer and Remicade (Infliximab) for the treatment of rheumatoid arthritis.

See: *acetylsalicylic acid; antibodies*

1. Ganellin, C. R., R. Jefferis and S. Roberts (eds). *Introduction to Biological and Small Molecule Drug Research and Development: Theory and Case Studies.* London: Elsevier, 2013.

bioplastics

Throughout the 20th century plastics have been prepared mainly from petroleum feedstocks. Bioplastics are derived from renewable sources and some are biodegradable such as polylactic acid. Genetically engineered *Escherichia coli* was used to convert glucose to muconic acid that could be hydrogenated to adipic acid.[1] The latter is conventionally derived from petroleum on an industrial scale and converted to nylon. The global production capacity in 2011 for bioplastics was just over 1 million tonnes compared to 280 million tonnes for all plastics.

See: *biodegradable plastics; plastics; polylactic acid*

1. Niu, W., K. M. Draths and J. W. Frost. Benzene-free synthesis of adipic acid. *Biotechnology Progress* (2002) 18(2), 201–11.

biopolymers

Biopolymers are naturally occurring polymers and include proteins such as insulin, cellulose, nucleic acids, for example deoxyribonucleic acid (DNA) and bioplastics.[1] The latter, unlike plastics derived from petrochemical sources, are biocompatible and biodegradable. Bioplastics are usually produced in a fermentation process on a sugar-based medium to produce monomers that can be condensed and polymerised.

See: *poly-3-hydroxybutyrate; polymers*

1. Ebnesajjad, S. (ed). *Handbook of Biopolymers and Biodegradable Plastics*. London: Elsevier, 2013.

bioprinting

The success of organ transplantation is limited by the availability of donors. Bioprinting is an emerging technology that controls the distribution of living cells within a biodegradable matrix material to build up a tissue structure layer by layer to replace diseased organs, for example livers or kidneys.[1,2] Three-dimensional printing has the potential to construct artificial organs when combined with living cells and matrix material as the materials to be printed. The matrix can be a biodegradable hydrogel printed as a suspension containing living cells. Examples of matrix material include collagen, fibrin, chitosan and hyaluronic acid.

See: *hydrogels; scaffolds; stereolithography; three-dimensional printing*

1. S. K. Hasan. Three-dimensional tissue generation. United States Patent Application 2011/0250688, published on 13 October 2011.
2. Challener, C. Cell prints. *Chemistry & Industry* (Feb. 2014) 78, 24–7.

bio-1,3-propanediol

Bio-1,3-propanediol ($HOCH_2CH_2CH_2OH$) is an example of how genetically engineered organisms can produce higher-value products from biomass.[1] Glucose extracted from corn is digested by such organisms in a vessel containing a growth medium. The glucose is metabolized, yielding 1,3-propanediol, which can be reacted with terephthalic acid ($COOHC_6H_4COOH$) to produce a polymer that can be spun into fibres for clothing and carpets. The trade name for this polymer is Sorona.

See: *bio-based chemicals; biomas; green chemistry*

1. Armitage, Y. The industrial biotechnology opportunity. *Materials World* (June 2014), 22, 45–7.

bio-succinic acid

An example of the versatility of genetic engineering is illustrated by the use of fermentation of corn glucose by genetically engineered yeast or *Escherichia coli* to yield succinic acid ($HOOCCH_2CH_2COOH$).[1] Succinic acid is a commodity chemical and it is an intermediate for the manufacture of solvents such as tetrahydrofuran and maleic anhydride, which is used for production of resins and polymers. It is also an intermediate for manufacture of polyesters and polyimides.

See: *bio-based chemicals; green chemistry; recombinant DNA technology; yeast*

1. Johnson, E. Sweeter feedstock. *Chemistry and Industry* (Nov. 2014) 78, 42–5.

bleaching agents

Peracids, for example peracetic acid (CH_3COOOH), are used as mild bleaching agents in detergent products;[1,2] peracids are also used in bleaching wood pulp in the paper-making process. The current method to produce peracids involves reaction of a carboxylic acid with hydrogen peroxide (H_2O_2) in the presence of a strong acid such as concentrated sulphuric acid. This process produces toxic waste products. In a greener reaction peracids have been produced enzymatically by replacement of sulphuric acid with Bacillus subtilis 31954. For example, treatment of triacetin with H_2O_2 at pH 6.5 in the presence of this strain at 25°C produced a concentration of approximately 3500 ppm of peracetic acid after 10 minutes.

See: *green chemistry*

1. Dicosimo, R. Production of peracids using an enzyme having perhydrolysis activity. United States Patent 8,329,441, granted on 11 December 2012.
2. DeRosa, T. F. *Engineering Green Chemical Processes: Renewable and Sustainable Design.* New York: McGraw-Hill, 2015.

bone china

While many materials are used for their functional properties, others are admired for their aesthetic properties. Bone china was first produced in the mid-18th century in England and is associated with English tableware.[1,2] Its name originates from the use of calcined bone ash (from animal bones) and a feldspar (e.g. $(K, Na)_2O;Al_2O_3;6SiO_2$) that acts as a liquid-phase sintering aid. The first firing involves kiln temperatures up to 1280°C. Glazes are used to seal pores in bone china products. A detailed account of the materials and processes involved in producing tableware such as bone china is given in reference (2).

See: *porcelain*

1. Hennicke, H. W., and A. Hesse. Traditional ceramics. In *Concise Encyclopaedia of Advanced Ceramic Materials*, p. 489. Oxford: Pergamon Press, 1991.
2. Miller, J. (Gen. Ed.). *Miller's Antiques Encyclopedia.* London: Octopus Publishing, 2013.

Brazilian bioethanol

Bioethanol has been produced economically on a large scale in Brazil for some years.[1] The feedstock is sugarcane that can be grown at a large tonnage per hectare because of the favourable climate. Sugar is extracted from the sugarcane and converted to ethanol by fermentation using yeast; the remaining sugarcane bagasse can be burnt, supplying energy for purification of ethanol by distillation.

See: *bioethanol*

1. Northen, T. R. Biofuels and biomaterials from microbes. In D. S. Ginley and D. Cahen (eds), *Fundamentals of Materials for Energy and Environmental Sustainability*, p. 318. Cambridge: Cambridge University Press, 2012.

breathable garments

The design, colours and comfort of clothing all affect the choices consumers make when making purchases. Breathable waterproof clothing has been produced by incorporating a porous layer of, for example, Gore-Tex or similar material as the pores are too small to let raindrops through but large enough to let sweat molecules through.[1] Thus, the porous layer acts as a membrane. Certain footwear ranges including those marketed under the trade name GEOX have pores of a controlled size that act as a breathable membrane.

See: *Gore-Tex*

1. Tyrell, J. A. *Fundamentals of Industrial Chemistry: Pharmaceuticals, Polymers and Business*, p. 128. New York: Wiley, 2014.

brominated flame retardants

Brominated compounds, for example hexabromocyclododecane, are used as fire retardants, particularly in textiles, where they are incorporated into organic polymers in a melt-processing step.[1] Bromine free radicals in the gas phase scavenge highly energetic oxygen and hydroxyl free radicals derived from combustion products of the polymer, hence limiting flame propagation by free radicals. There are concerns over the possible toxicity of low molecular weight brominated compounds and their ability to accumulate in body tissue.

1. Moran, N. Phasing out fire retardants. *Chemistry World* (Aug. 2013), 10, 50–3.

bullet-proof vests

Bullet-proof vests are made out of a synthetic polymer, Kevlar, and can be considered to be made out of a plastic.[1] Kevlar has the chemical name of poly (p-phenylene terephthalamide) and solutions of the polymer can be spun into fibres. Kevlar is lightweight with a low density, high stiffness and high tensile strength. Kevlar has a strength and stiffness greater than steel; bullet-proof vests are made of cross-woven Kevlar fibres which absorb energy on bending under impact and dissipate the energy over a wide area. Kevlar is an example of an aramid fibre.

See: *aramids; Kevlar*

1. Dodgson, M., and D. Gann. *Innovation: A Very Short Introduction*, pp. 51–4. Oxford: Oxford University Press, 2010.

candy floss

Foodstuffs can be considered as materials and their processing can have similarities with the processing of other materials. Candy floss (also called cotton candy) is familiar to generations of children and their parents during visits to funfairs or the seaside. A patent was issued for a machine to prepare candy floss in 1899 and the general method used nowadays involves dropping flavoured granulated sugar onto a spinning disc with small holes at the

Fig. G.3 *Cotton candy is made by melting flavoured, dyed sugar until it turns liquid and then spinning it by machine into fine threads.*
(Photo courtesy of Shutterstock.)

edges; the spinning disc is a spinerette (Fig. G.3).[1] A heating element melts the sugar that streams out of the holes in the spinning disc. The streams of liquid sugar (i.e. the floss) solidify in the colder air into a sugar glass, namely the candy floss. There are similarities in the way candy floss and Kevlar is produced by spinning. In the 20th century glass fibres have been prepared by extruding molten glass through small holes in a spinerette.[2] A more detailed description of candy floss is given in reference (3).

See: *fibreglass; Kevlar*

1. Morrison, W. J., and J. C. Wharton. Candy machine. United States Patent 618428, granted on 31 January 1899.
2. Slayter, G. Method and apparatus for making glass wool. United States Patent 2,133,235, 1938.
3. Goldstein, D. (ed.). *The Oxford Companion to Sugar and Sweets*, pp. 190–1. Oxford: Oxford University Press, 2015.

carbohydrates

Carbohydrates are compounds with the general formula $C_6(H_2O)_y$ and include the simple sugars (monosaccharides) that cannot be broken down into smaller units.[1] Examples of monosaccharides are glucose ($C_6H_{12}O_6$), fructose ($C_6H_{12}O_6$), which is stereoisomeric with glucose, and ribose ($C_5H_{10}O_5$), a constituent of ribonucleic acid (RNA); deoxyribose ($C_5H_{10}O_4$) is a constituent of deoxyribonucleic acid (DNA). Sucrose is a carbohydrate consisting of a molecule of glucose chemically linked to a molecule of fructose.

See: *bio-based chemicals; polysaccharides; sugary foods*

1. Gratzer, W. *Giant Molecules: From Nylon to Nanotubes.* Oxford: Oxford University Press, 2013.

carbon aerogels

Carbon aerogels are obtained when organic aerogels are pyrolysed at high temperature, around 1000°C.[1,2] High surface areas that can be obtained in these materials, on the order of 1000 m^2 g^{-1}, are potentially useful for battery electrodes and catalyst supports.

See: *organic aerogels*

1. Mayer, S. T., J. L. Kaschmitter and R. W. Pekala. Carbon aerogel electrodes for direct energy conversion, International Patent Application WO 95/20246, 1995.
2. Segal, D. *Exploring Materials through Patent Information*, p. 114. Cambridge: Royal Society of Chemistry, 2014.

carbon capture

The removal of carbon dioxide produced from the combustion of coal and fossil fuels in power stations is a challenge for the 21st century. This removal from flue gases is known as carbon capture, or carbon sequestration.[1] Materials that have been considered for carbon capture include organic amines in a liquid or solid form that can be regenerated. Inorganic adsorbents are also under consideration.

See: *industrial effluents; carbon dioxide*

1. Frost, S. Capturing the bogeyman. *Materials World* (Sept. 2015) 23, 10–11.

carbon fibre

Carbon fibres for use as lamp filaments had been produced by Thomas Edison in the 19th century from carbonised cotton and bamboo.[1] Since the 20th century continuous carbon fibres have been produced from synthetic polymers, in particular by carbonisation of wet-spun polyacrylonitrile fibres. Carbonisation involves a heat treatment of the spun fibres in an inert atmosphere at temperatures in the range 1000–1500°C. An alternative approach involves carbonisation of wet-spun mesophase pitches derived from thermal polymerisation of petroleum- or coal-tar-based pitches; these mesophase sources contain liquid crystals. Carbon fibres have an as-prepared diameter in the range 5–15 μm and can be woven into mats and felts. Examples of carbon fibres are AS-4 (Hexel), T300 (Toray) and CN60

(Nippon Graphite). The fibre has a graphite-type structure in layers along the length of the fibre with strong covalent bonding between carbon atoms along the layer and weak van der Waals forces between the layers. Carbon fibres are characterised by (i) stiffness as measured by Young's modulus up to around 900 GPa, (ii) being lightweight with a density approximately 2.0 g cm^{-3}, (iii) tensile strength of more than 5 GPa and (iv) thermal conductivity of around 1000 W m^{-1} K^{-1}. In a carbon–carbon composite the carbon matrix can be built up by, for example, vapour phase infiltration of carbon fibre mats. Carbon–carbon composites are stable at high temperature over 3000°C. Carbon fibre-reinforced composites have applications where lightweight, strong and tough structural components are required, particularly for space, military and civilian aircraft: for example, components for aircraft (Boeing 777, Airbus 380 and the Boeing Dreamliner), brake discs for luxury cars such as Porsche and Audi models, rocket nozzles and re-entry parts for missiles. Carbon–carbon composites have low oxidation stability above 400°C. This lack of stability can be overcome by coating the fibres with silicon carbide. In addition, carbon fibre-reinforced silicon matrices have been used in propulsion systems in spacecraft and satellite structures such as telescope components where the low thermal expansion of carbon–silicon carbide composites is desirable.

See: *acrylic fibre; composite materials; liquid crystals*

1. Bansal, N. P., and J. Lamon (eds). *Ceramic Matrix Composites: Materials, Modeling and Technology.* New York: Wiley, 2015.

carbon fibre composites

Continuous carbon fibre can be fabricated into carbon–carbon composites by infiltration of a mat of the fibre, for example from the vapour phase. Carbon fibre-reinforced composites have applications where lightweight, strong and tough structural components are required, particularly for space, military and civilian aircraft:[1] for example, components for aircraft (Boeing 777, Airbus 380 and the Boeing Dreamliner), brake discs for luxury cars such as Porsche and Audi models, rocket nozzles and re-entry parts for missiles. The addition of fibres to a matrix increases the toughness of the component, that is the resistance to fracture.

See: *carbon fibre*

1. Bansal, N. P., and J. Lamon (eds). *Ceramic Matrix Composites: Materials, Modeling and Technology.* New York: Wiley, 2015.

carbon nanotubes

A carbon nanotube is an allotrope of carbon and consists of an ordered sheet of carbon atoms arranged in a hexagonal array rolled into a tube; the sheet of carbon atoms corresponds to a sheet of graphene (Fig. G.4). Carbon nanotubes were first identified in 1991 by Sumio Iijima, who passed an electric arc discharge between graphite rods.[1] Other preparative methods include laser evaporation, catalytic hydrocarbon decomposition and chemical vapour deposition.[2–4] Carbon nanotubes can be single-walled or multi-walled, where the latter consist of concentric layers of single-walled nanotubes. Carbon nanotubes have diameters from several angstroms up to several nanometres and lengths up to several thousand times larger than the diameter. These nanotubes have high current capacity (up

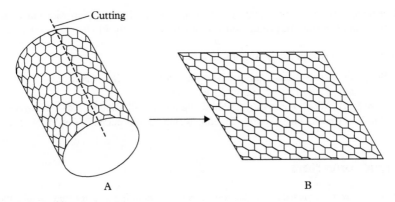

Fig. G.4 *Conversion of a carbon nanotube (A) to monolayer graphene (B).*
(B. Z. Zhang, et al. United States Patent Application 2005/0271574, 2005.)

to 10^{10} amp cm^{-2}), mechanical flexibility, high tensile strength (up to 100 GPa) and chemical stability. Potential applications for carbon nanotubes include hydrogen storage media, supercapacitors, structural materials, fillers for polymer composites and transparent electrodes for flat panel displays.

1. Iijima, S. Helical microtubes of graphitic carbon. *Nature* (1991) 354 (7), 56–8.
2. Silva, S. R., S. Haq and B. O. Boskovic. Production of carbon nanotubes. United States Patent 8,715,790, granted on 6 May 2014.
3. Talapatra, S., S. Kar, S. Pal, R. Vajtai and P. Ajayan. Carbon nanotube growth on metallic substrate using vapour phase catalyst delivery. United States Patent 8,207,658, granted on 26 June 2012.
4. Predtechensky, M. R., O. M. Tukhto and I. Y. Koval. United States Patent 8,137, 653, granted on 20 March 2012.

carbon/SiC braking systems

There are many examples of industrial equipment that incorporate braking systems either for routine operation or in emergency braking systems:[1] for example, in lifts, cranes, funfairs and high-performance lifts in tall buildings. During braking, kinetic energy is converted into heat and materials which are used for brake discs require a high coefficient of friction that is maintained at high temperatures produced on contact between the brake pads; temperatures as high as 1200°C can be reached. Carbon fibre-reinforced silicon carbide composites fulfil these requirements.

1. Bansal, N. P., and J. Lamon (eds). *Ceramic Matrix Composites: Materials, Modeling and Technology.* New York: Wiley, 2015.

catalysts

Catalysts are used to direct the selectivity of chemical reactions and to speed up the rate of reaction.[1,2] The term catalysis was first proposed by Jons Jakob Berzilius in 1835 and covers heterogeneous, homogeneous and enzymatic catalysis. An example from everyday life involves three-way automotive catalysts for the treatment of exhaust emissions in vehicles. A further example relates the production of synthetic diesel in the Fischer–Tropsch reaction.

See: *three-way automotive catalysts; synthetic diesel*

1. Busca, G. *Heterogeneous Catalytic Materials: Solid State Behaviour, Surface Chemistry and Catalytic Behaviour*. London: Elsevier, 2014.
2. Cornils, B., W. A. Hermann, C.-H. Wong and H.-W. Zantoff (eds). *Catalysis from A to Z: A Concise Encyclopedia*, 4th edn. Weinhem, Germany: Wiley-VCH, 2013.

catalytic converters

A catalytic converter consists of a stainless steel can containing a supported metal catalyst that is fitted to the exhaust pipe of vehicles.[1] The catalyst is supported on a high surface area alumina coating on a low thermal expansion ceramic honeycomb, usually made out of cordierite. The catalyst converts carbon monoxide, unburned hydrocarbons and nitrogen oxides in the hot exhaust gases, which are around 1000°C, to carbon dioxide, nitrogen and water. Catalysts include platinum (or palladium) and rhodium compounds. Cerium oxide is sometimes added to the catalyst to aid catalytic activity under cold-start conditions.

See: *three-way automotive catalysts*

1. Lucas, J., P. Lucas, T. Le Mercier, A. Rollat and W. Davenport. *Rare Earths: Science, Technology, Production and Use*, p.142–58. London: Elsevier, 2015.

catalytic cracking

Catalytic cracking takes place in petroleum refineries and converts large hydrocarbon molecules into smaller molecules of industrial use by vaporising the larger molecular fraction and passing the vapour phase over a powdered zeolite catalyst such as faujasite in a fluidised bed.[1] Products include petrol, diesel and fuel oil as examples. Lanthanum and cerium are added to petroleum-cracking catalysts, which prevent sintering and collapse of the porous zeolite structure.

1. Lucas, J., P. Lucas, T. Le Mercier, A. Rollat and W. Davenport. *Rare Earths: Science, Technology, Production and Use*, p.158–9. London: Elsevier, 2015.

cellophane

Cellulose can be dissolved in a mixture of carbon disulphide and sodium hydroxide, after which the resulting cellulose xanthate solution is extruded through a narrow slit to produce sheets of cellophane used as a wrapping material although it is flammable.[1]

See: *cellulose*

1. Brandenberger, J. E. Composite cellulose film. United States Patent 1,266,766, granted on 21 May 1918.

cellulase enzymes

The preparation of biofuels such as bioethanol from lignocellulose requires enzymatic hydrolysis with cellulase enzymes to degrade cellulose and hemicellulose polymers to monomeric sugar molecules with a low molecular weight, for example C_5 and C_6 sugars that are then fermented in the presence of yeasts or bacteria to ethanol. Cellulase enzymes have been obtained from fungi such as *Aspergillus* and *Trichoderma*, which express and secrete these soluble enzymes.[1] Examples of cellulase enzymes are endogluconases, cellobiohydrolases and β-glucosidase and these enzymes are used in the production of faded cotton jeans.

See: *faded jeans*

1. Bellof, M., and R. Maas. Enzymes on display. *Chemistry and Industry* (Mar. 2014) 78, 36–9.

celluloid

Celluloid, a highly flammable thermoplastic, was produced by mixing cellulose nitrate and a binder, camphor; the latter acted as a plasticiser. It had applications in the film industry in the form of tape that could be wound onto a spool but was eventually discontinued in this application due to its flammability. Developments on celluloid were made by John Hyatt Jr in 1869 who prepared it on a commercial scale and celluloid was used as a substitute for ivory billiard balls.[1] In 1863 the increasing demand for billiard balls has driven the African elephant to the edge of extinction with 12,000 animals being killed each year.[2] This synthetic resin had various applications, for example as dentures, spectacle frames, fountain pens, combs, buttons and dolls, but its flammability limited its use. John Hyatt and his younger brother Isaiah developed an injection moulding machine in 1872 that allowed bars or sheets of celluloid to be produced that were then worked into the final product.[3]

See: *cellulose; cellulose nitrate; Parkesine*

1. Hyatt, J. W. Jr. Improved molding composition to imitate ivory and other substances. United States Patent 88,633, granted on 6 April 1869.
2. Donald, G. *The Accidental Scientist: The Role of Chance and Luck in Scientific Discovery*, pp. 32–8. London: Michael O'Mara Books Limited, 2013.
3. Chaline, E. *Fifty Machines that Changed the Course of History*, pp. 42–3. London: Apple Press, 2012.

cellulose

Cellulose is the most abundant biorenewable material on earth and this crystalline polysaccharide consists of linear polymeric chains formed by repeated connection of β-D-glucose building blocks through a 1-4 glycoside linkage (Fig. G.5).[1–3] Cellulose is insoluble in water and common organic solvents due to the close-packed polymer chains that are linked through hydrogen bonding and van der Waals forces; it is found in the cell walls of plants, providing rigidity to the cell walls. Hence it is a major component of trees, vegetable fibre,

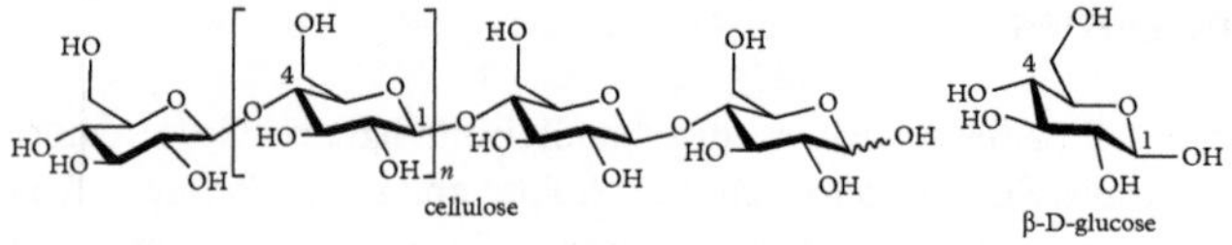

Fig. G.5 *Schematic diagram of the structure of cellulose and β-D glucose.*
(J. Holbrey, et al. United States Patent 8,574,368, 2013.)

bamboo, flax and hemp, as examples. Polymer chains are oriented in the direction of the vertical axis in trees and this orientation adds to the strength of the trunk. Cellulose (e.g. wood pulp) can be converted to viscose rayon, a fibre used in textiles by dissolution in a mixture of carbon disulphide and sodium hydroxide, after which the resulting aqueous cellulose xanthate solution is converted to fibres of regenerated cellulose (i.e. rayon) by passage through fine nozzles into sulphuric acid. Cellulose xanthate is extruded through a narrow slit to produce sheets of cellophane used as a wrapping material although it is flammable.[4] Cellophane is impermeable to moisture, flexible and transparent. Microcrystalline cellulose powders are used in pharmaceutical applications. Cellulose is broken down by hydrolysis using enzymes from micro-organisms into simple sugars that can be fermented into ethanol for use as biofuels.

See: *bioethanol; biomass; biopolymers; cellophane; lignocellulose; polysaccharides*

1. Walton, D., and P. Lorimer. *Polymers*, p. 16. Oxford: Oxford University Press, 2005.
2. Buchanan, C. M., and N. L. Buchanan. Cellulose esters and their production in halogenated ionic liquids. United States Patent 8,273,872, granted on 25 September 2012.
3. Askeland, D. K., and W. J. Wright. *The Science and Engineering of Materials*, 7th edn, p. 660. Boston, MA: Cengage Learning, 2011.
4. Brandenberger, J. E. Composite cellulose film. United States Patent 1,266,766, granted on 21 May 1918.

cellulose acetate

Cellulose acetate is obtained by reaction of cellulose with glacial acetic acid and acetic anhydride in the presence of sulphuric acid and was first prepared towards the end of the 19th century.[1] It can be spun into fibres and applications include lacquers and shatterproof glass for windscreens and these fibres are known as acetate rayon. Cellulose acetate can be formed into sheets that as a thermoplastic can be shaped by pressing a hot sheet into a mold. Plastics and fibres derived from cellulose are known as cellulosics.

See: *cellulose*

1. Fried, J. R. *Polymer Science and Technology*, 3rd edn, p. 350. Upper Saddle River, NJ: Prentice Hall, 2014.

cellulose-derived hydrocarbons

Sugars derived from hydrolysis of cellulose can be catalytically converted to hydrocarbons.[1] For example, glucose can be hydrogenated to sorbitol ($CH_2OH(CHOH)_4CH_2OH$).

Sorbitol is then deoxygenated to hexane (C_6H_{14}) by using hydrogen and a palladium or platinum catalyst supported on silica or alumina at approximately 250°C.

1. Shi, J., Q. Qing, T. Zhang, C. E. Wyman and T. A. Lloyd. Biofuels from cellulosic biomass via aqueous processing. In D. S. Ginley and D. Cahen (eds), *Fundamentals of Materials for Energy and Environmental Sustainability*, pp. 336–48. Cambridge: Cambridge University Press, 2012.

cellulose nitrate

Cellulose nitrate, or nitrocellulose, was first obtained in 1846 by Frederick Schonbein, a professor of chemistry at the University of Basle, by reaction of cellulose in the form of cotton or wood pulp with nitric acid;[1] nitrocellulose is an ester ($-CONO_2$) and not a nitro compound ($-C-NO_2$). Cellulose nitrate has been used as a shock-sensitive explosive in the form of gun cotton and is highly flammable. It is cellulose trinitrate in contrast to collodion, which is cellulose dinitrate.[2] When saturated with water, nitrocellulose can be transported relatively safely as a pulp that is detonated by a small quantity of dry nitrocellulose.[3]

See: *celluloid; cellulose; collodion; Parkesine*

1. Gratzer, W. Giant Molecules: *From Nylon to Nanotubes.* Oxford: Oxford University Press, 2013.
2. Walton, D., and P. Lorimer. *Polymers*, p. 17. Oxford: Oxford University Press, 2005.
3. Akhaven, J. *The Chemistry of Explosives*, 3rd edn, p.4. Cambridge: Royal Society of Chemistry, 2011.

cellulosic ethanol

Bioethanol obtained from lignocellulose in agricultural waste such as corn stover, wheat straw and sugar cane bagasse is referred to as a second-generation biofuel. Process development is at an early stage but this route to bioethanol could become important as the 21st century progresses. In the process lignin is separated from cellulose and hemicellulose in a lignocellulosic feedstock, after which cellulose and hemicellulose undergo enzymatic hydrolysis with cellulase enzymes that break down sugar polymers to smaller molecules with five or six carbon atoms.[1] The C_6 and C_5 sugars are then fermented by bacteria or yeasts to ethanol.

See: *bioethanol; cellulase enzymes*

1. Bellof, M., and R. Maas. Enzymes on display. *Chemistry and Industry* (Mar. 2014) 78, 36–9.

ceramic glazes

Conventional ceramic articles such as tableware (e.g. cups, saucers, plates and jugs) are manufactured from a mixture of primarily clays and a flux such as feldspar in which the mixture can have the consistency of 'dough' that is shaped to form, for example, plates. Alternatively, larger articles such as sinks and lavatory basins are shaped by slip casting of a slurry. The shaped article is fired in a kiln at about 1100°C, during which the feldspar or similar sintering aid encourages liquid-phase sintering between reactants. In the case of tableware the fired bodies are porous and known as 'biscuit' and pores are sealed by dipping the article into a slurry of glass powder to produce a coating on the article, which is then fired again in a kiln at a lower temperature to yield a hard glaze coating. Decoration is applied on top of the glaze coating and a further firing step in the kiln bonds the decoration to the article surface. Hand-decorated ware refers to applying by hand a transfer or decal containing the required pattern to the glaze surface. An account of the different types of glazes (e.g. salt glazes, tin glazes, transparent glazes) is given in reference (1).

See: *ceramic stains*

1. Miller, J. (Gen. Ed.). *Miller's Antiques Encyclopedia*. London: Octopus Publishing, 2013.

ceramic honeycombs

Concerns over pollution emanating from the exhausts of petrol and diesel vehicles has led to legislation to limit the pollutants. Extruded ceramic honeycombs are often made out of cordierite, a magnesium aluminium silicate ($2MgO;2Al_2O_3;5SiO_2$) that is a glass ceramic with a low thermal expansion able to withstand thermal cycling in hot exhausts without cracking.[1] Porous walls of the channels in the honeycombs are coated with a catalyst supported on a high surface area alumina for conversion of carbon monoxide to carbon dioxide, hydrocarbons to water and carbon dioxide and nitrogen oxides to nitrogen. The channel density is around 200 per square inch of cross-section of the honeycomb. For diesel engines the honeycomb acts as a particulate filter or trap to remove carbon particulates.

See: *diesel particulates; glass ceramics; three-way automotive catalysts*

1. Liu, P. S., and G. F. Chen. *Porous Materials*, pp. 295–6. London: Elsevier, 2014.

ceramic materials

Ceramics are non-metallic inorganic oxides.[1,2] The word ceramic is of Greek origin and its translation (keramos) means potter's earth. Traditional ceramics are derived from naturally occurring raw materials and include clay-based products such as tableware and sanitary-ware as well as structural claywares, for example bricks and pipes. Raw materials include clay minerals such as kaolinite, also known as kaolin ($Al_2Si_2O_5(OH)_4$), silica sand and a feldspar (e.g. $(K, Na)_2O;Al_2O_3;6SiO_2$) that acts as a liquid-phase sintering aid; clay minerals disperse in an aqueous environment to form colloidal systems. The fired micro-structure contains glassy phases, pores and crystalline phases. Traditional ceramics also include glasses, cements and refractories. The latter include chrome-magnesite refractories used in the steel-making industry. Advanced ceramics are derived from chemical synthetic routes or from naturally occurring materials that have been highly refined. Examples of advanced ceramics include yttria-stabilised zirconia, silicon carbide fibres for composites, nanoparticles of titanium dioxide for sunscreens, synthetic hydroxyapatite for coated hip implants, silica aerogels, barium titanate for capacitors and tin oxide gas sensors.

See: *electroceramics; engineering ceramics; ceramic matrix composites*

1. Somiya, S. (ed.). *Handbook of Advanced Ceramics*. San Diego, CA: Academic Press, 2013.
2. Verdeja, L. F., J. P. Sancho, A. Ballester and R. Gonzalez. *Refractory and Ceramic Materials*. Madrid: Editorial Sintesis, 2014.

ceramic matrix composites

Ceramic matrix composites (CMC) consist of a component in which ceramic reinforcement such as fibre or powder is dispersed throughout a ceramic matrix so that the reinforcement increases the toughness of the composite, that is increases its resistance to fracture. Examples of CMCs are silicon carbide fibre as reinforcement for a silicon carbide matrix or alumina fibre reinforcement in an alumina matrix.[1]

See: *ceramic materials; silicon carbide fibre composites*

1. Bansal, N. P., and J. Lamon (eds). *Ceramic Matrix Composites: Materials, Modeling and Technology*. New York: Wiley, 2015.

ceramic precursors

Ceramic precursors are usually associated with polymeric species that can be converted to ceramic materials, particularly non-oxide ceramics by decomposition at high temperature. Examples of these precursors are polycarbosilanes and polysilazanes, which can be converted to powders, fibres, coating or monolithic shapes of silicon carbide or silicon nitride.[1]

See: *polycarbosilanes; polysilazanes*

1. Riedel, R. Advanced ceramics from inorganic polymers. In R. J. Brook (ed.), *Materials Science and Technology*, vol. 17B, *Processing of Ceramics Part II*, pp. 1–50. Weinheim, Germany: VCH Publishers, 1996.

ceramic stains

The price of conventional ceramic articles such as tableware (e.g. cups, saucers, plates and jugs) is affected by its design and the quality of the decoration. The decoration contains ceramic pigments known as stains that are typically multicomponent oxides or sulphides that are prepared by solid-state reactions between raw materials at high temperature, after which grinding is used to produce powders with particle sizes around 10–30 μm; ceramic stains are also known as ceramic colours. Intricate coloured designs can be prepared on a transfer or decal that is immersed in water to remove a backing paper and then applied to a glazed ceramic article. This process is to known as 'hand decoration', in contrast to hand painting. Calcination of the article in a kiln removes residues from the decal, after which the stain reacts with the glaze to form an adherent decoration on the tableware. Examples of ceramic stains are praseodymium zircon (yellow), chrome tin (pink) and cobalt aluminate (blue). In the case of wall tiles, screen printing of ceramic stain particles has been conventionally used to print decorations onto the unfired tiles (known as 'green bodies'), which due to their fragility are prone to break. Screen-printed decorations do not extend to the edge of the tiles. Ink-jet printing of ceramic stains has been explored for printing decorations up to the edge of wall tiles and as it is a non-contact process, breakages of tiles are reduced. An account of the materials used in different types of ware, such as earthenware, stoneware and porcelains, stains and glazes, their preparation and uses, is given in reference (1).

See: *ceramic glazes*

1. Miller, J. (Gen. Ed.). *Miller's Antiques Encyclopedia*. London: Octopus Publishing, 2013.

chemical mechanical planarisation

There are requirements for polishing certain surfaces to a high degree in order to produce level surfaces: for example, lenses in binoculars, telescopes, microscopes and spectacle lenses. In addition, silicon wafers must be polished to produce high-quality surfaces before their use in integrated circuits. This polishing process is known as chemical mechanical planarisation (CMP).[1] Nanopowders of cerium oxide and silica are used for CMP.

1. Lee, J.-H., J.-H. Cho, J.-S. Choi and D.-J. Lee. Spin-on glass composition and method of forming silicon oxide layer in semiconductor manufacturing process using the same. United States Patent 7,270,886, granted on 18 September 2007.

chewing gum

While natural and synthetic rubbers are associated with tyre compositions, the major components of gum base in chewing gum are rubbers including butyl rubber, styrene-butadiene, polyisobutylene and low molecular weight polyvinyl acetate, although the materials do not, of course, have the same composition as materials used in car tyres.[1,2] Additives to the base are used to modify the physical properties of the gum, such as the texture and chewiness although the exact compositions are closely guarded trade secrets. A distinction between chewing gum and bubble gum arises because the polymer molecules in bubble gum have a higher molecular weight, stretch more easily and maintain the integrity of the bubble.

See: *elastomers*

1. Hartel, R. W., and A. Hartel. *Candy Bites: The Science of Sweets*. New York: Springer, 2014.
2. Emsley, J. *Chemistry at Home: Exploring the Ingredients in Everyday Products*, pp. 125–6. Cambridge: Royal Society of Chemistry, 2015.

chinese porcelain

The oldest fired, shaped clay artefacts date back approximately 8000 years while animal reliefs were made with coloured glazed bricks in Egypt around 1600 BC. Chinese porcelain was developed after 600 AD and continued during the Sung (960–1127) and Ming periods (1368–1644),[1,2] and these soft porcelains are highly valued in the 21st century for their aesthetic appeal. Soft porcelains have lower kaolin contents than hard porcelains, characteristic of European products such as Meissen porcelain. A detailed historical account of the development of Chinese porcelain is given in reference (2).

See: *porcelain*

1. Hennicke, H. W., and A. Hesse. Traditional ceramics. In *Concise Encyclopaedia of Advanced Ceramic Materials*, p. 489. Oxford: Pergamon Press, 1991.
2. Miller, J. (Gen. Ed.). *Miller's Antiques Encyclopedia*. London: Octopus Publishing, 2013.

chitin

Chitin is a naturally occurring polysaccharide and is found in the exoskeletons (i.e. shells) of shellfish and insects where it strengthens these structures.[1–3] It is a derivative of glucose with a chemical structure differing from that of cellulose because of the substitution of a hydroxyl group with an acetyl group. It consists of chains of *N*-acetyl-D-glucosamine (2-acetamido-2-deoxy-D-glucose).

See: *chitosan; iridescent organisms*

1. Daintith, J. (ed.). *A Dictionary of Chemistry*, 6th edn. Oxford: Oxford University Press, 2008.
2. Fried, J. R. *Polymer Science and Technology*, 3rd edn, p. 343. Upper Saddle River, NJ: Prentice Hall, 2014.
3. Atkins, P. *Atkins' Molecules*, 2nd edn, p. 119. Cambridge: Cambridge University Press, 2003.

chitosan

Chitosan is a derivative of the natural polysaccharide chitin obtained by *N*-deacetylation and has free amino groups.[1,2] It is a copolymer of *N*-acetyl-glucosamine and glucosamine, is biocompatible and biodegradable and is a linear polyamine. It is soluble in acidic media and has applications in pharmaceuticals, cosmetics and food technology, as examples. Deacetylation removes the acetyl functional group ($COCH_3$) from chitin and the resulting chitosan behaves as a linear cationic polymer.[3] Removal of acetyl groups allows for tighter packing of polymer chains and an increase in crystallinity. Chitosan can be processed into fibres, films, gels and three-dimensional porous structures for application as polymer scaffolds.

See: *chitin*

1. Borbely, J. Additives for cosmetic products and the like. United States Patent Application 2008/0070993, published on 20 March 2008.

2. Fried, J. R. *Polymer Science and Technology*, 3rd edn, p. 343. Upper Saddle River, NJ: Prentice Hall, 2014.
3. Agrawal, C. M., J. L. Ong, M. R. Appleford and G. Mani. *Introduction to Biomaterials: Basic Theory with Engineering Applications*. Cambridge: Cambridge University Press, 2014.

cholesteric phases

Liquid crystals can form cholesteric phases in which rod-shaped molecules are arranged in nematic-type layers where the axes of the molecules lie in the plane of the layers and are aligned. However, the layers are rotated relative to each other so that the direction of molecules in one layer are not parallel to the direction of molecules in the adjacent layer and are slightly twisted relative to the direction of molecules in the layer below.[1] After a number of rotations the direction of molecules in the upper plane is the same as the molecules in the lower plane so that the layers have a helical screw-like arrangement. This alignment occurs after a 360° twist of the planes. The separation of the upper plane from the lower plane is known as the pitch (Fig. G.6). Cholesteric phases have also been called chiral nematic phases and here chiral means that the helical structure is not superimposable on its mirror image.

See: *liquid crystals; nematic liquid crystals; smectic phases*

1. Kawamoto, H. The history of liquid crystal displays. *Proceedings of the IEEE* (2002) 90, 460–500.

chondroitin sulphate

Chondroitin sulphate is a glycosaminoglycan found in cartilage and in the cornea as well as in arteries, skin and bone and acts as a lubricant.[1] It is an unbranched polysaccharide and is

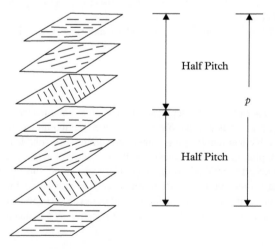

Fig. G.6 *Schematic representation of cholesteric liquid crystals illustrating the pitch and half-pitch. (H. Kawamoto. The history of liquid crystal displays.* Proceedings of the IEEE *(2002) 90, 460–500.)*

important for maintaining the structural integrity of tissue such as cartilage. Chondroitin sulphate is water-soluble and can be cross-linked with other polymers such as chitosan, gelatin and collagen.[2] It is marketed as a nutritional supplement for osteoarthritis. As a negatively charged polymer it can, when cross-linked with collagen, be used for the treatment of burns and in reconstructive surgery. Examples of commercial products are Viscoat and Integra.

See: *hyaluronic acid*

1. Gratzer, W. *Giant Molecules: From Nylon to Nanotubes*, p. 82. Oxford: Oxford University Press, 2013.
2. Agrawal, C. M., J. L. Ong, M. R. Appleford and G. Mani. *Introduction to Biomaterials: Basic Theory with Engineering Applications*. Cambridge: Cambridge University Press, 2014.

classification of porosity

Porosity is an important property of many materials including catalysts, zeolites, membranes and photonic materials. Pore sizes are divided into three regions.[1] Widths less than 2 nm are characteristic of micropores, those between 2 and 50 nm correspond with mesopores while macropores have widths greater than 50 nm.

See: *photonic materials; zeolites*

1. Sing, K. S. W., D. H. Everett, R. A. W. Haul, L. Moscou, R. A. Pierotti, J. Rouquerol and T. Siemieniwska. Reporting physisorption data for gas/solid systems with special reference to the determination of surface area and porosity. *Pure and Applied Chemistry* (1985) 57, 603–19.

coal gasification

Carbon-containing materials such as coal react with mixtures of steam and air (or oxygen) at elevated temperatures and pressures to form synthesis gas. This process is known as coal gasification.[1]

See: *hydrogen; synthesis gas*

1. Morreale, B. D., C. A. Powell and D. R. Luebke. Advancing coal conversion technologies: materials challenges. In D. S. Ginley and D. Cahen (eds), *Fundamentals of Materials for Energy and Environmental Sustainability*, p. 120. Cambridge: Cambridge University Press, 2012.

collagen

Collagen is a fibrous protein that makes up about 30% of the body protein.[1,2] The polypeptide backbone that contains amino acids is made up predominantly of glycine and proline. Three of the chains are bound together to form helical coils that are bound together to form thick fibres that are very rigid and strong. Collagen is found in tendons, muscles, arterial walls and hair and is bound up with the protein elastin, which has elastic properties. The protein is part of the connective tissue in the skin that helps to create firmness. Collagen is a component of bones and forms a natural composite with hydroxyapatite.

See: *anti-ageing skin treatments; elastin; hydroxyapatite*

1. Atkins, P. *Atkins' Molecules*, 2nd edn, p. 99. Cambridge: Cambridge University Press, 2003.
2. Gratzer, W. *Giant Molecules: From Nylon to Nanotubes*. Oxford: Oxford University Press, 2013.

collodion

Early work on polymers in the 19th century concentrated on derivatives of cellulose.[1] Collodion is a solution of cellulose nitrate (nitrocellulose) in ethanol and was used as a wound dressing. Evaporation of alcohol from coatings of the solution produced a hard, flammable sheet of cellulose nitrate. Collodion is based on cellulose dinitrate and is not an explosive unlike cellulose trinitrate.

See: *cellulose nitrate*

1. Walton, D., and P. Lorimer. *Polymers*, p. 17. Oxford: Oxford University Press, 2005.

colloidal crystals

Colloidal crystals are assemblies of colloidal particles around 1 μm in diameter where the particles have an ordered structure.[1] Opal is a naturally occurring colloidal crystal comprising of packed silica spheres. Synthetic colloidal crystals can form photonic crystals that contain periodic structures due to packing of the colloidal particles where the periodicity is on the order of the wavelength of visible light. Optical diffraction occurs within photonic crystals and the periodic structure is responsible for the colours in butterfly wings and peacock feathers.

See: *colloidal systems; photonic crystals*

1. Ozin, G. A., A. C. Arsenault and L. Cademartiri. *Nanochemistry: A Chemical Approach to Nanomaterials*, pp. 147–9. Cambridge: Royal Society of Chemistry, 2009.

colloidal systems

In 1861 Thomas Graham used the word colloid to describe 'glue-like' material prepared by dialysis of silicic acid made by acidifying silicate solutions and also organic species such as gums, caramel, tannin and albumen.[1] These systems did not crystallise. Nowadays colloidal systems are defined as comprising a disperse phase with at least one dimension between 1 nm and 1 μm in a dispersion medium. Colloidal systems include dispersions of particles where the colloidal dimension refers to particle diameter, emulsions, aerosols and foams with a film thickness in this size range. These systems also include nanofibres. Colloidal systems have a significant percentage of atoms in their surface regions compared to non-colloidal systems.[2–4]

See: *aerosol-derived powders; nanofibres; nanotechnology; quantum dots; sol-gel processes*

1. Graham, T. *Philosophical Transactions of the Royal Society* (1861) 151, 183–224.
2. Buzak, S., and D. Rende. *Colloid and Surface Chemistry: A Laboratory Guide for Exploration of the Nano World*. Boca Raton, FL: CRC Press, 2014.
3. Shaw, D. *Introduction to Colloid and Surface Chemistry*, 4th edn. Oxford: Butterworth-Heinemann, 1992.
4. Jones, R. A. L. *Soft Condensed Matter*. Oxford: Oxford University Press, 2013.

coloured diamonds

Diamonds have many applications, including in drill bits, that rely on their hardness and as heat sinks for semiconductors to conduct heat due to its high thermal conductivity. Diamonds are also widely admired in jewellery. Diamond has a large band gap (5.5 eV), which means that it does not absorb radiation from the ultraviolet to infrared wavelengths and thus remains transparent. Coloured diamonds can be produced by introducing impurities, for example by ion implantation although natural diamonds can contain impurities. For example, natural diamond containing nitrogen has luminescent defect centres due to unbound electrons that can give rise to a red appearance. Nitrogen concentrations of around 100 ppm give rise to a yellow appearance. Absorption levels in the band structure of diamond are introduced on doping so that electrons can absorb some components of visible light.

See: *cubic zirconia*

1. Askeland, D. K., and W. J. Wright. *The Science and Engineering of Materials*, 7th edn, p. 539. Boston, MA: Cengage Learning, 2011.

commercial bioplastics

There is increasing interest in developing bioplastics from renewable sources. Those materials that are commercially available are polypropylene from ethanol, polyethylene from ethanol, polylactic acid, polyhydroxyalkanoate, polytrimethylene terephthalate, cellulose acetate, nylon 11 and thermoplastic starch.[1]

See: *bioplastics*

1. Ashby, M. F., D. F. Balas and J. S. Coral. *Materials and Sustainable Development*, p. 119. Oxford: Butterworth, 2016.

composite materials

Composite materials consist of a reinforcement phase or filler dispersed in a matrix phase.[1] The filler can be continuous or chopped-up fibre, often aligned, for example carbon fibre and glass fibres, while the matrix can be a thermoplastic or thermoset polymer, or ceramic or metal. Composites can be low weight compared to metals with attractive mechanical properties such as toughness and strength. Lightweight composites have applications as structural components for aircraft. Examples of other composites include silicon carbide whisker-reinforced alumina, silicon carbide fibre-reinforced titanium compressor discs and glass fibre-reinforced epoxy resins.

See: *bio-based resins; epoxy adhesives; nanocomposites*

  1. Bansal, N. P., and J. Lamon (eds). *Ceramic Matrix Composites: Materials, Modeling and Technology.* New York: Wiley, 2015.

concrete

While a mixture of cement, water and sand is known as mortar, concrete is a mixture of cement and aggregate that is a coarse stony material.[1] Concrete can be described as a composite material and as with cements is strong under compression but weak under tension. In reinforced concrete tensile forces are transmitted through steel rods embedded in the concrete.

See: *Portland cement*

  1. Bolton, W., and R. A. Higgins. *Materials for Engineers and Technicians*, 6th edn, pp. 364–7. London: Routledge, 2015.

conducting polymers

Polyacetylene, poly-(p-phenylenevinylene) (PPV), polyaniline, polypyrrole and polythiophene are examples of electrically conducting polymers.[1] A common feature of these polymers is a conjugated Π-electron bond structure with alternating single and double bonds along the polymer chain. The conductivity is modified by addition of additives, for example arsenic pentafluoride to polyacetylene. PPV is a semiconductor with potential applications in organic light-emitting diodes (OLEDs). Other applications and potential applications are as electrodes and solid electrolytes in batteries, conductive paints and electrochromic displays.

See: *organic light-emitting diodes; polyacetylene; poly-(p-phenylenevinylene)*

  1. Geoghegan, M., and G. Hadziioannou. *Polymer Electronics*. Oxford: Oxford University Press, 2013.

contact angle

When a drop of liquid is present on a surface then the contact angle is the angle measured from the surface through the drop to the tangent at the air–liquid–solid interface (Fig. G.7).[1]

See: *hydrophilic materials; hydrophobic materials; superhydrophobic materials*

1. Buzak, S., and D. Rende. *Colloid and Surface Chemistry: A Laboratory Guide for Exploration of the Nano World.* Boca Raton, FL: CRC Press, 2014.

contrast agents for MRI

Magnetic resonance imaging is widely used to highlight soft tissue in the body. Contrast agents are administered at the time of the examination and enhance the differences in signal intensity between neighbouring tissues.[1] An example of a contrast agent is based on gadolinium (a lanthanide element) known under the trade name Magnevist (Gadopentetat-Dimeglumin, Gd-DTPA). The Gd (III) ion is paramagnetic, which decreases the relaxation time of protons and is referred to as a positive contrast agent.

See: *superconducting metals and alloys*

1. Lucas, J., P. Lucas, T. Le Mercier, A. Rollat and W. Davenport. *Rare Earths: Science, Technology, Production and Use*, p. 311. London: Elsevier, 2015.

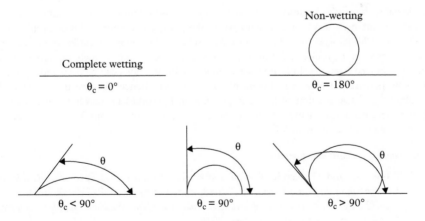

Fig. G.7 *Contact angles* $(\theta)_c$ *at a water–solid interface.*

(D. Segal. Exploring Materials through Patent Information, p. 220. Cambridge: Royal Society of Chemistry, 2014.)

copolymers

Copolymers, or heteropolymers, are made from two or more different monomers.[1] Styrene-butadiene, styrene-acrylonitrile-butadiene and polystyrene-polymethylmethacrylate are examples of heteropolymers. Both nylon and Kevlar are copolymers produced in a condensation reaction between monomers.

See: *polymers; tyre compositions*

> 1. Bolton, W., and R. A. Higgins. *Materials for Engineers and Technicians*, 6th edn, p. 272. London: Routledge, 2015.

copper indium gallium diselenide

The quest for renewable solutions to worldwide concerns over global warming has led to a search for materials that can be used in solar cell, that can be manufactured economically and that have high efficiencies. Copper indium gallium diselenide (CIGS), copper indium gallium disulphide and cadmium telluride are candidate materials for thin-film photovoltaic technology.[1]

See: *silicon-based solar cells*

> 1. Burke, M. Solar future looks bright. *Chemistry and Industry* (26 Sept. 2011) 19–21.

cosmetic formulations

Colloidal systems play an important role in cosmetic formulations.[1,2] For example, moisturisers are often oil-in-water emulsions where the particle diameter of the oil droplets lies within the colloidal size range. The emulsion must be stable, that is, not settle out into oil and water phases after purchase. Oil-in-water emulsions where water is the continuous phase can be removed from the skin with water. Emulsion formulations can be prepared as an aid to allow penetration of active ingredients into the skin; hence, the use of oil–water–oil emulsions or encapsulation of active ingredients in liposomes to deliver the ingredients. Active ingredients can include retinoids that are derivatives of vitamin A and antioxidants such as vitamin C and vitamin E.

See: *colloidal systems; moisturisers*

> 1. Buzak, S., and D. Rende. *Colloid and Surface Chemistry: A Laboratory Guide for Exploration of the Nano World*, pp. 226–8. Boca Raton, FL: CRC Press, 2014.
> 2. Ngo, C., and M. van de Voorde. *Nanotechnology in a Nutshell: From Simple to Complex Systems*. Amsterdam: Atlantis Press, 2014.

critical materials

Materials have been described as critical if access to them could be limited or they are essential for national security or important economically.[1] Lists of critical materials are produced by governments and critical materials on the U.S. list include antimony, beryllium, bismuth, cerium, chromium, cobalt, dysprosium, erbium, europium, gallium, germanium, indium, iridium, lanthanum, lithium, manganese, neodymium, osmium, palladium, platinum, praseodymium, rhodium, ruthenium, samarium, scandium, tantalum, tellurium, terbium, thulium, tin, tungsten and yttrium.

See: *smartphones*

1. Ashby, M. F., D. F. Balas and J. S. Coral. *Materials and Sustainable Development,* p. 11. Oxford: Butterworth, 2016.

cross-laminated timber

There is much interest in the second decade of the 21st century in the use of renewable and sustainable materials in all areas of everyday life. Prefabricated cross-laminated timber has potential as a construction material for walls, floors and ceilings in buildings.[1-3] Laminates are composite materials made up of two or more bonded layers. Cross-laminated structures are, by definition, orthogonal angle-ply laminates and cross-lamination enhances the strength and toughness of the material.

1. Askeland, D. K., and W. J. Wright. *The Science and Engineering of Materials,* 7th edn, p. 666. Boston, MA: Cengage Learning, 2011.
2. Coulson, J. *Wood in Construction.* Chichester: Wiley-Blackwell, 2012.
3. Wilson, P. Defence can be a passive activity. *Materials World* (2016) 24(12), 37–40.

cubic zirconia

The naturally occurring materials zircon sands (zirconium silicate) and baddeleyite can be purified to produce zirconium dioxide (zirconia) that can be stabilised in the cubic crystalline phase by addition of 8 wt% of yttrium oxide (yttria). Unstabilised zirconia does not exist as the cubic phase at room temperatures. Cubic zirconia is an ionic conductor and used as an oxygen sensor in automotive engines and to measure dissolved oxygen in molten steel and also used as an oxygen gas sensor.[1] However, cubic zirconia is best known among the general public for its use as a cheaper alternative to diamond in jewellery.

See: *coloured diamonds*

1. Askeland, D. K., and W. J. Wright. *The Science and Engineering of Materials,* 7th edn, p. 538. Boston, MA: Cengage Learning, 2011.

cyanobiphenyl liquid crystals

Cyanobiphenyl liquid crystals were developed by George Gray and co-workers at the University of Hull in the 1970s.[1-3] The significance of these materials is that they have stable nematic phases at or near ambient temperature such that liquid crystal displays can operate at these temperatures. A key feature of cyanobiphenyl liquid crystals is the presence of a cyano group (CN) in the structure, for example in $C_5H_{11}-C_6H_4.C_6H_4-CN$ (pentyl-cyanobiphenyl). The two aromatic rings in the structure are directly connected and this connection confers stability against decomposition. The flexible alkyl chain (C_5H_{11}) stabilises formation of the liquid crystal phase around room temperature. A mixture of similar compounds broadens the temperature range of the liquid crystal phase to between -10 and $60°C$. Cyanobiphenyls have a positive dielectric anisotropy characteristic of the twisted nematic liquid crystal phase.

See: *liquid crystals*

1. Andrews, B. M., N. Carr, G. W. Gray and C. Hogg. Liquid crystal compounds. European Patent 0097033 B, 1990.
2. Kawamoto, H. The history of liquid crystal displays. *Proceedings of the IEEE* (2002) 90, 460–500.
3. Dunmur, D., and T. Sluckin. *Soap, Science, and Flat-Screen TVs: A History of Liquid Crystals.* Oxford: Oxford University Press, 2014.

decaffeinated coffee

Foods can be considered as materials and labels often highlight the protein and carbohydrate contents. Materials are often used in food processing. While carbon dioxide will be familiar to the general public in connection with greenhouse gases and global warming, the phrase supercritical carbon dioxide is largely unknown to many people. A supercritical fluid is a substance above its critical temperature where it remains as a single fluid phase. This means that a supercritical fluid in a container does not have a meniscus that separates a liquid from a vapour phase. The critical temperature and pressure for carbon dioxide are $31°C$ and 7.38 MPa, respectively. Supercritical fluids have been used for preparing aerogels but they also act as very good solvents and the low critical temperature is a benefit for large-scale use. A major use for supercritical carbon dioxide is in the decaffeination of coffee.[1]

See: *aerogels*

1. Downie, N. A. *The Ultimate Book of Saturday Science*, pp. 75–6. Princeton, NJ: Princeton University Press, 2012.

density functional theory

Materials modelling is the development and use of mathematical models for describing and predicting electronic and atomic properties of materials and is an important part of activities on materials science in the 21st century.[1] The model is developed from first principles and this approach is known as an *ab initio* model after a Latin phrase meaning 'from the beginning'; no empirical data are required for the calculations. Ever-increasing computer power available at an economic price has encouraged computational modeling of materials. Density functional theory is based on quantum mechanics, in particular at finding approximate solutions of the many-body Schrodinger equation for the equilibrium structures of materials consisting of many atoms as represented by a gas of electrons. Approximate because while the Schrodinger equation can be solved for a single-electron system (e.g. H, He^+, Li^{2+}) it cannot be solved exactly for electron–electron interactions when there is more than one electron as the Heisenberg uncertainty principle prevents knowledge of the position of electrons in multi-electron systems. The key feature of density functional theory is that the ground state energy of the system (i.e. the lowest energy) is a *functional* of only the electron density, that is the electron gas, and not on the wave function that depends on the atomic positions; a functional may be considered as a function of a function. The method has applications to molecules, solids, surfaces and interfaces. For example, a characteristic feature of superconducting materials is a sudden jump in the heat capacity versus tempera-ture plot at the superconducting critical temperature. This jump was predicted for magne-sium diboride that has a critical temperature of 39 K. Other examples include the phase diagram for iron and the brittle fracture of silicon. Density functional theory can be used to predict the properties of materials that have not been prepared previously such as catalysts for hydrogen production.

See: *superconducting metals and alloys*

1. Giustino, F. *Materials Modeling Using Density Functional Theory*. Oxford: Oxford University Press, 2014.

diabetes monitoring

Diabetes is a major healthcare problem for governments in advanced economies due to the financial costs for healthcare systems.[1] Electrochemical methods for determining sugar concentrations in blood have been developed. The principle of these electrochemical sensors involves enzymatic oxidation of glucose on one electrode that facilitates electron transfer to a mediator, a ferrocene derivative. The reduced mediator is then oxidised at the other electrode, generating a current whose magnitude can be related to the sugar concentration. Methods used for semiconductor fabrication can be used to produce compact sensors for measuring blood glucose concentration. Electrochemical sensors can also be used for measuring sugar concentrations in exhaled breath.

1. Clarke, S. F., and J. R. Foster. A history of blood glucose meters and their role in self-monitoring of diabetes mellitus. *British Journal of Biomedical Science* (2012) 69(2), 83–93.

diesel particulates

There is concern among governments about levels of air pollution and the contribution that exhausts from diesel engines make to these levels. Carbonaceous particulates, that is diesel particulates in diesel exhausts, are primarily of carbon (soot) but include hydrocarbons such as soluble organic fractions that can be adsorbed on the surface of soot particles; the soot particles can be small, of the order of nanometres.[1,2] Diesel particulate filters also known as wallflow filters are used as particulate traps for diesel exhausts. A representative wallflow filter dimension is 5.66 inches diameter, 6 inches high with 100 cells per square inch. The wall thickness separating parallel rectangular channels is around 17 thousandths of an inch. The walls are porous with a mean pore diameter around 10 μm. Alternate channels are blocked off at the ends so that the exhaust passes through the walls that trap particulates. Examples of these filters made of cordierite have been sold under the trade name Celcor (Corning Inc.) although silicon carbide wallflow has been produced. Carbon is removed from the filters by use of catalysts and/or heating the filter to burn off the soot.

See: *ceramic honeycombs; three-way automotive catalysts*

1. Hall, S. I., J. T. Shawcross, R. McAdams, D. L. Segal, M. Inman, K. Morris, D. Raybone and J. Stedman. Non-thermal plasma reactor. United States Patent Application 2004/0208804, published on 21 October 2004.
2. Lucas, J., P. Lucas, T. Le Mercier, A. Rollat and W. Davenport. *Rare Earths: Science, Technology, Production and Use*, pp. 155–6. London: Elsevier, 2015.

direct band gap semiconductors

Gallium arsenide (GaAs) and its solid solutions are an example of a direct band gap semiconductor. Electroluminescence occurs when an excited electron falls back from the conduction band into the valence band and combines with a hole which results in the emission of light. In a direct band gap semiconductor there is no loss of momentum of the electron in this transition.[1]

See: *indirect band gap semiconductors*

1. Tilley, R. J. D. *Understanding Solids: The Science of Materials*, p. 453. Chichester: Wiley, 2013.

disposable nappies

Disposable nappies are made not from paper but from highly absorbent polymers known as hydrogels.[1] Thus, these nappies are made out of plastic. The hydrogels are often used in the form of fine powders or granules that absorb body fluids and they can absorb over ten times their dry weight as liquids. Hydrogels are three-dimensional networks of hydrophilic homopolymers or copolymers that are swollen by absorption and retention of water. Although the polymers are themselves water-soluble, they are rendered insoluble in the hydrogel through cross-linking. Hydrogels can be prepared by, for example, polymerisation of acrylic acid.

See: *hydrogels*

1. Walton, D., and P. Lorimer. *Polymers*, p. 125. Oxford: Oxford University Press, 2005.

DNA

DNA (deoxyribonucleic acid), a nucleic acid, is found in the chromosomes (23 pairs) within the cell nucleus, is responsible for expressing (i.e. synthesis) proteins in cells and replicates itself on cell division. DNA is a heteropolymer and the backbone of the polymer chain consists of the sugar deoxyribose linked by phosphate bridges (Fig. G.8).[1–3] One of four bases, adenine, cytosine, guanine and thymine, designated by the letters A, C, G and T, is attached to a deoxyribose molecule. A unit of base, sugar molecule and phosphate is known as a nucleotide and a chain or strand of DNA can contain millions of nucleotides. A unit of base and sugar molecule is known as a nucleoside. Two chains in DNA are wrapped around each other in a double helix so that hydrogen bonds link A to T and C to G and the linkages are contained within the helical structure. A is complementary to T and C is complementary to G. The three-dimensional structure of DNA based on the double helix was proposed by Francis Crick and James Watson, who used X-ray diffraction data obtained by Raymond Gosling (a Ph. D student) and Rosalind Franklin, a crystallographer working at King's College in London in the early 1950s.[1,2] An x-ray pattern in the form of an 'X' that is characteristic of a helical structure was present in what is referred to

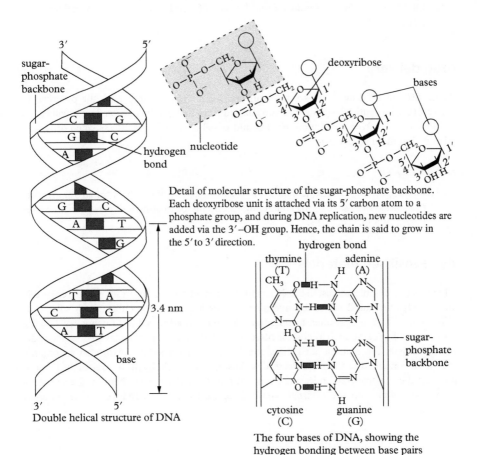

Detail of molecular structure of the sugar-phosphate backbone. Each deoxyribose unit is attached via its 5' carbon atom to a phosphate group, and during DNA replication, new nucleotides are added via the 3'–OH group. Hence, the chain is said to grow in the 5' to 3' direction.

Double helical structure of DNA

The four bases of DNA, showing the hydrogen bonding between base pairs

Fig. G.8 *Molecular structure of DNA.*

(R. S. Hine (ed.). Oxford Dictionary of Biology, *7th edn, p. 175. Oxford: Oxford University Press, 2015.)*

as photograph 51 in Franklin's work.[3] The A to T and C to G linkages are known as base pairs and the full characteristic sequence of these base pairs in all of the chromosomes represents the human genome and genetic code of the organism; there are about 3 billion base pairs in the human genome. A sequence of three bases known as a codon encodes a specific amino acid. Regions of base pairs along the chromosome constitute genes, which are separated from other genes by regions of what has been referred to as junk DNA as its role is unclear.

See: *genes; polymer; polymerase chain reaction*

1. Crick, F., and Watson, J. Molecular structure of nucleic acids. *Nature* 1953 717, 737–9.
2. Barber, J., and C. Rostron (eds). *Pharmaceutical Chemistry*, pp. 242–8. Oxford: Oxford University Press, 2013.
3. Glynn, J. *My Sister Rosalind Franklin*, p. 126. Oxford: Oxford University Press, 2012.

drug delivery

About 80% of all drug doses are taken in oral form.[1] Ionic liquids have the potential to convert pharmaceutical compounds into a liquid form that can be delivered externally such as transdermally from a patch on the skin and possibly reduce side effects when drugs are adsorbed in the digestive tract.

See: *ionic liquids*

1. Palmer, P., T. Schreck, L. Hamel, S. Tzannis and A. Poutiatine. Small volume oral transmucosal dosage forms containing sufentanil for treatment of pain. United States Patent, 8,535,714, 2013.

dye-sensitised materials

There is considerable interest in the 21st century in identifying efficient sources of renewable energy. Dye-sensitised materials have potential for increasing the efficiency of solar cells (Fig. G.9).[1] Here, a dye, often a ruthenium complex, is adsorbed onto nanoparticles of a semiconducting oxide such as anatase titanium dioxide that is in contact with one electrode. An electrolyte, for example an iodine/tetraalkylammonium iodide mixture in an organic solvent, is contained between the electrodes. Absorption of light by the dye raises the latter to an excited electronic state, releasing an electron into the conduction band of the semiconductor and then into the external circuit. The oxidised dye is reduced at the counter electrode.

See: *Perovskite-based solar cells*

1. Christie, R. M. *Colour Chemistry*, 2nd edn. Cambridge: Royal Society of Chemistry, 2015.

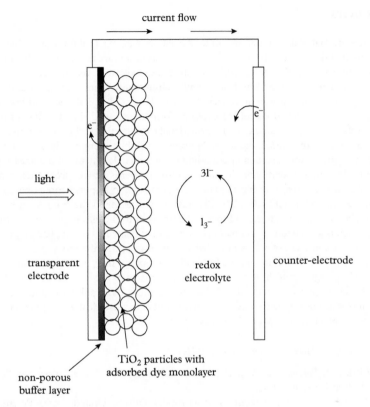

current flow

light

transparent
electrode

redox
electrolyte

counter-electrode

$3I^-$

I_3^-

e^-

e^-

TiO$_2$ particles with
adsorbed dye monolayer

non-porous
buffer layer

Fig. G.9 *Schematic representation of a dye-sensitized solar cell.*
(R. M. Christie. Colour Chemistry, *2nd edn, p. 284. Cambridge: Royal Society of Chemistry, 2015.)*

elastin

Elastin is a fibrous protein that is rich in proline and glycine and forms flexible random coils
where polypeptide chains from neighbouring coils are linked at intervals to each other.[1] This
protein is often present in the body with collagen and is an elastic solid that is particularly
found in the cartilage, blood vessel walls and in the heart. The lung is made of 30% elastin,
whereas skin is made up of 2–4% elastin.[2] Elastic ligaments and large arteries are made of
70 and 50% elastin, respectively. While molecules of collagen are close-packed, elastin is
relatively loose with unstructured polypeptide chains cross-linked to form an elastic network.

See: *collagen*

1. Gratzer, W. *Giant Molecules: From Nylon to Nanotubes*, pp. 44–5. Oxford: Oxford
   University Press, 2013.
2. Agrawal, C. M., J. L. Ong, M. R. Appleford and G. Mani. *Introduction to Biomaterials:
   Basic Theory with Engineering Applications*. Cambridge: Cambridge University Press,
   2014.

elastomers

Elastomers are materials that deform under the action of an external force and which return to their original shape when the force is removed;[1] elastomers are associated with natural and synthetic rubbers and also polyurethanes. Charles Goodyear developed the vulcanisation process in which latex gum was heated with sulphur and white lead at around 180°C, producing a material that was no longer sticky, set to a hard and elastic solid that did not soften on heating.[2] The cured rubber could be formed into solid sheets of India rubber or used for vehicle tyres or coated onto a fabric to make it waterproof; the sulphur content was around 3 wt% of the rubber content. Nowadays sulphur compounds are used in the vulcanisation process that results in cross-linking of molecular chains of the latex liquid by sulphur bridges, hence solidifying the gum. Natural rubber is derived from latex gum from the tree *Hevea brasiliensis* and has the chemical composition of polyisoprene based on the monomer *cis*-isoprene $(CH_2=C(CH_3)CH=CH_2)$ in contrast to the *trans* form known as gutta percha. The latter is hard and inelastic compared to the elastic properties of natural rubber, which is an elastomer and had been obtained initially by cooking latex tapped from indigenous trees in Malaysia.[3] Its potential use in surgical instruments was recognised by William Montgomerie working for the East India Company Army in Singapore in the 1840s but a breakthrough application was as a durable insulator for long-distance underwater telegraph cables in the latter half of the 19th century. Vulcanisation and the role of accelerators is discussed in reference (4). An account of the historical development of rubber technology is given in reference (5) by Loadman.

See: *chewing gum; latex; neoprene; natural rubber*

1. Fried, J. R. *Polymer Science and Technology*, 3rd edn, pp. 374–9. Upper Saddle River, NJ: Prentice Hall, 2014.
2. Goodyear, C. Improvement in India-rubber fabrics. United States Patent 3633, granted on 15 June 1844.
3. Weightman, G. *Eureka: How Invention Happens*, pp. 61–2. New Haven, CT: Yale University Press, 2015.
4. Walton, D., and P. Lorimer. *Polymers*, p. 120. Oxford: Oxford University Press, 2005.
5. Loadman, J. *Tears of the Tree: The Story of Rubber—A Modern Marvel*. Oxford: Oxford University Press, 2014.

electrical transmission lines

Overhead cables on pylons need to be mechanically strong and also electrically conducting.[1] This can be achieved by use of a core of steel fibres for strength that has fibres of conducting aluminium twisted around it. Alternative materials for consideration include a composite core of aluminium-containing Nextel fibres with the objective of increasing the current-carrying capacity of the conductor.

1. Ginley, D. S., and D. Cahen (eds). *Fundamentals of Materials for Energy and Environmental Sustainability*. Cambridge: Cambridge University Press, 2012.

electroceramics

Electroceramics, or electronic ceramics, are a class of advanced ceramic materials and their use depends primarily on their electrical or magnetic properties rather than on their mechanical behaviour.[1,2] Optical properties are important for opto-electronic devices and materials with these properties fall within the class of electroceramics. Examples of electroceramics include lithium niobate for electro-optic devices, lead zirconate titanate for piezoelectric materials and barium titanate for capacitors.

See: *ceramic materials*

1. Somiya, S. (ed.). *Handbook of Advanced Ceramics*. San Diego, CA: Academic Press, 2013.
2. Moulson, A. J., and J. M. Herbert. *Electroceramics: Materials, Properties, Applications*. London: Chapman and Hall, 1990.

electrochromic materials

Electrochromic materials change colour when a voltage or electric current is applied.[1] An application is in window glass and the effect is reversed when the direction of the voltage is reversed. Examples of materials that exhibit the electrochromic effect include tungsten oxides (WO_3), which exhibits a colour change in a thin film from pale yellow to blue when a voltage is applied due to reduction of $W(VI)$ to $W(V)$.

See: *smart materials*

1. Monk, P. M. S., R. J. Mortimer and D. R. Rosseinsky. *Electrochromism and Electrochromic Devices*. Cambridge: Cambridge University Press, 2007.

electroluminescence

Electroluminescence is the process whereby light is emitted from a material when subject to an applied electrical voltage.[1] Electroluminescence is the process responsible for light emission from light-emitting diodes and organic light-emitting diodes. Electroluminescence was first observed by Henry Round in 1907 on applying a voltage across regions of the semiconductor silicon carbide. It is associated with recombination of electrons and holes at a junction region in a semiconductor material between a *p*-type region and *n*-type region: for example, at the junction of a *p*-type gallium arsenic phosphide (GaAsP) layer and *n*-type GaAsP layer. At this time there was considerable interest in radio detectors, such as the cat's whisker (see Chapter 3) in which a rectifier could be made by placing a metal wire in contact with a semiconductor such as galena (lead sulphide), and Henry Round had interests in radio detection.[2]

See: *phosphors; light-emitting diodes*

1. Tilley, R. J. D. *Understanding Solids: The Science of Materials*. Chichester: Wiley, 2013.
2. Ploszajski, A. Silicon carbide. *Materials World* (Jan. 2016) 24(1), 61–3.

electronic ink

Electronic books can be downloaded from websites and read on an e-book reader and it is quite common to see people on public transport or in coffee shops reading books in this way in the second decade of the 21st century.[1] An integral part of these readers is electronic ink, which consists of millions of small capsules filled with a dark dye that float in a liquid. The capsules contain negatively charged white chips that move up and down in response to a positive charge applied to the back of the screen. This motion produces the words on the screen.

1. van Dulken, S. *Inventing the 21st Century*, p.10–13. London: British Library, 2014.

electrospinning

Electrospinning is a technique for producing fibres, including nanofibers, in which a polymer solution or melt is ejected as an electrically charged filament from a nozzle onto a grounded electrode.[1] A mat of fibres can be produced in this spinning process. Polymers that have been electrospun include polylactic acid, polystyrene and nylon. Fibre diameters of 1 μm can be obtained by electrospinning.

See: *nanofibres*

1. Mitchell, G. R. (ed.). *Electrospinning: Principles, Practice and Possibilities*. Cambridge: Royal Society of Chemistry, 2015.

emulsion polymerisation

Emulsion polymerisation is used to prepare polymer latices that are, for example, used in paints.[1] In the method an emulsion of monomer droplets is formed in an aqueous environment containing micelles and a water-soluble initiator; potassium persulphate has been used as an initiator. Monomer and initiator diffuse into the micelles where polymerisation occurs to form latex particles.

See: *latex; micellar systems*

1. Buzak, S., and D. Rende. *Colloid and Surface Chemistry: A Laboratory Guide for Exploration of the Nano World*. Boca Raton, FL: CRC Press, 2014.

engineering ceramics

Engineering ceramics are a class of advanced ceramic materials and their use in structural components depends on their mechanical properties or their refractory properties, that is chemical resistance to the working environment.[1,2] Examples of engineering ceramics include yttria-stabilised zirconia, sintered silicon nitride, silicon carbide fibres for use in composites and hard boron nitride components.

See: *ceramic materials*

1. Tilley, R. J. D. *Understanding Solids: The Science of Materials*, p. 147. Chichester: Wiley, 2013.

2. Somiya, S. (ed.). *Handbook of Advanced Ceramics*. San Diego, CA: Academic Press, 2013.

engineering fabrics

Engineering fabrics are textiles or nonwoven products that have a dual role particularly in the area of cleaning products.[1] Examples are 'wet wipes' that combine the usefulness of a household spray with the convenience of a wipe. Alternatively, such fabrics can have an antistatic activity or antibacterial role.

See: *biodegradable hygiene products*

1. Burke, M. Smart clothes. *Chemistry and Industry* (Apr. 2013) 77, 21–3.

engineering plastics

Engineering plastics are distinct from commodity polymers used for high-volume, low-value products such as textiles by having applications in niche markets.[1] The properties of engineering plastics such as mechanical strength, high-temperature resistance, high impact strength and chemical resistance are important for their applications. Examples of engineering plastics are styrene-acetonitrile-butadiene (ABS), aramids such as Kevlar, polyphenylene oxide, fluoropolymers and polycarbonate.

1. Fried, J. R. *Polymer Science and Technology*, 3rd edn. Upper Saddle River, NJ: Prentice Hall, 2014.

epoxy adhesives

Epoxy resins are made by copolymerising phenols (e.g. bisphenol A, $HOC_6H_4CH_3CCH_3$. C_6H_4OH) with an epoxide such as epichlorohydrin, CH_2OCHCH_2Cl; epoxide molecules contain oxygen atoms that are part of a three-membered ring.[1] The initial reaction product is a viscous liquid that is hardened by the addition of amines that cause cross-linkages. Epoxy resins are thermosetting polymers, are used as adhesives and in composite materials and have high chemical resistance.

See: *bioadhesives; bio-based resins; composite materials; hairy adhesives; Post-it notes; superglue*

1. Bolton, W., and R. A. Higgins. *Materials for Engineers and Technicians*, 6th edn, p. 292. London: Routledge, 2015.

excimer lasers

Excimer lasers, in particular the argon fluoride laser (ArF), is the major light source used in conjunction with photolithography to prepare silicon chips and integrated circuits.[1] In an excimer laser the active lasing material consists of argon and fluorine atoms bound together as ArF in an excited state. The output is ultraviolet radiation with a wavelength of 193 nm.

See: *photoresists*

1. Solymar, L., D. Walsh and R. R. A. Syms. *Electrical Properties of Materials*, 9th edn, p. 307. Oxford: Oxford University Press, 2014.

exfoliating creams

Exfoliating creams are used in cosmetic preparations to help remove dead cells from the skin.[1] Glycolic acid (CH_2=C(OH)COOH) has been used as an ingredient of exfoliating creams. It can be made synthetically and also occurs naturally in sugar cane and sugar beet. Salicylic acid has also been used as a gentle skin peel to remove dead skin cells.

1. Libbert, L. Face wipes that are as good as a facial. *Daily Mail*, p. 42, 6 April 2015.

extrinsic semiconductors

An extrinsic semiconductor is one that contains impurity atoms that are specifically added to modify the electronic properties, in particular electrical conductivity.[1] Impurities can be introduced by ion implantation or diffusion. The distribution of charge carriers in respect to whether the majority are electrons (*n*-type semiconductors) or holes (*p*-type semiconductors) depends on the valency of the dopant atom.

See: *band gaps; intrinsic semiconductors*

1. Askeland, D. K., and W. J. Wright. *The Science and Engineering of Materials*, 7th edn, p. 692. Boston, MA: Cengage Learning, 2011.

fabric conditioners

Many household products contain fragrances, for example deodorants, fabric sprays and fabric conditioners.[1,2] The fragrance can be released if pressure is applied or by heat or by added water but how materials are used in these applications may not be obvious to consumers. One way fragrances are incorporated into fabric conditioners is to incorporate an emulsion of the fragrance in alcohol into microspheres (around 400 μm diameter) of gum arabic or a protein such as gelatin so that a free-flowing powder is produced. The latter is then added to the conditioner and the fragrance is released during the washing cycle.

1. McGee, T., and Sgaramela, R. P. Substrate care products. United States Patent Application, 2008/0234172, published on 25 September 2008.
2. Emsley, J. *Chemistry at Home: Exploring the Ingredients in Everyday Products*, p. 38. Cambridge: Royal Society of Chemistry, 2015.

faded jeans

'Faded' jeans, such as 'stonewashed' items, have been popular items of clothing in recent years.[1] The fungus *Trichoderma reesei* produces cellulase enzymes that are able to degrade cotton fabrics, reducing them to rags.[1] *T. reesei* enzymes are used commercially to produce faded cotton fabric. The fungus produces cellobiohydrolases and endoglucanases that work together to hydrolyse cellulose in the fading process.

See: *indigo*

1. Shi, J., Q. Qing, T. Zhang, C. E. Wyman and T. A. Lloyd. Biofuels from cellulosic biomass via aqueous processing. In D. S. Ginley and D. Cahen (eds), *Fundamentals of Materials for Energy and Environmental Sustainability*, p. 341. Cambridge: Cambridge University Press, 2012.

fats

There is much debate in governments on the causes of increasing obesity among populations and the role of fats and sugars in the causes of obesity. Animal fats are mixtures of lipids, mainly triglycerides that are esters of glycerol (also called glycerine), a trihydric alcohol $HOCH_2CH(OH)CH_2OH$;[1] esterification can take place at one, two or all three hydroxyl groups. Animal fats are solids at body temperatures and triglycerides are components of biodiesel.

See: *biodiesel; lipids*

1. Atkins, P. *Atkins' Molecules*, 2nd edn, p. 58. Cambridge: Cambridge University Press, 2003.

fermentation processes

Fermentation is often associated with the enzymatic conversion of sugars to ethyl alcohol by yeasts in the production of beer.[1,2] However, in the 21st century fermentation processes using yeasts (e.g. *Saccharomyces cerevisiae*) or micro-organisms such as *Escherichia coli* and *Clostridia* are under active investigation for the production of bio-based chemicals from renewable sources.

See: *bio-based chemicals; recombinant DNA technology; yeast*

1. Northen, T. R. Biofuels and biomaterials from microbes. In D. S. Ginley and D. Cahen (eds), *Fundamentals of Materials for Energy and Environmental Sustainability*, p. 322. Cambridge: Cambridge University Press, 2012.
2. Bamforth, C. W., and R. E. Ward. *The Oxford Handbook of Food Fermentation*. Oxford: Oxford University Press, 2014.

ferroelectric materials

All ferroelectric materials are pyroelectrics as they both have a spontaneous polarisation but only in the case of ferroelectrics can the direction of spontaneous polarisation be reversed in an electric field.[1] Examples of ferroelectric materials include lithium niobate

(LiNbO$_3$), lithium tantalate (LiTaO$_3$) and Rochelle salt (potassium sodium tartrate tetrahydrate).

See: *pyroelectric materials*

    1. Tilley, R. J. D. *Understanding Solids: The Science of Materials.* Chichester: Wiley, 2013.

ferromagnetic materials

Ferromagnetic materials are associated with permanent magnets and include SmCo$_5$, Nd$_2$Fe$_{14}$B, iron, cobalt and nickel.[1] As for paramagnetic atoms and ions, ferromagnetic materials have atoms and ions with unpaired electrons so that the atoms have oriented magnetic dipoles (magnetic moments) that remain oriented in the absence of applied magnetic fields. In paramagnetic materials (e.g. lanthanide atoms) the alignment breaks down when the magnetic field is removed. Ferromagnetic materials contain small magnetised regions or domains.

See: *neodymium-based magnets; samarium-based magnets*

    1. Tilley, R. J. D. *Understanding Solids: The Science of Materials.* Chichester: Wiley, 2013.

fibreglass

Fibreglass is a lightweight and tough composite of glass fibres in a polymer matrix, in particular in a polyester resin.[1,2] It has applications as lightweight structures such as in bathtubs, the hulls of small boats and as panels in automobiles and lorries.

See: *glass epoxy composites*

    1. Gordon, J. E. *The New Science of Strong Materials,* 2nd edn. London: Penguin, 1991.
    2. Walton, D., and P. Lorimer. *Polymers,* p. 116. Oxford: Oxford University Press, 2005.

fibre-reinforced plastic composites

Ceramic matrix composites such as silicon carbide fibre-reinforced silicon carbide matrices have specialist applications.[1] However, fibre-reinforced plastics have more general applications. An early example of such a composite was the addition of fibres to phenol-formaldehyde resins (Bakelite) to improve the toughness. Improved toughness arises because when an advancing crack encounters a fibre the crack tip spreads outward and the energy of the crack is dissipated. Fibreglass is a trade name for a composite of glass fibres in a plastic, in particular a polyester matrix, and has been used for lightweight structures such as in bathtubs and the hulls of small boats. Other examples of fibre-reinforced plastics use include Kevlar fibre and carbon fibre while materials for the matrix include epoxy resins. Particulates can also be used to reinforce plastic matrices. Graphene flakes can be incorporated into polymer blends to increase the compressive strength when the composite is used as a structural component of a tennis racket.

See: *epoxy adhesives; glass epoxy composites; silicon carbide fibre composites*

1. Bansal, N. P., and J. Lamon (eds). *Ceramic Matrix Composites: Materials, Modeling and Technology*. Chichester: Wiley, 2015.

flax fibre

Flax fibre is a by-product of the linen industry and has potential uses for insulating commercial and residential buildings, hence a replacement for synthetic insulation such as foamed products with a natural product.[1] Flax fibres in the form of a mat have been formed into a biocomposite with a sugar-based bio-resin, which acts as a binder for the fibres, for use in seatbacks in cars (EcoTechnilin) while a composite of flax fibres, a bioresin and basalt known as FibreRock has been used for the galley trolleys in aircraft aisles.

See: *biocomposites*

1. Lawler, R. Natural insulation materials: which is best? *Materials World* (Feb. 2014) 22, 32–3.

float glass

Sheets of glass such as window glass are made by the float glass process that was developed in the 1950s by Sir Alistair Pilkington and Kenneth Bickerstaff of the Pilkington Glass Company in England.[1,2] In this process a continuous strip of glass (at about 1000°C) from the furnace floats on a bath of molten tin (melting temperature 232°C). At the exit from the bath, the temperature is around 600°C, at which point the tin is still molten but the glass is rigid enough to be removed to the annealing stage. Both the tin and glass remain horizontal so that the surfaces of the glass remain parallel and the flat surface of the metal produces a smooth and undistorted glass surface. An earlier process for making plate glass involved flowing glass from the furnace through rollers followed by grinding and polishing the resulting rough surfaces.

1. Bolton, W., and R. A. Higgins. *Materials for Engineers and Technicians*, 6th edn, p. 340. London: Routledge, 2015.
2. Felice, M. Sheet glass. *Materials World* (July 2013) 21, 62–4.

fluorographene

Graphene is an electrical conductor and has been described as a 'wonder material' but to displace silicon as the main material used for the fabrication of integrated circuits and computer chips, doped graphene needs to be produced that has semi-conducting properties, that is has a band gap. There is much activity on developing doped graphenes. Fluorinated graphenes, or fluorographenes, are an example of doped graphenes.[1]

See: *graphene*

1. Wolf, E. L. *Graphene*. Oxford: Oxford University Press, 2013.

fluoropolymers

Fluoropolymers are polymers that have carbon–carbon linkages ($-C-C-$) in the polymer backbone and fluorine-based side groups.[1] They have been manufactured by the free radical polymerisation of fluorinated olefins; polytetrafluoroethylene (Teflon) is an example of a fluoropolymer as is poly (chlorotrifluoroethylene). The stable carbon–fluorine bond and strong carbon–carbon bond confer unique properties to these polymers. Replacement of hydrogen by fluorine in a polymer increases the fire resistance, elevates the maximum temperature at which the polymer can be used, produces a low coefficient of friction, lowers the surface energy useful for non-stick properties and improves electrical insulation. Applications for fluoropolymers are diverse and include gaskets, pipe liners, tubing, wire and cable insulation, seals and rings in automotive power steering, non-stick coatings in cookware, cardiovascular grafts and heart patches, coated fabrics for buildings and roofs, anti-dripping agents and wafer carriers in the semiconductor industry.

See: *polytetrafluoroethylene*

1. Smith, D. W. Jr, S. T. Iaccono and S. S. Iyer (eds). *Handbook of Fluoropolymer Science and Technology*. Chichester: Wiley, 2014.

flywheels

Flywheels store kinetic energy in a rotating wheel and the energy stored is proportional to the square of the rotational velocity of the wheel.[1] Flywheels can supply large pulses of current for starting generators and are under development for vehicle and stationary storage systems. Material properties, particularly mechanical properties, are important for the construction of flywheels in order to prevent failure. Fibre composites are an attractive option for their construction besides the use of metallic components. The Toyota Prius hybrid vehicle uses small composite-based flywheels in its regenerative braking system to store energy from braking and use it to boost acceleration of the vehicle.

1. McBride, T., B. Bollinger and D. Kepshire. Mechanical energy storage: pumped hydro, CAES, flywheels. In D. S. Ginley and D. Cahen (eds), *Fundamentals of Materials for Energy and Environmental Sustainability*, pp. 624–36. Cambridge: Cambridge University Press, 2012.

food emulsions

Colloidal systems are important for the preparation of processed food. For example, ice cream and mayonnaise are oil-in-water emulsions where the particle size of oil droplets lies within the colloidal size range. Butter, margarine and fat-based spreads are water-in-oil emulsions. Emulsions can be prepared by forcing the disperse phase (oil in an oil-in-water emulsion) through a membrane into the continuous phase.[1] Egg yolk, which contains the phospholipid lecithin, can act as an emulsifier for stabilising food emulsions.

See: *colloidal systems*

1. Buzak, S., and D. Rende. *Colloid and Surface Chemistry: A Laboratory Guide for Exploration of the Nano World*, pp. 221–2. Boca Raton, FL: CRC Press, 2014.

formica

A laminate consists of layers of materials joined together by an organic adhesive. Formica is an organic laminate derived from a formaldehyde-based resin and is used for decorative purposes such as tabletops.[1]

See: *cross-laminated timber; plywood*

1. Askeland, D. K., and W. J. Wright. *The Science and Engineering of Materials*, 7th edn, p. 647. Boston, MA: Cengage Learning, 2011.

fracking

Hydraulic fracturing, commonly referred to as fracking, is a process for releasing natural gas from shale deposits by injecting chemicals in water together with sand to break up the shale, often using horizontal drilling to cover a wide area of land from one well.[1–3]

See: *shale gas*

1. Hester, R. E., and R. M. Harrison (eds). *Fracking, Issues in Environmental Science and Technology*, vol. 39. Cambridge: Royal Society of Chemistry, 2015.
2. Bunger, J. W. Oil shale and tar sands. In D. S. Ginley and D. Cahen (eds), *Fundamentals of Materials for Energy and Environmental Sustainability*, pp. 127–36. Cambridge: Cambridge University Press, 2012.
3. Prud'homme, A. *Hydrofracking: What Everyone Needs to Know*. Oxford: Oxford University Press, 2014.

freeze-drying

Many materials, particularly foods, are produced in the form of powders. In freeze-drying a solution or slurry is first frozen by immersion in, for example, solid carbon dioxide, after which liquid is removed directly in a solid–vapour transition without going through a liquid phase.[1] Freeze-dried coffee is an example of a foodstuff prepared in this way and freeze-dried products tend to have a 'fluffy' rather than lumpy texture.

See: *spray-drying*

1. Buzak, S., and D. Rende. *Colloid and Surface Chemistry: A Laboratory Guide for Exploration of the Nano World*, p. 117. Boca Raton, FL: CRC Press, 2014.

fuel cell materials

Polymer exchange membrane (PEM) fuel cells convert chemical energy into electrical energy, usually using hydrogen or methanol as the fuel, producing water as a by-product. Fuel cells are an attractive option as a source of clean energy in the 21st century.[1,2] A representative fuel cell consists of two electrodes, often particulate platinum dispersed on carbon, that are separated by a solid polymer electrolyte membrane that allows transport of protons and water but not electrons. Hydrogen dissociates at the anode, allowing electrons to flow through the electrode and out of the cell, and protons migrate through the membrane to the cathode where they combine with oxygen and the returning electrons to form water. Nafion has an application as the solid polymer exchange membrane in fuel cells.

See: *electroceramics; Nafion*

1. Behling, N. H. *Fuel Cells—Current Technology Challenges and Future Needs*. London: Elsevier, 2013.
2. Steele, B. C. H. Electrical ceramics for fuel cells and high energy batteries. In B. C. H. Steele (edr), *Electronic Ceramics*, pp. 203–26. London: Elsevier, 1991.

fuel cells: high-temperature

Fuel cells can be classified into two types: those operating at (i) near ambient temperatures or (ii) high temperature.[1-3] The first category includes direct methanol fuel cells and proton exchange membrane cells. High-temperature fuel cells include solid oxide fuel cells, molten carbonate fuel cells, phosphoric acid fuel cells and alkaline fuel cells (Fig. G.10).[3]

1. Behling, N. H. *Fuel Cells—Current Technology Challenges and Future Needs*. London: Elsevier, 2013.
2. Steele, B. C. H. Electrical ceramics for fuel cells and high energy batteries. In B. C. H. Steele (edr), *Electronic Ceramics*, pp. 203–26. London: Elsevier, 1991.
3. Kocha, S., B. Pivovar and T. Gennett. Fuel cells. In D. S. Ginley and D. Cahen (eds), *Fundamentals of Materials for Energy and Environmental Sustainability*, p. 639. Cambridge: Cambridge University Press, 2012.

fullerenes

Fullerenes, which are also known as buckyballs, are carbon allotropes and were discovered in 1985.[1] They are named after the architect Buckminster Fuller. A fullerene molecule is a closed hollow aromatic carbon compound that is made up of twelve pentagonal faces and differing numbers of hexagonal faces with a surface pattern similar to that of a football and involves 60 carbon atoms bonded together.[2] Fullerenes are nanomaterials. An overview of fullerenes is given in reference (3).

See: *nanomaterials*

1. Kroto, W. H., J. R. Heath, S. C. O'Brien, R. F. Curl and R. F. Smalley. C_{60}: Buckminsterfullerene. *Nature* (1985) 328, 162–3.

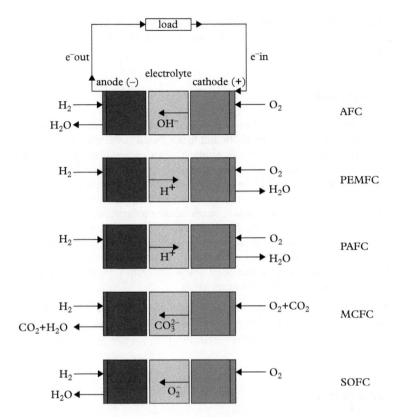

Fig. G.10 *Schematic diagrams for fuel cells: AFC, alkali fuel cell; PEMFC, proton exchange membrane fuel cell; PAFC, phosphoric acid fuel cell; MCFC, molten carbonate fuel cell; SOFC, solid oxide fuel cell.*

(R. J. D. Tilley. Understanding Solids, *2nd edn, p. 260. Chichester: Wiley, 2013.)*

2. Benavides, J. M. Method for manufacturing high quality carbon nanotubes. United States Patent 7,008,605, granted on 7 March 2006.
3. Yang, S., and C.-R. Wang. *Endohedral fullerenes: From Fundamentals to Applications.* London: Imperial College Press, 2014.

furan-2, 5-dicarboxylic acid

5-(Chloromethyl) furfural (CMF) has been extracted from lignocellulosic biomass by treatment with hydrochloric acid. CMF can be converted into furan-2, 5-dicarboxylic acid (FDCA), which can be polymerised with ethylene glycol to poly (furan-2, 5-dicarboxylic acid), also known as polyethylene furanoate (PEF), which is a potential replacement for polyethylene terephthalate when used in the production of plastic soft-drink bottles.[1,2]

See: *lignocellulose; PlantBottle; polyester*

1. Davies, E. Biomass bonanza. *Chemistry World* (Mar. 2013) 10, 40–3.
2. O'Driscoll, C. Next generation polyester beats PET. *Chemistry & Industry* (Dec. 2012) 76, 15.

fusion materials

Nuclear fusion of light elements, the energy source of the stars and the hydrogen bomb, has been under investigation since the 1950s and fusion-based power plants offer the potential of a relatively clean power source with minimal effect on greenhouse gases and nuclear proliferation if it can be successfully commercialised.[1] Heat is released from the fusion of deuterium (^2D) and tritium (^3T) ions at high temperature to overcome repulsive forces between the ions. Tritium is obtained by neutron bombardment of lithium, in particular of ^6Li or ^7Li. Reactions occur in a plasma at thermonuclear temperatures of millions of degrees kelvin and to exploit these reactions the plasma needs to be contained without contact with surrounding materials. This can be achieved, in principle, by magnetic confinement or by inertial confinement in which pellets of fuel remain stationary while its contents are rapidly heated by, for example, a laser to thermonuclear temperatures. Successful exploitation will depend in part on suitable confinement materials.

1. Fisch, N. J., J. L. Peterson and A. Cohen. Material requirements for controlled nuclear fusion. In D. S. Ginley and D. Cahen (eds), *Fundamentals of Materials for Energy and Environmental Sustainability*, pp. 196–8. Cambridge: Cambridge University Press, 2012.

gallium arsenide

Gallium arsenide (GaAs) is a III–V semiconductor that is a red light-emitter when used in light-emitting diodes and has applications as the active layer in semiconductor lasers. It has a band gap energy of 1.35 eV and has the zinc blende structure. Its early applications in the 1960s were for light-emitting diodes and this is still an application for the semiconductor. Solid solutions of GaAs with other elements such as aluminium allow the band gap energy to be tuned and hence the frequency and colour of the light emitted during electroluminescence to be varied.[1,2]

See: *band gaps; direct band gap semiconductors; gallium nitride; semiconductor lasers*

1. Solymar, L., D. Walsh and R. R. A. Syms. *Electrical Properties of Materials*, 9th edn, pp. 137–40. Oxford: Oxford University Press, 2014.
2. Askeland, D. K., and W. J. Wright. *The Science and Engineering of Materials*, 7th edn, p. 699. Boston, MA: Cengage Learning, 2011.

gallium nitride

Gallium nitride (GaN) is a crystalline semiconductor with a band gap energy of 3.34 eV and it is a key material for the practical exploitation of light-emitting diodes (LEDs) in domestic, industrial and street lighting as well as for the use of LEDs in displays. White light can be obtained by mixing light from the primary colours red, green and blue or by mixing yellow and blue light because yellow is a combination of the primary colours of red and green light. Red light-emitting and green light-emitting LEDs have been known since the 1960s, for example compositions based on gallium arsenic phosphide produces red light and gallium phosphide-based LEDs emits green light. However, it was not until the early 1990s that pioneering work by Shuji Nakamura and co-workers showed that doped GaN could be prepared by techniques such as metal organic vapour deposition, for example by using magnesium as the dopant and these thin layers of doped semiconductors emitted blue light (e.g. 450 nm wavelength) by electroluminescence.[1] Blue light from the LED can be converted to yellow light by use of a cerium-doped yttrium aluminium garnet phosphor. Nowadays blue-emitting LEDs in the visible spectrum are obtained by using alloys of indium-doped GaN.

See: *band gaps*

1. Nakamura, S., N. Iwasa and M. Senoh. Method of manufacturing p-type compound semiconductor. United States Patent 5,468,678, 1995.

gas hydrates

Gas hydrates, also known as hydrates or clathrate hydrates, are ice-like crystalline solids that are formed when water and gas are contacted at high pressures and low temperatures.[1] The hydrates are composed of hydrogen-bonded water molecules that form cages which trap small gas molecules. Methane clathrates are found in sediments under the permafrost at depths of around 1200 m below the surface and at around 500 m below the sea floor at depths of 1200 m. These hydrates are a potentially significant energy source.

1. Koh, C. A., E. D. Sloan, A. K. Sum and D. T. Wu. Unconventional energy sources: gas hydrates. In D. S. Ginley and D. Cahen (eds), *Fundamentals of Materials for Energy and Environmental Sustainability*, p. 138. Cambridge: Cambridge University Press, 2012.

generic drugs

Pharmaceutical compositions and processes for making them are often protected by patents. A patent gives the inventor a financial reward for a limited period (usually 20 years from the filing date) in exchange for making a public disclosure of the invention in the public domain. When a patent has expired, anyone can make the invention described in it and in the case of pharmaceutical compounds, the phrase generic drug is used when a medicine is made in this way.[1] Generic drugs tend to be considerably cheaper for governments to purchase for use in health services. However, the original composition may be protected by trademarks and designs if sold as tablets so that the generic drug cannot use the same marks.

See: *trademarks*

1. Greene, J. A. *Generic: The Unbranding of Modern Medicine*. Baltimore, MD: Johns Hopkins University Press, 2014.

genes

Regions of base pairs along the chromosome constitute genes, which are separated from other genes by regions of what has been referred to as junk DNA as its role is unclear.[1,2] Thus, a gene can be considered to be a length of DNA, hence a natural polymer. There are about 20,000 genes in the human body. Genes carry instructions on how the body works, how to express (i.e. synthesise) proteins and how to reproduce (i.e. copy) the sequence of base pairs. Genes encode the functions of the organism.

See: *DNA*

1. Rapley, R., and D. Whitehouse (eds). *Molecular Biology and Biotechnology*. Cambridge: Royal Society of Chemistry, 2015.
2. Northen, T. R. Biofuels and biomaterials from microbes. In D. S. Ginley and D. Cahen (eds), *Fundamentals of Materials for Energy and Environmental Sustainability*, pp. 326–8. Cambridge: Cambridge University Press, 2012.

genome

Regions of base pairs along the chromosome in a cell nucleus constitute genes, which are separated from other genes by regions of what has been referred to as junk DNA as its role is unclear. A gene can be considered to be a length of DNA, hence a natural polymer. The complete set of DNA, that is the set of base pairs and their order, is known as the genome.[1,2] The study of genomes is sometimes referred to as genomics, which has become a focus of attention of medical research in the early part of the 21st century as it has the potential to improve a population's health through application of personalised medicine. There are 3165×10^6 base pairs in humans, that is in the human genome.[3] This compares with 12×10^6 base pairs in fungi (e.g. *Saccharomyces cerevisiae*), 180×10^6 in fruit flies, 400×10^6 in rice and 3454×10^6 in mice.

See: *DNA; genes; personalised medicine*

1. Rapley, R., and D. Whitehouse (eds). *Molecular Biology and Biotechnology*. Cambridge: Royal Society of Chemistry, 2015.
2. Northen, T. R. Biofuels and biomaterials from microbes. In D. S. Ginley and D. Cahen (eds), *Fundamentals of Materials for Energy and Environmental Sustainability*, pp. 326–8. Cambridge: Cambridge University Press, 2012.
3. Lesk, A. M. *Introduction to Protein Science: Architecture, Function and Genomics*, 3rd edn, p. 15. Oxford: Oxford University Press, 2016.

glass ceramics

Glass ceramics are glasses in which fine crystals are grown by a controlled heat treatment.[1] They have higher strength, chemical durability and electrical resistivity than the parent glass. They can have low thermal expansions, producing very good resistance to thermal shock. Glass ceramics containing crystalline phases of β-spodumene (a lithium aluminium silicate) are particularly suitable for use in cookware and counter-top cooking surfaces because of their low thermal expansion, high strength and thermal stability. Pyroceram is a glass ceramic that has been used for cookware.

See: *ceramic honeycombs*

1. Grossman, D. G. Glass ceramics. In R. J. Brook (ed.), *Concise Encyclopaedia of Advanced Ceramic Materials*, p.170–6. Oxford: Pergamon Press, 1991.

glass epoxy composites

Glass fibre yarns, mats and felts when impregnated with unpolymerised epoxy resins are known as prepregs (pre-impregnated materials) and have been used in laminated structures as substitutes for heavier aluminium alloys for structural components (e.g. boat hulls); other resins can be used, for example phenolic-based resins. The mats can be combined in layers and fabricated into the desired shape during which polymerisation of the resin takes place. In aircraft, fibreglass epoxy composites have been used on aerodynamic control surfaces, fairings and trailing edge panels on the Boeing 747 while on the Boeing 757 and 767 these composites have been used on main landing- and nose-gear doors, engine cowls, wing spoilers and ailerons, rudder, elevators, fin and stabiliser tips.[1]

See: *composite materials; fiberglass; fibre-reinforced plastic composites; prepregs*

1. Bansal, N. P., and J. Lamon (eds). *Ceramic Matrix Composites: Materials, Modeling and Technology*. New York: Wiley, 2015.

glass fibres

Glass fibres are used in fibre-reinforced-plastic composites, such as fibre-reinforced poly-ester resins, and can be prepared in continuous form from a melt.[1] For example, S-glass is a high-strength magnesia-aluminosilicate composition with a tensile strength of 4.6 GPa and E-glass is a non-alkali metal borosilicate composition. Glass fibres are strong and cheap but not as stiff as carbon fibre.

1. Bolton, W., and R. A. Higgins. *Materials for Engineers and Technicians*, 6th edn. London: Routledge, 2015.

glass transition temperature

Synthetic polymers are an important class of materials, many of which were developed in the 20th century, and their use continues to be widespread in the 21st century. At room temperatures polymers can be soft and flexible and when cooled this type of polymer can become hard and glassy. The glass transition temperature is the temperature at which the transition from soft to glassy takes place.[1] Polystyrene has a glass transition temperature of 100°C and becomes soft and flexible when heated above this temperature. Polythene has a glass transition temperature of $-120°C$, so it remains soft and flexible at room temperature. Natural rubber has a glass transition temperature of $-73°C$.

1. Bolton, W., and R. A. Higgins. *Materials for Engineers and Technicians*, 6th edn, p. 302. London: Routledge, 2015.

glucosamine

Glucosamine is an amino-substituted (NH_2) glucose and it and its derivatives, such as glucosamine sulphate, have been available to the general public as dietary supplements.[1] Glycosaminoglycans are naturally occurring polymers containing glucosamine derivatives, for example hyaluronic acid, and act as lubricants in joints.

See: *hyaluronic acid*

1. Gratzer, W. *Giant Molecules: From Nylon to Nanotubes*. Oxford: Oxford University Press, 2013.

glycoproteins

Glycoproteins are proteins that are covalently linked to carbohydrates and include antibodies.[1] Glycoproteins reside on the surfaces of red blood cells and are responsible for the distinction between different blood groups between people. Surface chemistry is important when considering the role of glycoproteins. Hormone receptors on cell surfaces are often glycoproteins.[2] A receptor, here a glycoprotein, is a species that binds with a specific molecule, a hormone in the case of a hormone receptor, to bring about a change in the cell.

See: *antibodies; biopharmaceuticals*

1. Gratzer, W. *Giant Molecules: From Nylon to Nanotubes*, pp. 83–4. Oxford: Oxford University Press, 2013.
2. Hine, R. S. (ed.). *Oxford Dictionary of Biology*, 7th edn. Oxford: Oxford University Press, 2015.

gore-tex

A microporous form of polytetrafluoroethylene (PTFE) obtained on stretching and expanding layers of the material and known as Gore-Tex is used in waterproof clothing as the pores are too small to let raindrops through but large enough to let sweat molecules through;[1-3] expanded polytetrafluoroethylene has been referred to as ePTFE.

See: *polytetrafluoroethylene*

1. Gore, R. W. United States Patent 3,953,566, 1976.
2. Gore, R. W. United States Patent 4,187,390, 1980.
3. Gore, R. W., and S. B. Allen. United States Patent 4,194,041, 1980.

graphene

Graphene is a two-dimensional sheet of carbon that is one atomic layer thick in which the atoms are arranged in a hexagonal array in a honeycomb structure and it is the thinnest known material (see Fig. G.2 under 'carbon nanotubes').[1-3] The honeycomb structure is characteristic of graphite. Graphite was isolated in 2004 at the University of Manchester by Sir Andre Geim and Sir Konstantin Novoselov. Its high electrical conductivity, mechanical strength, flexibility, electron mobility, optical transparency and thermal conductivity offer many potential applications including flexible display devices, transparent electrodes, field effect transistors, protective coatings and fillers or reinforcements for composites.

See: *two-dimensional materials*

1. Wolf, E. L. *Graphene*. Oxford: Oxford University Press, 2013.
2. Warner, J. H., F. Schaffel, A. Bachmatiuk and M. H. Rummeli. *Graphene: Fundamentals and Emerging Applications*. London: Elsevier, 2013.
3. Torres, L. F., S. Roche and J.-C. Charlier. *Introduction to Graphene-Based Nanomaterials*. Cambridge: Cambridge University Press, 2014.

graphene oxide

Graphene oxide is an exfoliated form of graphite oxide.[1-3] 'Exfoliated' means 'deaggregated'. That is, an assembly of crystallites which are joined together is broken down into a suspension of individual crystallites or primary crystallites.

See: *graphene; graphite oxide*

1. Wolf, E. L. *Graphene*. Oxford: Oxford University Press, 2013.
2. Warner, J. H., F. Schaffel, A. Bachmatiuk and M. H. Rummeli. *Graphene: Fundamentals and Emerging Applications*. London: Elsevier, 2013.
3. Torres, L. F., S. Roche and J.-C. Charlier. *Introduction to Graphene-Based Nanomaterials*. Cambridge: Cambridge University Press, 2014.

graphite oxide

The terminology used to discuss graphene can be confusing. Graphite can be oxidised in a liquid medium to form graphite oxide, an electrical insulator containing surface epoxy groups (Fig. G.11). Graphene oxide can be considered to be an exfoliated or delaminated form of graphite oxide and can be reduced to graphene, for example by use of hydrazine monohydrate.[1-3]

See: *graphene; graphene oxide*

1. Wolf, E. L. *Graphene*. Oxford: Oxford University Press, 2013.
2. Warner, J. H., F. Schaffel, A. Bachmatiuk and M. H. Rummeli. *Graphene: Fundamentals and Emerging Applications*. London: Elsevier, 2013.

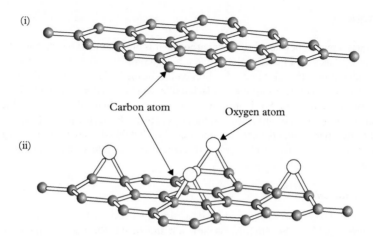

Fig. G.11 *Chemical structures for graphite (i) and graphite oxide (ii).*
(Y. Miyamoto. United States Patent Application 2014/0017440, 2014).

3. Torres, L. F., S. Roche and J.-C. Charlier. *Introduction to Graphene-Based Nanomaterials.* Cambridge: Cambridge University Press, 2014.

green chemistry

The use of genetically engineered organisms such as bacteria and yeast in the production of useful materials such as bio-polymers as replacement for synthetic polymers derived from petrochemical sources, the production of biodiesel from renewable sources, namely vegetable oils or animal fat, and ionic liquids as solvents with reduced environmental emissions fall within the area of green chemistry.[1,2] The latter involve processes that have a reduced detrimental effect on the environment compared to conventional processes for the production of materials.

See: *biodiesel; bio-1,3-propanediol; bio-succinic acid; ionic liquids; poly-3-hydroxybutyrate*

1. Kerton, F., and R. Marriot. *Alternative Solvents for Green Chemistry*, 2nd edn. Cambridge: Royal Society of Chemistry, 2013.
2. Bunzel, E., and R. A. Sayers. *Solvent Effects in Chemistry*, 2nd edn, pp. 164–76. Chichester: Wiley, 2016.

gutta percha

The 21st century is an age of digital communications. Smartphones, laptop and tablet computers proliferate and the use of social media is widespread. While some technologies are potentially disruptive technologies such as three-dimensional printing, others evolve over time and are built on technical advances made in previous ages. Gutta percha is a hard plastic, a naturally occurring polyisoprene, obtained by coagulation by heating of latex suspensions tapped from trees.[1,2] It is an isomer of polyisoprene, a different one of which forms elastic natural rubber. An application[3] for gutta percha in the second half of the 19th century was as a durable insulator for long-distance underwater telegraph cables and this allowed the expansion of communication technology at that time.

See: *elastomers; natural rubber*

1. Fried, J. R. *Polymer Science and Technology*, 3rd edn, p. 345. Upper Saddle River, NJ: Prentice Hall, 2014.
2. Weightman, G. *Eureka: How Invention Happens*, pp. 61–2. New Haven, CT: Yale University Press, 2015.
3. Loadman, J., and F. James. *The Hancocks of Marlborough*. Oxford: Oxford University Press, 2010.

hair dyeing: permanent

Permanent hair dyes are based on the chemical reaction between an ortho- or para-substituted aromatic diammine and a coupler (an aromatic compound such as certain naphthols) in the presence of an oxidising agent to form a coloured compound.[1] For example, p-phenylenediamine ($NH_2C_6H_4NH_2$) or p-aminophenol ($NH_2C_6H_4OH$) known as primary intermediates or developers react with an oxidising agent, usually hydrogen peroxide, under alkaline conditions for example in the presence of aqueous ammonia. The coupler modifies the colour produced on oxidation of the developer. Permanent hair colouring kits often consist of two components: (i) a mixture of dye precursors to produce the desired colour together with ammonia and (ii) a hydrogen peroxide solution. The dyeing process takes about 30 minutes. Hydrogen peroxide produces reactive electrophilic species on reaction with the developer that react with the coupler to form the coloured species. The latter are larger molecules than the precursors and are permanently bonded to the hair.

See: *hair dyeing; semi-permanent; hair dyeing; temporary*

1. Christie, R. M. *Colour Chemistry*, 2nd edn, p. 259. Cambridge: Royal Society of Chemistry, 2015.

hair dyeing: semi-permanent

Semi-permanent hair dyes are applied directly to the hair, do not involve an oxidative reaction to generate the colour and last roughly four to six washings.[1] The dyes are small non-ionic molecules that diffuse into the hair and are retained by van der Waals forces. Nitro dyes, for example, nitroanilines, nitroaminophenols and nitrophenylenediammines, are often used for semi-permanent dyes, producing a range of colours from yellow, orange, red-brown and violet.. Semi-permanent dyes are applied to the hair after a shampoo.

See: *hair dyeing; permanent*

1. Christie, R. M. *Colour Chemistry*, 2nd edn, p. 263. Cambridge: Royal Society of Chemistry, 2015.

hair dyeing: temporary

Temporary hair colours are designed to last from one shampoo to the next and have been referred to as colour rinses.[1] They can be used after permanent or semi-permanent hair dyes to modify the shade of colour. There is interest generally in obtaining colours from structures and nanoparticles that do not contain pigments, for example in coated mica flakes. Whether this approach is applied to hair colours in the 21st century remains to be seen.

See: *hair dyeing; permanent; iridescent pigments*

1. Christie, R. M. *Colour Chemistry*, 2nd edn, p. 264. Cambridge: Royal Society of Chemistry, 2015.

hairy adhesives (gecko lizard)

Not all adhesives are available in tubes, bottles or tins. The gecko is a lizard and is able to climb walls and even walk upside down along ceilings. Its ability to achieve this remarkable feat is subject to debate. However, its feet are covered with millions of densely packed thin hairs known as setae with diameters around 10 μm and length in the region of 100 μm.[1] The setae are made out of the protein β-keratin and the diameter can as much as ten times smaller than a human hair. The setae are arranged on pads in the lizard's feet. Each seta divides into between 100 and 1000 branches that are terminated by flat surfaces called spatulae. The tips of the spatulae are approximately 0.5 μm in length, 0.3 μm wide and 0.01 μm (10 nm) thick. As the foot presses against a surface an attractive van der Waals force comes into play between an individual hair and the surface. Although the force is very small, the magnitude of the force increases significantly when the individual forces for all of the hairs are added together. In a humid environment, capillary forces can increase the adhesion between the feet and damp surface. The gecko releases its grip on the surface by bending its feet which releases the attached pads.

See: *bioadhesives; epoxy adhesives; superglue*

1. Lee, M. (ed.). In *Remarkable, Natural Material Surfaces and Their Engineering Potential*, pp. 115–25. New York: Springer, 2014.

hemicelluloses

Hemicelluloses are branched polysaccharides consisting of simple sugar molecules including glucose, xylose, mannose and arabinose connected together.[1] Hemicellulose is a component of lignocellulose. An example of hemicellulose is xylan, which is found in plant cell walls and in algae.

See: *lignocellulose; lignin*

1. Askeland, D. K., and W. J. Wright. *The Science and Engineering of Materials*, 7th edn, p. 660. Bosont, MA: Cengage Learning, 2011.

heteropolymers

In homopolymers only one type of monomer is used; for example, polyethylene is derived from ethylene monomer. However, heteropolymers or copolymers are made from two or more different monomers.[1] Nylon and styrene-acrylonitrile-butadiene are examples of heteropolymers, as is DNA (deoxyribonucleic acid).

See: *DNA*

1. Daintith, J. (ed.). *A Dictionary of Chemistry*, 6th edn, p. 431. Oxford University Press, 2008.

heterostructures

Heterostructures consist of a thin layer (around 10 nm) of a semiconductor sandwiched between two layers of a different semiconductor that has a higher band gap than the well layer.[1] For example, a layer of GaAs (gallium arsenide) between layers of GaAlAs (gallium aluminium arsenide). This structure confines electrons to the layer with the smaller band gap and is an important component of solid-state semiconductor lasers that are used, for example, to read the stored information in a DVD (digital versatile disc). In the latter, a focused laser beam (405 nm in the blue region of the visible spectrum and used for the Blue-ray system) scans the surface of a metallised disc (polycarbonate), detecting the presence of etched pits in the surface.

See: *quantum wells*

1. Solymar, L., D. Walsh and R. R. A. Syms. *Electrical Properties of Materials*, 9th edn, pp. 194–8. Oxford: Oxford University Press, 2014.

hierarchical structures

Many plants have a hierarchical structure in their leaves.[1] For example, leaves of the lotus plant exhibit superhydrohobicity because of a hierarchical surface structure in the leaves which contains raised papillae that have a nanostructure of protruding tubular asperities of waxes. The combination of microscale and nanoscale features make up the hierarchic surface structure and which is essential for superhydrophobic properties. There is considerable interest in trying to mimic the structures of leaves in order to produce superhydrophobic structures in materials of interest.

See: *lotus leaves*

1. Lee, M. (editor). In *Remarkable, Natural Material Surfaces and Their Engineering Potential*. New York: Springer, 2014.

high-temperature ceramic composites

Zirconium diborides (ZrB_2) and hafnium diborides (HfB_2) are refractory transition metal diborides that have melting points above 3000°C.[1] Silicon carbide-reinforced ZrB_2

composites have been prepared using zirconium diboride powder for the matrix material using hot pressing and other densification methods, such as spark plasma sintering. ZrB_2- and HfB_2-based composites have high strength, high stiffness and good electrical and thermal conductivities among transition metal carbides, nitrides and diborides. Applications for diboride-based composites are limited but as the 21st century progresses these composites may find a niche market in extreme operating conditions.

1. Bansal, N. P., and Lamon, J. (eds). *Ceramic Matrix Composites: Materials, Modeling and Technology.* Chichester: Wiley, 2015.

hollow textile fibres

Lightweight hollow fibres have been developed for quick-drying fabrics including sheets and sportswear.[1] These materials are composite yarns made in a spinning process and consist of a core surrounded by a sheath. For example, filaments marketed under the trade name Wincall (Unitica) have a nylon sheath and a polyester core that can be eluted with sodium hydroxide solution. Other hollow textile fibres have the trade names Air-mint (Kuraray) and Lighton (Kanebo).

1. Kim, S., and H. A. Kim. Elution characteristics of high hollow filament nylon fabrics in dyeing and finishing processes and their physical properties for high emotional garments. In J. Fu (ed.), *Dyeing: Processes, Techniques and Applications*, p. 137. Hauppauge, NY: Nova Science Publishers, 2014.

humira

Humira (adalimubab) is a biopharmaceutical made by recombinant DNA technology. It has over 1,000 amino acids and a molecular weight around 40,000 Da and it is a monoclonal antibody. T cells (lymphocytes) release cytokines in the body that promote the destruction of tissues surrounding joints.[1] Tumor necrosis factor α (TNFα) is a cytokine that is blocked by adalimubab and is prescribed for the treatment of rheumatoid arthritis.

See: *biopharmaceuticals; recombinant DNA technology*

1. Tyrell, J. A. *Fundamentals of Industrial Chemistry: Pharmaceuticals, Polymers and Business.* New York: Wiley, 2014.

hyaluronic acid

Hyaluronic acid (also known as hyaluronan) is a linear polysaccharide of alternating disaccharide units of D-glucuronic acid and *N*-acetyl-D-glucosamine.[1-3] It belongs to the class of mucopolysaccharides although this term has now been replaced by the group of natural products known as glycosaminoglycans. The latter are so-called as they contain an amino sugar, here glucosamine. Hyaluronic acid is a lubricant in the joints and it is found in mammalian connective tissue such as in cartilage and the vitreous body of the eye as examples; it absorbs water to form a gel within joints. This natural biopolymer has a molecular weight in the range 10 to 6000 kDa and has a role in maintaining skin elasticity, transport of nutrients, cell adhesion, the movement of collagen and regulation of inflammation. Hydrogels are formed from hyaluronic acid by chemical modification or crosslinking and have applications in cosmetic products.

See: *anti-ageing skin treatments; glucosamine; polysaccharides*

1. Borbely, J. Additives for cosmetic products and the like. United States Patent Application 2008/0070993, published on 20 March 2008.
2. Lim, H. J., E. C. Cho, J. H. Lee and J. Kim. Chemically cross-linked hyaluronic acid hydrogel nanoparticles and the method for preparing thereof. International Patent Application WO 2008/100044, published on 21 August 2008.
3. Gratzer, W. *Giant Molecules: From Nylon to Nanotubes*, p. 82. Oxford: Oxford University Press, 2013.

hydrogels

Hydrogels are three-dimensional networks of hydrophilic homopolymers or copolymers that can be swollen by absorption and retention of water.[1] Hydrogels exhibit both liquid properties, because the major constituent is water, and solid properties, because of cross-linking during polymerisation. Hydrogels can absorb over 99.9% of water by weight based on the weight of the dry hydrogel (more than 10 times the dry weight) and can be prepared from natural or synthetic materials, for example crosslinked cellulose and acrylic acid. Applications for hydrogels include disposable nappies (diapers), soft contact lenses, wound dressings, scaffolds for tissue engineering and personal sanitary wear such as tampons.

See: *natural product hydrogels; superabsorbent polymers*

1. Segal, D. *Exploring Materials through Patent Information*. Cambridge: Royal Society of Chemistry, 2014.

hydrogen

Hydrogen is manufactured by steam reforming of methane in natural gas or by coal gasification, which yields a mixture of hydrogen and carbon monoxide.[1] More hydrogen can be obtained in the water–gas shift reaction in which carbon monoxide is reacted with steam over iron-, copper- or molybdenum-based catalysts that yields carbon dioxide and hydrogen.

See: *coal gasification*

1. Tyrell, J. A. *Fundamentals of Industrial Chemistry: Pharmaceuticals, Polymers and Business*, pp. 17–24. New York: Wiley, 2014.

hydrophilic materials

Hydrophilic materials can be defined in terms of contact angles and have contact angles from around 10° to 30° for water droplets.[1] An extremely wettable surface has a contact angle approaching zero. Hydrophilic surfaces are oleophobic; namely, oil droplets do not spread across their surface.

See: *contact angle*

1. Buzak, S., and D. Rende. *Colloid and Surface Chemistry: A Laboratory Guide for Exploration of the Nano World*. Boca Raton, FL: CRC Press, 2014.

hydrophilic polymers

Hydrophilic polymers are soluble in aqueous media and examples include polyethylene oxide, polyethylene glycol and polyacrylamide.[1] Hydrophobic polymers are insoluble in an aqueous medium and include polydimethylsiloxanes. The two polymer chains in block polymer have different degrees of hydrophilicity.

See: *polymers*

1. Buzak, S., and D. Rende. *Colloid and Surface Chemistry: A Laboratory Guide for Exploration of the Nano World*. Boca Raton, FL: CRC Press, 2014.

hydrophobic aerogels

Aerogels are ideal candidates for thermal insulation due to their very high porosity and low thermal conductivity. It is important in this application that the aerogel does not absorb water vapour as the pore structure will collapse. Hydrophobic aerogels that do not absorb water vapour have the potential to act as thermal insulation in buildings.[1]

See: *aerogels*

1. Segal, D. *Exploring Materials through Patent Information*. Cambridge: Royal Society of Chemistry, 2014.

hydrophobic ionic liquids

Hydrophobic ionic liquids are ionic liquids that do not absorb water vapour from the atmosphere.[1] Applications include polar solvents, phase transfer catalysts and liquid media for batteries.

See: *ionic liquids*

1. Kerton, F., and R. Marriot. *Alternative Solvents for Green Chemistry*, 2nd edn. Cambridge: Royal Society of Chemistry, 2013.

hydrophobic materials

Hydrophobic materials can be defined in terms of contact angles and have contact angles from around 70° to 90° and above for water droplets.[1] Teflon (polytetrafluoroethylene) is a hydrophobic material with a contact angle between 120° and 130°. Hydrophobic materials are oleophilic; namely, oil droplets can spread across their surface.

See: *contact angle; polytetrafluoroethylene*

1. Buzak, S., and D. Rende. *Colloid and Surface Chemistry: A Laboratory Guide for Exploration of the Nano World.* Boca Raton, FL: CRC Press, 2014.

hydroxyapatite

Hydroxyapatite, $Ca_{10}(PO_4)_6(OH)_2$ is a calcium phosphate and is found in bone as a composite with fibres of the protein collagen. Hydroxyapatite is used as a porous coating on medical implants for hip replacements to encourage growth of tissue into the pores.[1]

See: *biomimetic materials*

1. Prentice, T. C., K. T. Scott and D. L. Segal. Preparation of hydroxyapatite. United Kingdom Patent Application GB 2433257 A, published on 20 June 2007.

indigo

The blue textile vat dye indigo ($C_{16}H_{10}N_2O_2$) occurs in the leaves of a number of plant species including *Indigofera* and *Isatis*. Knotweed and woad are well-known indigo sources and indigo can be extracted from precursors such as indican and isatan in the leaves. The synthetic indigo industry based on aniline derivatives dates from the end of the 19th century and indigo is the dye used to impart the blue colour in denim. Indigo is insoluble in water and in use it is first converted by chemical reduction to a pale yellow soluble form, leuco indigo, by, for example, use of enzymes in a bacterial growth medium derived from natural products (e.g. molasses, dates), after which the fabric is soaked in the colourless solution and removed.[1] Leuco indigo is then oxidised back to indigo by air, producing the characteristic blue colour. A leuco dye is one that can exist in two forms, one of which, the leuco form, is colourless.

See: *faded jeans; textile printing*

1. Ratnapandian, S., S. Fergusson L. Wang and R Padhye. Indigo colouration. In J. Fu (ed.), *Dyeing: Processes, Techniques and Applications*, pp. 65–76. Hauppauge, NY: Nova Science Publishers, 2014.

indirect band gap semiconductors

Silicon and germanium are examples of indirect band gap semiconductors.[1] In this type of semiconductor electrons and holes combine to produce heat and the heat is dissipated throughout the semiconductor. This is because there is a change in momentum of the electron in the transition that prevents emission of light.

See: *direct band gap semiconductors; silicon carbide LEDs*

1. Tilley, R. J. D. *Understanding Solids: The Science of Materials*, p. 453. Chichester: Wiley, 2013.

indium tin oxide

Portable electronic devices—whether smartphones, laptop and tablet computers—are ubiquitous in the second decade of the 21st century. Indium tin oxide has a very important role in all of these devices as it is a transparent oxide that is electrically conducting and it is used as the electrode in electronic displays often as a coating on the screens.[1]

1. Ozin, G. A., A. C. Arsenault and L. Cademartiri. *Nanochemistry: A Chemical Approach to Nanomaterials*, p.120. Cambridge: Royal Society of Chemistry, 2009.

industrial effluents: acids

The Basil process (biphasic acid scavenging utilising ionic liquids) is an industrial process that uses ionic liquids.[1] Acids produced in chemical reactions have been removed by addition of amines, producing crystalline salts that are removed by filtration but this procedure can block up equipment. In the Basil process acids are removed from reaction solvents by interaction with a base in the form of an ionic liquid and the resulting liquid, rather than solid crystals, could be separated from the reaction product by decanting.

See: *ionic liquids*

1. Maase, M., K. Massonne, K. Halbritter, R. Noe, M. Bartsch, W Siegel, V. Stegmann, M. Flores, O. Huttenloch and M. Becker. International Patent Application WO 03/062171, published on 31 July 2003.

industrial effluents: carbon dioxide

Industrial processes often produce effluent streams and there is increasing interest in the 21st century on developing environmentally friendly processes. Traditional methods for removing carbon dioxide from exhaust gases involve chemical reaction with an amine-based solution, after which CO_2 is stripped from the solution at high temperature. Amidium-based ionic liquids have the potential for absorption of carbon dioxide at around 313 K and at pressures up to 2 MPa.[1] The ionic liquid can be regenerated by releasing the reactor pressure, a less-energy intensive process than traditional methods.

See: *carbon capture; ionic liquids*

1. Choi, S. J., J. Palgunadi, J. E. Kang, H. S. Kim and S. Y. Chung. Amidium-based ionic liquids for carbon dioxide absorption. United States Patent 8,613,865, 2013.

inkjet printing inks

Inkjet printing is a non-contact printing method and two methods are used in the process, the bubblejet system and the piezoelectric jet.[1] Small jets squirt tiny ink droplets onto the paper. In the bubblejet system, a heating element in each jet tube vaporises solvent in the ink, creating a bubble that expels the ink droplet from the jet. In the piezoelectric method, a piezoelectric material contracts on application of an electrical voltage and expels the ink droplet. Pigments that are used in the inks include carbon black, azo-based materials for yellow and magenta dyes and copper phthalocyanines for cyan inks.

1. Christie, R. M. *Colour Chemistry*, 2nd edn. Cambridge: Royal Society of Chemistry, 2015.

insulating fibres

Oxide fibres such as glass or aluminosilicates can be drawn from a melt.[1] In the case of aluminosilicates, the limitation of forming as-blown fibres is the melt viscosity. Kaowool is an example of an aluminosilicate fibre. Some fibres, particularly high-alumina-containing materials, can be spun from solutions of precursors. Nextel is an example of a fibre obtained from solution precursors. Fibres are used as thermal insulation. As an example, the US space shuttle was covered in insulated tiles containing oxide fibres that protect the aluminium skin from heat generated on re-entering the Earth's atmosphere. The fibres reduce radiation and conduction of heat from the outer surface to the skin. Silica- and alumina-containing fibres with high oxide contents can be prepared by sol-gel processes.

See: *sol-gel processes*

1. Birchall, J. D. Insulating fibres. In R. J. Brook (ed.), *Concise Encyclopaedia of Advanced Ceramic Materials*, pp. 236–8. Oxford: Pergamon Press, 1991.

insulin

In 1921 Frederick Banting and co-workers identified insulin, the pancreatic hormone that is deficient in diabetes. Insulin was obtained in a pure form from the pancreatic glands of cattle and shown to reduce the blood sugar concentration for the treatment of diabetes.[1] At the beginning of the 21st century there is concern in countries with public healthcare systems over rising levels of obesity and the associated Type 2 diabetes and the cost of treatment. Insulin is thus an important pharmaceutical nearly a century after its isolation and nowadays it is prepared through genetic engineering techniques.[2]

See: *yeast*

1. Banting, F. G., and C. H. Best. Extract obtainable from the mammalian pancreas or from related glands of fishes. Canadian Patent Application 234336 A, published 18 September 1923.
2. Barber, J., and C. Rostron (eds). *Pharmaceutical Chemistry*, pp. 13–14. Oxford: Oxford University Press, 2013.

integrated circuits

An integrated circuit, also known as a microchip, consists of a piece of semiconducting silicon, namely a silicon chip, that has been doped with elemental impurities and which has an intricate surface pattern of interconnected components (i.e. transistors, resistors, diodes and capacitors) that form the basis for carrying out logic operations in computers (Fig. G.12).[1,2] The integrated circuit has dimensions of about 1 mm in thickness with diameters up to 30 cm. The components are produced by photolithographic techniques using photoresists that are standard processes in the semiconductor industry. Interconnects are used to connect different parts of the circuit. In an integrated circuit the transistors act as switches rather than amplifiers and deal with pulse inputs and outputs while capacitors store electric charge and store data in a digital format. Integrated circuits are usually mounted in a plastic or ceramic package about the size of a postage stamp for ease of handling. Note that the engineer Geoffrey Dummer suggested in 1952 the idea that it should be possible to make electronic equipment in the form of a solid block with no connecting wires. This idea, the basic concept of an integrated circuit, was put into practice a few years later by Jack Kilby.

See: *interconnects; silicon chips; semiconducting-grade silicon*

1. Askeland, D. K., and W. J. Wright. *The Science and Engineering of Materials*, 7th edn, pp. 702–4. Boston, MA: Cengage Learning, 2011.
2. Thackray, A., D. C. Brock and R. Jones. *Moore's Law*. New York: Basic Books, 2015.

interconnects

Interconnects are thin, narrow electrically conducting metallic strips that connect different devices on an integrated circuit.[1] They are deposited from the gas phase, for example by sputtering onto exposed regions of silicon on a silicon wafer.

1. Askeland, D. K., and W. J. Wright. *The Science and Engineering of Materials*, 7th edn, p. 703. Boston, MA: Cengage Learning, 2011.

intrinsic semiconductors

An intrinsic semiconductor is one where the concentration of charge carriers (electrons and holes) is a characteristic of the material itself and not a characteristic of impurity atoms.[1] Thermal energies cause electrons to jump from the valence band to the conduction band, a process that leaves a hole behind in the valence band. An intrinsic semiconductor has an equal number of different charge carriers.

See: *band gaps; extrinsic semiconductors*

1. Askeland, D. K., and W. J. Wright. *The Science and Engineering of Materials*, 7th edn, p. 692. Boston, MA: Cengage Learning, 2011.

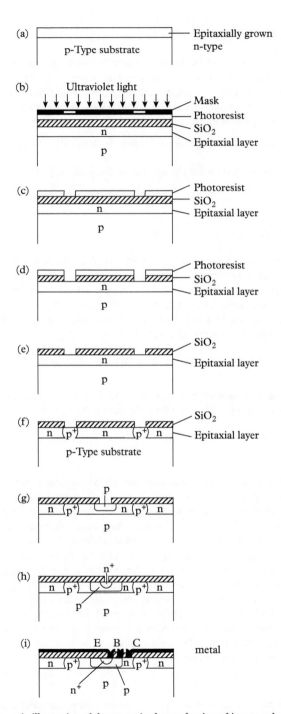

Fig. G.12 *Schematic illustration of the stages in the production of integrated circuits.*
(L. Solymar, et. al. Electrical Properties of Materials, *9th edn, p. 208. Oxford: Oxford University Press, 2014.)*

intumescent coatings

Intumescent coatings are used as fire retardants in which they act as a heat barrier, protecting the underlying substrates in the event of a fire.[1,2] During a fire the coating expands to form a solid foam on the underlying substrate and the coatings are used to protect internal walls of buildings including timber panels. Ammonium polyphosphate has been used as a component of intumescent coatings.

1. Ward, D. A. Fire retardant coatings, United Kingdom Patent Application 2,433,938 A, 2007.
2. Segal, D. *Exploring Materials through Patent Information*, p. 158. Cambridge: Royal Society of Chemistry, 2014.

ion implantation

The production of *n*-type semiconductors, *p*-type semiconductors, light-emitting diodes, silicon chips and integrated circuits relies on the ability to add controlled amounts of dopant atoms such as phosphorus or boron to pure silicon. A number of techniques can be used to introduce atoms into a lattice of a semiconductor to modify its electronic properties. Ion implantation is one such method and other methods are molecular beam epitaxy and metal-organic chemical vapour deposition (MOCVD).[1]

See: *integrated circuits; light-emitting diodes; n-type semiconductors; p-type semiconductors; silicon chips*

1. Page, T. Ion implantation. In Brook, R. J. (ed.), *Concise Encyclopaedia of Advanced Ceramic Materials*, pp. 252–7. Oxford: Pergamon Press, 1991.

ionic liquids

Ionic liquids are composed entirely of cations and anions from compounds that either have high or low melting points.[1–5] When ionic liquids have melting points at or below 303 K (30°C) they are referred to as room temperature ionic liquids. The latter usually consist of a bulky organic cation and organic anion and the large cation prevents the formation of an ordered crystalline solid structure. An example of an ionic liquid is 1-butyl 3-methylimidazolium acetate. Ionic liquids have low vapour pressure (i.e. they do not smell or have faint odours) and are environmentally friendly or 'green' alternatives to conventional organic solvents such as acetone and dichloromethane. Examples of their versatility include dissolution of cellulose for extraction of renewable chemicals such as sugars, a solvent for the preparation of biodiesel and use in separation processes such as removal of mercury vapours from oil and gas streams. An indication of their versatility as solvents is represented by the ability of ionic liquids to dissolve elemental sulphur, phosphorus, selenium and tellurium prior to purification of these elements.

See: *cellulose*

1. Segal, D. *Exploring Materials through Patent Information*. Cambridge: Royal Society of Chemistry, 2014.

2. Kerton, F., and R. Marriot. *Alternative Solvents for Green Chemistry*, 2nd edn. Cambridge: Royal Society of Chemistry, 2013.
3. Reade, L. Charged with success. *Chemistry and Industry* (Oct. 2013) 77, 42–5.
4. Paul, B. K., and S. P. Moulik (eds). *Ionic-Based Surfactant Science*. New York: Wiley, 2015.
5. Visser, A. E. (ed.). *Ionic Liquids: Science and Applications*, American Chemical Society Series 1117. Washington, DC: American Chemical Society, 2014.

iridescent organisms

The colours in soap bubbles arise from interference between the components of white light which are reflected from the front and back faces of soap films and interfere with each other. Also, the colours of butterfly wings and peacock feathers result not from the presence of pigments but from optical effects arising from scattering of the components of white light from the structures in the wings of the insects and birds.[1,2] The scattered components (i.e. wavelengths) interfere with each other generating the apparent colours. In some butterfly wings colours arise from interference of diffracted light from three-dimensional arrays of close-packed air holes in a chitin matrix so that the periodicity of the holes constitutes a diffraction grating. Pearl lustre effects that are often combined with coloured iridescence in pearls, shells and fish scales arise from optical effects due to the interaction of light with the materials.

See: *biomaterials; chitin; iridescent pigments; microstructures; photonic materials*

1. Ozin, G. A., A. C. Arsenault and L. Cademartiri. *Nanochemistry: A Chemical Approach to Nanomaterials*. Cambridge: Royal Society of Chemistry, 2009.
2. Luohong, S. Butterfly wings: nature's fluttering kaleidoscope. In M. Lee (ed.), *Remarkable Natural Material Surfaces and Their Engineering Potential*, p. 129. New York: Springer, 2014.

iridescent pigments

Conventional pigments absorb some wavelengths of light so that others are reflected and are seen as specific colours. The optical effect due to light scattering has been utilised in iridescent or pearlescent pigments.[1] Mica is a layered silicate mineral and can be broken into thin flakes. When the thin flakes are coated with titanium dioxide and then deposited as a coating, for example on the bodywork of cars, a white pearlescent appearance with a bluish colour on reflection is produced. Other colours can be produced from mica effect pigments including reds and oranges for metallic shades. Iridescent pigments are also used in cosmetics such as in nail varnish and in eye shadow. Coated mica can be produced with layers of other oxides including zirconium dioxide, chromium dioxide and iron oxide and multiple layers can be assembled. Mica has a lower refractive index than the oxide coatings and refraction and reflection can occur besides light scattering.

See: *iridescent organisms; mica*

1. Challener, C. Any colour you like. *Chemistry and Industry* (Dec. 2013) 77, 36–9.

keratin fibres

Keratin is a fibrous protein, a natural polymer that occurs as the α-form in hair, nails and feathers while the β-form is the protein of birds' beaks and claws.[1] The α-form consists of coiled helical polypeptide chains and several coils twist around each other to produce filaments with greater stiffness and strength. Coils are held together by disulphide linkages (-S-S-) between adjacent cysteine amino acids that give the hair its elasticity. Hair 'perms' involve chemically breaking the disulphide bonds through a reduction process producing –SH- groups, shaping the hair with clamps and rollers and reforming the disulphide linkages but between different pairs of –SH- groups. Keratin is a negatively charged polymer and many conditioners and shampoos contain or aim to contain positively charged polymers that bond to the hair and modify its texture.

1. Atkins, P. *Atkins' Molecules*, 2nd edn, pp. 107–8. Cambridge: Cambridge University Press, 2003.

Kevlar

Stephanie Kwolek is credited as the inventor of Kevlar (DuPont), a crystalline aromatic polyamide. The polymer is obtained by an interfacial or solvent polymerisation at temperatures less than $100°C$.[1,2] This polymerisation is an example of a condensation reaction. Reactants include as examples aromatic diamines and aromatic dichlorides. An example of a polymer composition is poly (p-phenylene terephthalamide), which corresponds to the chemical composition of Kevlar; the repeating unit is $-(NHC_6H_4NHCOC_6H_4CO)-$. Aromatic polyamides such as Kevlar are referred to as aramids while aliphatic polyamides are nylons. Solutions of the polymer (e.g. in concentrated sulphuric acid) can be spun by using a spinerette (a rotating disc with many holes in it) into continuous fibres that contain rod-shaped polymers in the nematic liquid crystalline phase. Rotation of the spinerette twists the filaments into a yarn. The spinning process is similar to that used for producing candy floss. The fibres are heated while under tension and this processing step produces a continuous fibre that is strong along the fibre axis and which can be woven into fabric. This aramid cannot be processed by melt-spinning as it decomposes below its melting point. Properties of Kevlar include low density, high stiffness (Young's modulus) and high tensile strength and the melting points for aramids are up to $500°C$. Kevlar has a strength and modulus greater than steel and is particularly associated with its use in lightweight body armour and helmets. Bullet-proof vests are made of cross-woven Kevlar fibres, which absorb energy on bending under impact and dissipating the energy over a wide area.[3] Other applications include conveyor belts, high-temperature electrical insulation and a substitute for steel in radial tyres and in the hulls of small boats. Nomex (DuPont) is another aramid that is used for flame retardant protective clothing.

See: *aramids; bullet-proof vests; candy floss; nematic liquid crystals; microstructures; nylon*

1. Hill, H. W. Jr, S. L. Kwolek and W. Sweeny. Aromatic polyamides. United States Patent 3,380,969, granted on 30 April 1968.
2. Kwolek, S. L. Optically anisotropic aromatic polyamide dopes. United States Patent 3,671,542, granted on 20 June 1972.

3. Gratzer, W. *Giant Molecules: From Nylon to Nanotubes*. Oxford: Oxford University Press, 2013.

laminated safety glass

Chipped and cracked windscreens are familiar to drivers of cars and other vehicles. Windscreens are made from a laminated safety glass.[1] Two glass pieces are laminated with a sheet of the plastic adhesive polyvinyl butyral between the pieces. The glass sheets are annealed; that is, they undergo a controlled heating and cooling cycle to produce glass that is nearly stress-free before the bonding process. When the windscreen is hit by an object, the plastic sheet holds the pieces of glass together, reducing risk of injury to the driver and passengers.

See: *plywood*

1. Askeland, D. R., and W. J. Wright. *The Science and Engineering of Materials*, 7th edn, p. 282. Boston, MA: Cengage Learning, 2011.

lanthanide-based magnesium alloys

Magnesium is the lightest structural material and is the third most common structural material after iron and aluminium.[1,2] Lanthanides are alloyed with magnesium to improve resistance to combustion, increase corrosion resistance, enhance high-temperature mechanical strength and improve resistance to creep. The alloys have applications as helicopter gearboxes, tail rotor housings, components in racing cars and high-performance vehicles such as transmission casings and components in aircraft engines.

1. Lucas, J., P. Lucas, T. Le Mercier, A. Rollat and W. Davenport. *Rare Earths: Science, Technology, Production and Use*, p. 185. London: Elsevier, 2015.
2. Pekguleryuz, M., K. Kainer and A. Kaya (eds). *Fundamentals of Magnesium Alloy Metallurgy*. Cambridge: Woodhead, 2013.

lanthanide dopants for lasers

The lanthanide elements (also referred to as the rare earths) are used as dopants for the optical medium used to produce lasing action:[1] for example, as dopants in single crystals, glasses or transparent ceramics. Dopant concentrations of the lanthanide are small, 0.1 at.% or less to avoid non-radiative relaxation between lanthanide ions instead of stimulated emission. Examples include neodymium-doped crystalline yttrium aluminium oxide ($Y_3Al_5O_{12}$, also known as yttrium aluminium garnet after the mineral garnet) with a wavelength around 1 μm, ytterbium-doped silicate glasses with a wavelength of around 1.1 μm, praseodymium-doped fluoride glass with a wavelength around 1.3 μm, holmium-doped yttrium aluminium garnet with a wavelength around 2.1 μm and cerium-doped fluoride crystals with a wavelength around 0.28 μm.

See: *laser-based materials*

1. Tilley, R. J. D. *Understanding Solids: The Science of Materials*, p. 458. Chichester: Wiley, 2013.

laser-based materials

Adsorption of a photon can excite an atom or molecule from a lower electronic energy state (or ground state) to a higher electronic energy state. The atom or molecule can drop back to the lower energy state by spontaneous emission of an identical photon whose energy is the difference in electronic energy between the lower and higher states. Light emitted by spontaneous emission is incoherent and unpolarised. Albert Einstein suggested in 1917 that there was another mode of emission, namely stimulated emission. Here a photon with energy exactly equal to the difference in electronic energy of the ground and excited states can interact with an atom in the excited state, causing it to return to the ground state and emit an identical photon. Thus, the light output is amplified. In stimulated emission, the emitted light is coherent and polarised. A laser is a device that produces a monochromatic, coherent beam of light that has little divergence and can travel over large distances while remaining collimated. It is an acronym for light amplification by stimulated emission of radiation. Emitted light can be produced continuously or in pulses with duration from microseconds to femtoseconds (1 femtosecond = 10^{-15} seconds). Materials have played a crucial role in the development of lasers and two key requirements for producing lasing action are (i) obtaining a population inversion between ground and excited states in a non-equilibrium system and (ii) maintaining the excited state to ensure emission is by stimulated and not by spontaneous emission.[1]

See: *ruby laser*

1. Lucas, J., P. Lucas, T. Le Mercier, A. Rollat and W. Davenport. *Rare Earths: Science, Technology, Production and Use*, pp. 319–28. London: Elsevier, 2015.

laser printing inks

Subtractive printing inks are used in laser printing in which powders are based on the three subtractive primary colours, magenta, cyan and yellow.[1] The yellows are based on diazo pigments, the magenta is represented by quinacridones and the cyan is based on copper phthalocyanine; carbon black powders are used for black.

See: *subtractive colour mixing*

1. Christie, R. M. *Colour Chemistry*, 2nd edn, p. 293. Cambridge: Royal Society of Chemistry, 2015.

latex

A latex is a suspension of polymer particles in a liquid medium, frequently an aqueous medium.[1] For example, natural rubber is derived from the sap of a tree and the sap consists of an aqueous suspension of microglobules. Polystyrene lattices are prepared in an emulsion polymerisation process. Polymer latices are used in paints and coatings.

See: *elastomers; emulsion polymerization; polystyrene*

1. Buzak, S., and D. Rende. *Colloid and Surface Chemistry: A Laboratory Guide for Exploration of the Nano World.* Boca Raton, FL: CRC Press, 2014.

LED televisions

Consumers have a wide range of televisions to choose from with screen sizes up to around 60 inches. The televisions are often lightweight and thin.[1] The screens are based on liquid crystal displays that have a back illumination from hundreds of light-emitting diodes (LEDs). Use of LEDs avoids the requirement for colour filters if a backlight based on a white light source is used and produce striking colours with improved clarity.

See: *light-emitting diodes; liquid crystals*

1. Kawamoto, H. The history of liquid crystal displays. *Proceedings of the IEEE* (Apr. 2002) 90, 460–500.

light-emitting diodes

Light-emitting diodes (LEDs) are based on the *p–n* junction diode.[1,2] Under the influence of a forward bias, electrons and holes recombine in the junction region, resulting in light emission (Fig. G.13). This process is known as electroluminescence. The band gap energy determines the wavelength (i.e. colour) of emitted light. Light emission does not occur for the semiconductors silicon and germanium as their band gap energies do not promote direct transition from the valence to conduction bands. In contrast, gallium arsenide (GaAs) emits in the infrared region of the electromagnetic spectrum. Red emissions in the visible spectrum

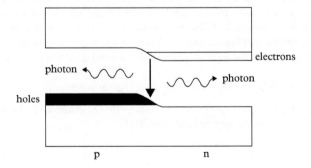

Fig. G.13 *The principle of LED operation: under a forward bias, electrons and holes recombine in the junction region between* p- *and* n-*type semiconductors and emit radiation.*
(R. J. D. Tilley. Understanding Solids, *2nd edn, p. 453. Chichester: Wiley, 2013.)*

were obtained by Holonyak in the early 1960s for alloys based on GaAs such as $GaAs_{0.24}$ $P_{0.76}$. LEDs are increasingly used for energy-efficient solid-state lighting, communication lasers used in fibre-optic transmission, solid-state lasers and displays for electronic devices such as smartphones and televisions. About 30×10^9 LED chips are produced each year. LEDs are inorganic materials (e.g. gallium nitride) in contrast to organic materials used in organic light-emitting diodes. An example of a composition for a LED involves a *p*-type gallium arsenic phosphide (GaAsP) layer on a *n*-type GaAsP layer supported on a gallium arsenide substrate.

See: *band gaps; organic light-emitting diodes; p–n junction diode*

1. Tilley, R. J. D. *Understanding Solids: The Science of Materials*. Chichester: Wiley, 2013.
2. Sinclair, I. *Electronics Simplified*, 3rd edn. Oxford: Newnes, 2011.

lignin

Lignin is a cross-linked polyphenolic viscous resin that is intimately mixed with the two other components of lignocellulose, namely cellulose and hemicellulose.[1]

See: *lignocellulose*

1. Askeland, D. K., and W. J. Wright. *The Science and Engineering of Materials*, 7th edn, p. 660. Boston, MA: Cengage Learning, 2011.

lignocellulose

Lignocellulose has three main components: 30–40 wt% cellulose, 20–30 wt% hemicellulose and 5–30 wt% lignin.[1] It is a major structural component of plants, is a form of biomass and is found widely in natural materials including sawdust, straw and bagasse (sugar cane residues). Lignocellulose fibres in the form of wood pulp are used for paper making, spun to cotton fibres and chemically modified; for example, treatment with nitric acid produces nitrocellulose, an ester used as the explosive guncotton. Bioethanol manufactured by enzymatic hydrolysis of sugar in a fermentation reactor is known as first-generation biofuel. However, the conversion of lignocellulose to bioethanol is at an earlier stage of development and ethanol made from lignocellulose is known as a second-generation biofuel.

See: *cellulose; hemicellulose; lignin; biomass; furan-2, 5-dicarboxylic acid*

1. Askeland, D. K., and W. J. Wright. *The Science and Engineering of Materials*, 7th edn, p. 320. Boston, MA: Cengage Learning, 2011.

lipids

Lipids are compounds found in living organisms, insoluble in water but soluble in organic solvents.[1,2] Complex lipids are esters of long-chain fatty acids and include glycerides that are the fats and oils of animals and plants; complex lipids also include phospholipids found in liposomes. Phospholipids are major components of cell membranes. Simple lipids include steroids and terpenes but not fatty acids. Steroids include the bile acids essential for the digestion of fats and terpenes, unsaturated hydrocarbons that are the essential oils with a distinctive scent in plants.

See: *lipidosomes*

1. Daintith, J. (ed.). *A Dictionary of Chemistry*, 6th edn. Oxford: Oxford University Press, 2008.
2. Barber, J., and C. Rostron (eds). *Pharmaceutical Chemistry*, pp. 261–8. Oxford: Oxford University Press, 2013.

liposomes

Liposomes are assemblies of phospholipids consisting of a lipid bilayer that acts as a permeable barrier or membrane to specific species and are known as phospholipid vesicles.[1] This bilayer is analogous to the lipid bilayers associated with cell membranes. The aqueous interior of liposomes can entrap or encapsulate materials such as hydrogel nanoparticles around 100 nm in diameter and pharmaceutical compounds; encapsulated hydrogels have been referred to as lipobeads. Potential applications of liposomes are in biomedical areas, including drug delivery and targeting of drugs, as well as in cosmetic preparations.

See: *hydrogels; lipids; nanoparticles*

1. Kazakov, S., M. Kaholek and K. Levon. Lipobeads and their production. United States Patent 7,618,565, granted on 17 November 2009.

lipsticks

Lipsticks are dispersions of colourants in a base of a blend of oils, fats and waxes.[1,2] Hydrophobic properties of lipsticks are important for run-free applications. Some lipsticks known as 'mood lipsticks' change colour when applied to the lips due to changes in pH. These lipsticks can contain indicators analogous to acid–base indicators used in chemical analysis. An example of a 'mood lipstick' has been produced by Lipstick Queen in which it is applied as a green layer, after which the colour transforms to a pink shade characteristic for each user.[3] In general, some inorganic pigments based on iron oxide are in lipstick formulations for dark colours such as brown while organic pigments (e.g. azo pigments) are used for bright colours, hence for reds and violets. Organic pigments are often present in formulations as 'lakes', that is precipitated insoluble forms of water-soluble dyes on a substrate, for example aluminium hydroxide.

1. Christie, R. M. *Colour Chemistry*, 2nd edn. Cambridge: Royal Society of Chemistry, 2015.
2. Emsley, J. *Chemistry at Home: Exploring the Ingredients in Everyday Products*, pp. 125–6. Cambridge: Royal Society of Chemistry, 2015.
3. Haslett, S. The green lippy that goes pink. *Daily Mail*, p. 35, 30 March 2015.

liquid crystal displays

In a typical liquid crystal display (1-3) two parallel glass substrates are coated with transparent electrodes of indium tin oxide on their inner surfaces. The electrodes are coated with an orientation layer made from an organic material (e.g. polyimide) and liquid crystals reside between the electrodes. The outside surfaces of the glass substrates are coated with polarising filters that are perpendicularly aligned. When an electric field is absent, light from a backlight passes through a vertical polarising filter. Polarised light that emerges from the filter then passes through the liquid crystal layer that rotates the angle of polarisation by 90°. Horizontal polarised light then passes through a horizontal polarising filter so that the glass screen appears to be bright to an observer (Fig. G.14). When an electric field is applied across the layer of liquid crystals, vertically polarised light emerging from the first polariser does not undergo a change in rotation in the angle of polarisation because of a change in orientation of the liquid crystals and cannot pass through the horizontally polarised filter. The glass screen then appears dark to an observer.

See: *liquid crystals*

1. German Flat Panel Display Forum. *European Technology: Flat Panel Displays*, 6th edn. VDMA, Verlag GmbH, 2008.
2. Platt, C., and F. Jansson. *Encyclopaedia of Electronic Components*, vol. 2. San Francisco, CA: Maker Media, 2014.
3. Kawamoto, H. The history of liquid crystal displays. *Proceedings of the IEEE* (Apr. 2002) 90, 460–500.

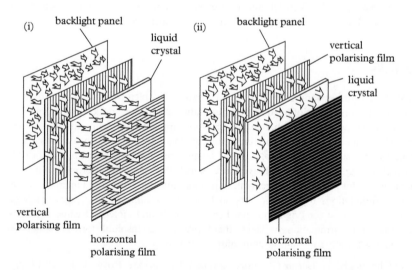

Fig. G.14 *Effect of an applied voltage on the appearance of a LCD display. (i) No voltage, the display appears transparent; and (ii) applied voltage and the display appears dark.*

(C. Platt and F. Jansson. Encyclopedia of Electronic Components, *vol. 2, p. 160. San Francisco, Ca: Maker Media, 2015.)*

liquid crystalline polymers

Cholesterol derivatives that form liquid crystals are small molecules. Liquid crystalline polymers, in particular liquid crystalline polyesters such as Xydar (Amoco), have been prepared.[1] The polymer chains with an aromatic backbone are ordered into liquid crystal phases and have higher tensile strengths and stiffness than polyesters that do not form liquid crystal phases. The aromatic polyamide Kevlar also forms liquid crystal phases when polymer molecules are aligned during the spinning process for fibres. Formation of liquid crystalline polymers represents a balance between decreasing entropy on forming ordered structures and favourable hydrogen bonding between close-packed polymer molecules that lowers the free energy of the system.

See: *liquid crystals*

1. Askeland, D. K., and W. J. Wright. *The Science and Engineering of Materials*, 7th edn, p. 587. Boston, MA: Cengage Learning, 2011.

liquid crystals

Liquid crystals are materials that form phases which are intermediate between crystalline solid phases and the fully disordered liquid phase and are associated with long-range molecular ordering.[1,2] The phases flow like liquids with low viscosity and have crystal-like properties such as anisotropy, and liquid crystalline phases are sometimes referred to as mesophases. Liquid crystals were discovered by Friedrich Reinitzer in 1888, who noticed that cholesterol benzoate became cloudy and viscous on melting at 145°C, after which it became isotropic (disordered like a liquid) at 179°C. The liquid crystalline phase, also referred to as a mesophase, lies in the temperature range between the first melting point temperature and the temperature at which isotropic liquid is formed. Liquid crystals are used in flat displays such as television screens and in mobile phones. Unlike light-emitting diodes and organic light-emitting diodes, liquid crystals do not emit light by electroluminescence and devices incorporating them have a backlight. Liquid crystals are birefringent and cause a rotation of polarisation in the incident light of about 90°. Examples of liquid crystalline phases are those formed from p-azoxyanisole, anisaldazine and dibenzaldbenzidine.

See: *cholesteric phases; nematic liquid crystals; smectic phases*

1. Kawamoto, H. The history of liquid crystal displays. *Proceedings of the IEEE* (Apr. 2002) 90, 460–500.
2. Dunmur, D., and T. Sluckin. *Soap, Science, and Flat-Screen TVs: A History of Liquid crystals*. Oxford University Press, 2014.

lithium–air batteries

Lithium–air batteries are under development for electric vehicles but are not yet commercially available.[1,2] The large free energy change for the reaction of lithium with oxygen is greater than for lithium-ion systems and makes the specific energy for lithium–air chemistry

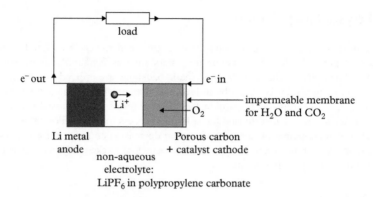

Fig. G.15 *Schematic diagram for lithium-air battery.*
(R. J. D. Tilley. Understanding Solids, *2nd edn, p. 259. Chichester: Wiley, 2013.)*

greater than for lithium-ion chemistry. The anode is metallic lithium and the cathode is air where the electrolyte is similar to that used in a lithium-ion battery, hence $LiPF_6$ in a polar organic solvent, for example a mixture of ethylene carbonate and dimethylcarbonate (Fig. G.15). Porous carbon also forms part of the cathode structure. The compound Li_2O_2 forms at the cathode and as this reacts with water vapour in the air to form lithium hydroxide, the cathode must be separated from the atmosphere by a membrane that only allows oxygen to pass through it.

1. Tilley, R. J. D. *Understanding Solids: The Science of Materials*, p. 259. Chichester: Wiley, 2013.
2. Ball, P. Up in the air. *Chemistry World* (Dec. 2012) 60–3.

lithium batteries

Combinations of different materials are important in the development of battery systems. In non-rechargeable lithium batteries the anode is metallic lithium and the cathode usually manganese dioxide (MnO_2).[1] The electrodes are separated by an electrolyte consisting of a lithium salt in a polar organic solvent. As lithium reacts violently with water the electrolyte must be a non-aqueous liquid. The output is about 3 V and these batteries are often produced in the form of flat discs for watches and other devices.

1. Ginley, D. S., and D. Cahen (eds). *Fundamentals of Materials for Energy and Environmental Sustainability*. Cambridge: Cambridge University Press, 2012.

lithium-ion batteries

Rechargeable lithium-ion (Li-ion) batteries are widely used in portable electronic devices such as mobile phones, laptop and tablet computers as they have a high energy density and can be manufactured in different shapes.[1] Li-ion batteries consist of a carbon anode and an oxide cathode, frequently lithium cobalt oxide $(LiCoO_2)$, a porous separator containing a

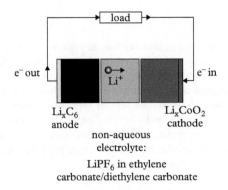

Fig. G.16 *A lithium-ion cell in discharge operation (schematic).*
(R. J. D. Tilley. Understanding Solids, *2nd edn, p. 258. Chichester: Wiley, 2013.)*

solution of a lithium salt such as $LiPF_6$ in a polar organic solvent, for example a mixture of ethylene carbonate and dimethylcarbonate (Fig. G.16). The oxide is an intercalation compound and lithium ions are transferred between the electrodes enabling the current to be utilised.

See: *nanoparticle-based batteries*

1. Tilley, R. J. D. *Understanding Solids: The Science of Materials*, p. 257. Chichester: Wiley, 2013.

living polymers

The 20th century has seen the development of a range of synthetic polymers (i.e. plastics) that continue to play an important role in everyday life in the 21st century. Preparation of a polymer from smaller monomer units often requires an initiator and terminating molecular additive. In the absence of a terminator, the molecular weight of the polymer, that is its size, continues to grow until all of the monomer is converted to polymer. The absence of a terminator results in a 'living polymerisation', where a narrow range of molecular weights and hence polymer sizes are produced.[1] Monodisperse living polymer particles such as those of polystyrene are used as standards in certain analytical techniques, including gel permeation chromatography. An example of a living polymer is styrene–butadiene. Styrene can undergo anionic polymerisation until all of the monomer is consumed. Addition of butadiene yields a styrene–butadiene copolymer, after which addition of more styrene monomer makes a styrene–butadiene–styrene triblock polymer.

See: *tyre compositions*

1. Tyrell, J. A. *Fundamentals of Industrial Chemistry: Pharmaceuticals, Polymers and Business*, p. 95. New York: Wiley, 2014.

lotus leaves

The lotus plant grows in marshes and lagoons in regions of the world.[1] The leaves are superhydrophobic so that water on them rolls up into droplets and falls off, removing dirt from the leaves. Superhydrohobicity arises from a hierarchical surface structure in which raised asperities of epidermal cells known as papillae produce an uneven surface and each papilla has a nanostructure of protruding tubular asperities of waxes. The combination of microscale and nanoscale features make up the hierarchic structure that prevents contact and adhesion between water droplets and the surface of the leaf. The phrase 'lotus effect' was coined by botanist Wilhelm Barthlott in 1997 who predicted that a synthetic material could mimic the waxy rough surface of the lotus leaf.

See: *superhydrophobic materials; self-cleaning glass; stain-resistant clothing*

1. Lee, M. Lotus leaves: humble beauties. In M. Lee (ed.), *Remarkable Natural Material Surfaces and Their Engineering Potential*, pp. 53–63. New York: Springer, 2014.

lotusan

Lotusan is a paint for external surfaces of buildings that exploits the 'lotus effect' to form a superhydrophobic self-cleaning coating.[1]

See: *lotus leaves; self-cleaning paint*

1. Lee, M. Lotus leaves: humble beauties. In M. Lee (ed.), *Remarkable Natural Material Surfaces and Their Engineering Potential*, p. 61. New York: Springer, 2014.

lyocell

Lyocell is a textile fibre derived from cellulose and its trade name is Tencel.[1] Cellulose, for example wood pulp, is dissolved in a solvent, N-methylmorpholine N-oxide, and the solution is spun into a water bath that regenerates the cellulose as continuous fibres. Tencel has a silk-like texture.

See: *rayon*

1. Christie, R. M. *Colour Chemistry*, 2nd edn. Cambridge: Royal Society of Chemistry, 2015.

macro-defect-free cement

Portland cement is porous and while its use depends on its compressive strength it is weak under tension. However, reducing the fraction of voids (i.e. the pores) produces cement that can be formed into a coiled spring rather like the springs used for suspensions in vehicle. This macro-defect-free cement can be prepared by blending Portland cement with a distribution of particle sizes with a polyacrylamide gel in water which produces a dough that can be shaped under pressure.[1]

See: *Portland cement*

1. Bolton, W., and R. A. Higgins. *Materials for Engineers and Technicians*, 6th edn, p. 333. London: Routledge, 2015.

magnetic resonance imaging

Magnetic resonance imaging (MRI) is widely used in hospital environments for obtaining high-quality images of soft tissue of varying composition. The principle of the technique is identical to that of nuclear magnetic resonance (NMR) that has been used in research environments since the 1950s for determining the chemical structure of molecules often in solution but because the word nuclear is associated with radioactivity the name has been changed when the method is applied to medical diagnostics.[1] In the technique, the nuclear spin of protons in water molecules is aligned in a strong magnetic field, after which they are perturbed by electromagnetic radiation in the radio spectrum. Emitted signals as the spins return to their random alignments are detected and combined to build up the image. In practice the patient lies on a table inside a double-walled cylinder. A coil of a superconducting alloy (niobium-germanium) is wrapped around the cylinder inside the cavity that is filled with liquid helium and a strong magnetic field (around 2 T) is generated when an electric current is passed through it.

See: *superconducting metals and alloys*

1. Woolfson, M. M. *Resonance: Applications in Physical Science*, pp. 87–103. London: Imperial College Press, 2014.

mauveine dye

The discovery of a purple dye by Sir William Henry Perkin in 1856 laid the foundation for the synthetic dye industry and organic chemicals industry.[1–3] The dye known as mauveine (aniline purple or Perkin's mauve) was initially prepared by oxidation of aniline with potassium dichromate. The dye was prepared on the large scale and required industrial production of aniline by reduction of nitrobenzene and this large-scale production encouraged expansion of the organic chemicals industry (Fig. G.17). Mauveine was a commercial success after launch in 1857 and its distinctive purple colour on textiles such as dresses was very popular. Nowadays colours are abundant in everyday life, from textiles to cosmetics; paints and coloured plastics as examples.

1. Perkin, W. H. *Journal of the Chemical Society Transactions* (1979) 717.
2. Christie, R. M. *Colour Chemistry*, 2nd edn, pp. 6–7. Cambridge: Royal Society of Chemistry, 2015.
3. Perkin, W. H. Producing a new colouring matter for the dyeing with a lilac or purple color stuffs of silk, cotton, wool or other material. United Kingdom Patent GB1984, granted 24 February 1857.

Fig. G.17 *An early sample of mauveine and a shawl dyed with an original sample.*
(J. Barber and C. Rostron (eds). Pharmaceutical Chemistry, *p. 180. Oxford: Oxford University Press, 2013.)*

metal foams

Metal foams are stiff and can be made into complex shapes without deformation and have been made from a variety of metals including titanium, tantalum and stainless steel.[1] Potential applications include medical implants, lightweight components (compared to dense metal) in vehicles and porous supports for electrodes in fuel cells. Pores in a metallic form can be interconnected in an open-cell foam or sealed bubbles in a closed-cell foam.[2] Closed-cell foams can be prepared by injecting gas into a molten metal or adding a blowing agent such as titanium hydride (TiH_2), which decomposes in the melt, producing gas bubbles. The blowing agent can be incorporated into a powder compact and the latter heated close to the melting point of the metal, causing decomposition of the agent and release of gas bubbles. Closed-cell aluminium foam has good strength and vibration dampening properties and acts as a construction metal when sandwiched between aluminium sheets. Metal foams also act as materials for absorbing energy on impact between vehicles. Open-cell metal foams have potential applications as diesel particulate filters and catalytic converters. Fabrication of open-cell foams involves using a template around which the molten metal solidifies.

1. Bolton, W., and R. A. Higgins. *Materials for Engineers and Technicians*, 6th edn, p. 429. London: Routledge, 2015.
2. Ploszajski, A. Metal foams. *Materials World* (2016) 24(2), 59–61.

metallic nanoparticles

Nanotechnology has attracted much attention in the early part of the 21st century and activity in this area often involves nanoparticles. Examples of the latter are gold nanoparticles, which exhibit optical properties different to the bulk material.[1] Whereas bulk gold has a yellow lustre, colloidal dispersions of gold can exhibit a range of colours when illuminated by visible light where the colour depends on the particle size, around 10 nm or less. A significant percentage of atoms reside in the surface of small particles compared to the number in the bulk material. Light that is incident on the particle surfaces induces oscillations in electron charge distribution known as plasmons which modify the light reflected from the surface due to distinct resonance frequencies. Such effects are seen when colloidal metals are used in stained-glass windows and decorations in glassware.

See: *colloidal systems; nanoparticles; nanotechnology*

1. Louis, C., and O. Pluchery. *Gold Nanoparticles for Physics, Chemistry and Biology*. London: Imperial College Press, 2013.

metamaterials

Metamaterials have a negative refractive index and have the ability to guide electromagnetic radiation, visible light or other wavelengths, for example microwaves, around an object, rendering the object invisible.[1,2] Metamaterials have an engineered microstructure or nanostructure where the structure does not occur in nature and the structure has a periodicity. Hence metamaterials may be referred to as artificial materials. An example of a metamaterial that operates in the microwave region was constructed from a network of straight wires that mimic a free-electron response to an electric field and millimetre-sized split-ring resonators that are responsible for a magnetic resonance.[1] An 'invisibility cloak' can be produced in theory and has been widely publicised by the fictional character Harry Potter, but such cloaks have not been made in practice.

1. Novotny, L., and B. Hecht. *Principles of Nano-optics*, 2nd edn. Cambridge: Cambridge University Press, 2006.
2. Solymar, L., D. Walsh and R. R. A. Syms. *Electrical Properties of Materials*, 9th edn. Oxford: Oxford University Press, 2014.

methotrexate

Throughout the 20th century most drugs have been based on small synthetic molecules. For example, acetylsalicylic acid has a molecular weight of 74. An example of the synthetic route is given for methotrexate that is used in chemotherapy and in the treatment of rheumatoid arthritis where it inhibits the formation of folic acid.[1] The synthesis of drugs often involves a series of steps but methotrexate has been prepared in a one-step synthesis. Here it was prepared by reaction of tetraaminopyrimidine, p-(*N*-methylamino)-benzoyl glutamic acid and dibromopropionaldehyde.[2] A more recent drug used for treatment of rheumatoid arthritis is Humira, a biopharmaceutical.

See: *acetylsalicylic acid; Humira*

1. Tyrell, J. A. *Fundamentals of Industrial Chemistry: Pharmaceuticals, Polymers and Business*, p. 171. New York: Wiley, 2014.
2. Singh, B., and F. C. Schaefer. United States Patent 4,374,987, 1983.

mica

Mica is a naturally occurring aluminosilicate mineral and is characterised by having very thin sheets less than around 0.025 mm thick (25 μm) that can be cleaved easily from the bulk material because the bonding forces between the sheets are small. Coated mica has applications as iridescent pigments. Mica is an 'effect' pigment for metallic car paints where the mica flakes have a thickness around 300–600 nm although thin glass flakes known as Glassflake with a thickness around 350 nm when coated with metal oxide exhibit a range of colours in effect pigments. Fluorine-containing mica crystals have been used in a family of non-porous machinable glass ceramics sold under the trade name MACOR.[2]

See: *iridescent pigments*

1. Perkins, J. 10 minutes with Simon Brigham. *Materials World* (Sept. 2015) 23, 19.
2. Grossman, D. G. Glass ceramics. In *Concise Encyclopaedia of Advanced Ceramic Materials*, pp. 170–6. Oxford: Pergamon Press, 1991.

micellar systems

Surface-active agents or surfactants are molecules that are amphiphilic, that is have both hydrophilic and hydrophobic regions. They are often described as having a hydrocarbon 'tail' and a hydrophilic 'headgroup'. The headgroup can be cationic, anionic or non-ionic although zwitterionic surfactants have both cationic and anionic headgroups. An example of a surfactant is sodium dodecyl sulphate. On dissolution in water, surfactants adsorb at the air–water interface at the surface, lowering the surface tension, and once a monolayer is formed at the surface the surfactant molecules aggregate or self-assembly on increasing concentration to form spherical micelles in the aqueous medium.[1,2] Micelle formation takes place above a critical micelle formation. In these micelles the headgroups point out into the aqueous medium and the hydrocarbon chains point inwards. Surfactants are found in many household products, including washing-up liquids and toothpastes and are responsible for the foaming properties of these products. Micelles can solubilise dirt and grease which migrate from surfaces into the core of the micelles. Facial wipes are a popular consumer product and some are described as 'micellar facial wipes'.

1. Buzak, S., and D. Rende. *Colloid and Surface Chemistry: A Laboratory Guide for Exploration of the Nano World*. Boca Raton, FL: CRC Press, 2014.
2. Jones, R. A. L. *Soft Condensed Matter*, pp. 136–58. Oxford: Oxford University Press, 2014.

microelectromechanical systems

Microelectromechanical systems (MEMS) are fabricated by methods widely used for the fabrication of integrated circuits: for example, through the use of silicon wafers, photoresists, photolithography and etching.[1] These devices can be used as sensors such as accelerometers in determining when airbags in vehicles should be inflated or in optical systems where rotatable mirrors are required. MEMS also include gyroscopes, magnetometers and actuators.

See: *wearable technology*

1. Solymar, L., D. Walsh and R. R. A. Syms. *Electrical Properties of Materials*, 9th edn, p. 215. Oxford: Oxford University Press, 2014.

microlaminates

Microlaminates consist of alternating layers of different material in particular layers of aluminium sheet and fibre-reinforced polymer. For example, Arall consists of a prepreg of an aramid fibre such as Kevlar impregnated with an adhesive and laminated between sheets of an aluminium alloy.[1] The laminate is lightweight, stiff and strong with good corrosion resistance. Glass-aluminium microlaminates have been used in the fuselage of the Airbus 380. Microlaminates have good lightning resistance, are machinable and are relatively easy to repair.

See: *laminated safety glass; prepregs*

1. Askeland, D. R., and W. J. Wright. *The Science and Engineering of Materials*, 7th edn. Boston, MA: Cengage Learning, 2011.

microstructures

The properties of a material are determined by its microstructure, that is the nature, quantity and distribution of structural elements or phases that make up the material.[1] These properties include optical properties such as the colours observed in the wings of butterflies; mechanical properties such as strength, toughness, stiffness (Young's modulus), hardness, elasticity (e.g. for rubber); and electronic properties such as electrical conductivity in superconductors. The strength of Kevlar depends on oriented liquid crystalline phases of a polyamide in its microstructure. The microstructure in DNA may be considered to be the sequence of nucleotide bases in the polymer chain. The term microstructure is a technical one but in everyday life the word structure can be used interchangeably with the term 'microstructure'.

1. Brook, R. J. Advanced ceramic materials: An overview. In R. J. Brook (ed.), *Concise Encyclopaedia of Advanced Ceramic Materials*, pp.1–8. Oxford: Pergamon Press, 1991.

moisturisers

Moisturisers are widely used in cosmetics and applied to the skin.[1] An example of a moisturiser is γ-polyglutamic acid which can be derived in a fermentation process from L-glutamic acid and converted to a hydrogel using cross-linking agents or by irradiation with electron beams or gamma rays.

See: *cosmetic formulations; hydrogels*

1. Segal, D. *Exploring Materials through Patent Information*. Cambridge: Royal Society of Chemistry, 2014.

monoclonal antibodies

Monoclonal antibodies are antibodies from identical cells derived from a parent cell.[1] The parent cell is derived by fusion of a lymphocyte cell, which produces antibodies, and a mouse cell, after which the hybrid cell multiplies quickly, producing the antibody. Monoclonal antibodies form the basis of biopharmaceuticals also known as biologic drugs. Humira, a biologic drug, is an immunoglobulin and a key difference between biologic drugs compared to conventional drugs such as aspirin is that they have molecular weights in the tens of thousands rather than on the order of several hundreds or less. The conventional drug aspirin has a molecular weight of 76.

See: *antibodies; biopharmaceuticals*

1. Rapley, R., and D. Whitehouse. *Molecular Biology and Biotechnology*. Cambridge: Royal Society of Chemistry, 2015.

monosaccharides

Monosaccharides or simple sugars are carbohydrates that are not broken down (i.e. hydrolysed) into simpler compounds by the action of dilute acids.[1] They include glucose, also referred to as an aldohexose as this molecule contains six carbon atoms and an aldehyde group ($-CH{=}O$) and fructose, a ketohexose containing six carbon atoms and a ketone group ($-C{=}O$). A monosaccharide such as glucose is a monomer that forms the polymer backbone of cellulose.

See: *biomass; carbohydrates; cellulose; polysaccharides*

1. Barber, J., and C. Rostron (eds). *Pharmaceutical Chemistry*, p. 256. Oxford: Oxford University Press, 2013.

moore's law

Gordon Moore, the founder and president of Intel Corporation, suggested in 1965 that the number of components such as transistors on a silicon chip in an integrated circuit would increase exponentially over time; that is, the number would double for a fixed interval of time for example every 18 months or every two years. This suggestion is known as Moore's law.[1] Since this suggestion was made, this law has been followed and the number of transistors that have been fabricated on a silicon chip has doubled nearly every two years.

See: *silicon chips*

1. Thackray, A., D. C. Brock and R. Jones. *Moore's Law*. New York: Basic Books, 2015.

MOSFET

MOSFET refers to a metal oxide semiconductor field effect transistor and is widely used in integrated circuits.[1] As with all transistors, MOSFETs have three terminals, the drain, the source and gate. The latter is insulated from a p-type substrate and the gate–substrate acts as a capacitor. The aim of the device is to control the current flow from source to drain and this is achieved by varying the voltage applied to the gate. When a voltage is applied between the gate and source, electrons flow into a channel between drain and source, the channel conductivity increases and current flows between the source and drain. The device is switched on. No current flows in the absence of an applied voltage. Hence, the MOSFET acts as a switch. MOSFETs have low power consumption, low operating voltages and high processing speeds due to the short channel length.

See: *transistors*

1. Solymar, L., D. Walsh and R. R. A. Syms. *Electrical Properties of Materials*, 9th edn, pp. 206–10. Oxford: Oxford University Press, 2014.

nacre

Molluscs are animals belonging to the phylum Mollusca that have soft bodies and hard shells. Species include conch shells, abalone, clams, limpets, mussels and oysters as examples. The shells consist of an outer layer of brittle aragonite crystals (calcium carbonate) that has an inner layer of nacre, a composite of aragonite crystals and a small amount (< 5%) of soft protein that acts as ceramic armour.[1] Nacre is also known as mother-of-pearl that exhibits lustrous colours and pearls are made out of nacre. The composite has a layered structure analogous to a brick wall where the aragonite represents the bricks and the protein is the mortar. Nacre is a tough material and can deform when subjected to attack by a predator, preventing cracks spreading throughout the composite.

See: *biomaterials*

1. Lee, M. (ed.). In *Remarkable, Natural Material Surfaces and Their Engineering Potential*. London: Springer, 2014.

Nafion

The perfluorinated polymer Nafion is a perfluorosulphonic acid ionomer, a copolymer of tetrafluoroethylene with trifluorovinyl ether containing a trifluorovinyl ether group which is hydrolysed to the hydrophilic sulphonic acid group after copolymerisation;[1] the backbone of Nafion corresponds to polytetrafluoroethylene. It is used as a thin membrane in polymer electrolyte membrane (PEM) fuel cells operating below 100°C.

See: *fuel cell materials*

1. Smith, D. W. Jr, S. T. Iaccono and S. S. Iyer (eds). *Handbook of Fluoropolymer Science and Technology*, p. 125. Chichester: Wiley, 2014.

nanocellulose

Nanocellulose is a renewable, biodegradable fibre with an aspect ratio (length:diameter) of around 50 and is a component of lignocellulose but embedded in a matrix of hemicellulose and lignin.[1] It can be extracted by mechanical processing in contrast to the use of chemical methods and has a Young's modulus of around 100 GPa; that is, the fibres are stiff. The tensile strength is greater than that of steel. These stiff fibres have a range of potential applications: reinforcements in composite materials such as in plastics and concrete, drug delivery and wound dressings.

See: *lignocellulose; cellulose*

1. Gross, M. Nature's building blocks. *Chemistry and Industry* (Dec. 2014) 78, 18–21.

nanoclays

Clays are aluminosilicate minerals. Nanoclays refer to these minerals when they have a particle size less than 50 nm and have applications as flame retardants when used as fillers in plastic components.[1]

1. Bolton, W., and R. A. Higgins. *Materials for Engineers and Technicians*, 6th edn, p. 428. London: Routledge, 2015.

nanocomposites

Nanocomposites consist of a nanomaterial dispersed as a reinforcing material in a matrix.[1,2] Examples of matrix materials include epoxy resins and thermoplastic resins, for example polyester, polyetherimide and polyurethane elastomers.[1] Use of nanomaterials as a filler can improve flame resistance, wear resistance, thermal stability, strength and Young's modulus (stiffness) of the composite compared to the use of filler with a larger particle size.

See: *composite materials; nanomaterials; polyetherimide*

1. Simmons, M., and J. Cawse. Composite materials. United States Patent Application 2014/0047710, published on 20 February 2014.

2. Bolton, W., and R. A. Higgins. *Materials for Engineers and Technicians*, 6th edn, p. 428. London: Routledge, 2015.

nanofibres

Nanofibres have been described as having a diameter of less than 1 µm and an aspect ratio (length:diameter) greater than 50.[1] Nanofibres can be considered to be colloidal systems and have been prepared by electrospinning in which a polymer solution such as polylactic acid is ejected as an electrically charged filament from a nozzle. Potential applications for nanofibres include fillers for composites and biomimetic scaffolds for cell growth.

See: *colloidal systems*

1. Stevens, B., B. Park and F. Sartain. Nanofibres to nanocomposites. *Materials World* (Apr. 2013) 21, 26–7.

nanofiltration membranes

Membranes are porous barriers that are selective in the type of species allowed to pass through them.[1] Nanofiltration membranes have pore sizes around 2 nm in diameter with molecular weight cut-offs around 200 and have been prepared from polymers and ceramics such as porous zirconia. Applications for nanofiltration membranes include gas separations, separation of sugars and separations in the pharmaceutical and water-treatment industries.

See: *ultrafiltration membranes*

1. Basile, A., and S. P. Nunes. *Advanced Membrane Science and Technology for Sustainable Energy and Environmental Applications*. Cambridge: Woodhead, 2011.

nanomaterials

Nanomaterials can be considered to be any material that has at least one dimension on the nanometre scale.[1-3] The word is often used indiscriminately and nanomaterials are also described as nanoparticles. Examples of nanomaterials include particles of nanoclays such as those based on montmorillonite, graphene, graphene oxide, nanofibres for example cellulose nanofibres, nano-whiskers such as cellulosic nano-whiskers and nano-tubes including carbon nanotubes. Nanomaterials are used as reinforcing fillers for nanocomposites.

See: *fullerenes; nanocomposites; nanoparticles*

1. Palmese, G. R., and A. L. Watters. Room temperature ionic liquid-epoxy systems as dispersants and matrix materials for nanocomposites. United States Patent Application 2012/0296012, published on 22 November 2012.

2. Bolton, W., and R. A. Higgins. *Materials for Engineers and Technicians*, 6th edn, p. 428. London: Routledge, 2015.
3. Ozin, G. A., A. C. Arsenault and L. Cademartiri. *Nanochemistry: A Chemical Approach to Nanomaterials*. Cambridge: Royal Society of Chemistry, 2009.

nanoparticle-based art

Colours seen in stained glass windows dating from medieval times as well as colours in ancient glassware are admired for their beauty. These colours often arise from incorporation of nanoparticles such as gold into the glass. An example is the Lycurgus cup that was made by ancient Roman artists.[1–3] When illuminated by white light from behind, the cup shows a rich shade of colours ranging from deep green to bright red. The transmitted light is red while observing scattered light perpendicular to the direction of incident light indicates a greenish colour in the cup. Another example involves red and yellow colours seen in 'October: The Labours of the Months', a stained glass window dating from around 1480 in Norwich (England). The colours arise from the absorption and scattering of light by gold and silver nanoparticles around 50 nm in size that are embedded in the glass. Small gold particles absorb green and blue light, producing a red colour, whereas larger particles scatter mainly in the green, rendering a greenish colour.

1. Novotny, L., and B. Hecht. *Principles of Nano-optics*, 2nd edn. Cambridge: Cambridge University Press, 2006.
2. Louis, C., and O. Pluchery. *Gold Nanoparticles for Physics, Chemistry and Biology*. London: Imperial College Press, 2013.
3. Extance, A. Plasmons with a purpose. *Chemistry World* (Sept. 2012) 56–9.

nanoparticle-based batteries

The growth in the worldwide use of portable devices (e.g. smartphones, tablet and laptop computers) with increasing functions requires batteries with a long battery life before recharging. Nanotechnology applications, in particular the evaluation of nanoparticles for electrode materials in rechargeable batteries to increase performance through for example faster charging rates than existing batteries, is an area of considerable commercial interest.[1] Examples of nanoparticles include graphene, nanotubes and fullerenes and lithium titanate for cathode materials.

See: *lithium-ion batteries*

1. Amin, S. M., and A. M. Giacomoni. Toward the smart grid: the US as a case study. In D. S. Ginley and D. Cahen (eds), *Fundamentals of Materials for Energy and Environmental Sustainability*, p. 590. Cambridge: Cambridge University Press, 2012.

nanoparticle-based medicines

There is interest in the early part of the 21st century on 'personalised medicine' in which the genome of an individual would be sequenced with the intention that this approach will lead to specific drugs that can be used for targeting diseases. Nanoparticle-based medicines fall within this category of 'personalised medicine'.[1] Here nanoparticles are used to incorporate a drug, for example either internally or on the surface as a coating and species (e.g. antibodies) that will recognise specific cells such as cancer cells will be attached to the nanoparticle so that once injected the nanoparticle will become attached to the diseased cells and hopefully release the drug.

See: *nanoparticles; personalised medicines*

1. Rathi, A. Seek and destroy. *Chemistry World* (Oct. 2013) 10, 56–7.

nanoparticles

In a colloidal system particles have, by definition, a size in the range 1 nm to 1 μm. The definition of nanoparticles is not well-defined but many particles with a diameter less than 100 nm are classified as nanoparticles and often such particles have a size less than 10 nm.[1,2] Hence nanoparticles are colloidal systems.

See: *colloidal systems; liposomes; nanomaterials; silver nanoparticles; sunscreens*

1. Hosokawa, M., K. Nogi and M. Naito. *Nanoparticle Technology Handbook*. London: Elsevier, 2012.
2. Ozin, G. A., A. C. Arsenault and L. Cademartiri. *Nanochemistry: A Chemical Approach to Nanomaterials*. Cambridge: Royal Society of Chemistry, 2009.

nanophosphors

Nanophosphors are fluorescent materials with diameters in the region of 2–10 nm in diameter and contain between 10 and 50 atoms; quantum dots are nanophosphors.[1] Phosphors have been prepared in conventional processes that involve mixing oxide components followed by milling and calcination to achieve a solid-state chemical reaction between the reactants. A composition made in this way is, as an example, yttrium aluminium oxide ($Y_3Al_5O_{12}$). Phosphors made in this way have particle sizes of the order of micrometres. Quantum dots with sizes in the range of nanometres can be prepared by nucleation and growth processes in the gaseous or liquid phase.

See: *quantum dots; phosphors*

1. Segal, D. *Exploring Materials through Patent Information*, p. 48. Cambridge: Royal Society of Chemistry, 2014.

nanotechnology

The phrase nanotechnology has attracted much attention in the media in the early years of the 21st century and is often promoted as promising a future based on materials with properties that were previously unattainable.[1,2] Nanotechnology is often associated with small particles, often referred to as nanoparticles, nanocrystals and nanomaterials. Small particles are often colloidal systems and include quantum dots and quantum wires. Small particles have a larger percentage of atoms in their surfaces than larger bulk materials that can generate properties not available in the bulk. Quantum dots are subject to quantum confinement.

See: *colloidal systems; nanoparticle-based medicines; nanoparticles; optical tweezers; quantum dots*

1. Ozin, G. A., A. C. Arsenault and L. Cademartiri (eds). *Nanochemistry: A Chemical Approach to Nanomaterials*. Cambridge: Royal Society of Chemistry, 2009.
2. Ngo, C., and M. van de Voorde. *Nanotechnology in a Nutshell: From Simple to Complex Systems*. Amsterdam: Atlantis Press, 2014.

nanowires

Nanowires are cylinders of material with diameter between around 10 and 100 nm, although there is not a rigorous size range in this definition.[1] Potential applications for nanowires include electrical conductors in electronic devices, electrodes in batteries and chemical sensors. The technique of vapour-liquid-solid (VLS) growth is often used to produce metallic nanowires. As an example, if a substrate containing a thin layer of gold is heated the gold forms solid nanoparticles in the form of hemispheres on the surface as the gold no longer wets the surface. If the gold is above the gold-silicon eutectic temperature of 636 K but below the melting point of gold then silicon vapour from the decomposition of silane gas flowing over the metal diffuses into the nanoparticles to form a liquid droplet with the nanoparticles. As the silicon content increases in the molten hemispheres, a solid silicon phase, the nanowire, forms and grows while silane gas is present. The nanowires have the same diameter as the droplet that sits on top of the wire as the latter grows upward from the substrate.

1. Ozin, G. A., A. C. Arsenault and L. Cademartiri. *Nanochemistry: A Chemical Approach to Nanomaterials*. Cambridge: Royal Society of Chemistry, 2009.

natural product hydrogels

Hydrogels can be prepared from natural products, in particular from natural polymers:[1] for example, by cross-linking the polysaccharide, hyaluronic acid, and proteins such as elastin for potential use as injectable dermal fillers.

See: *hydrogels*

1. Segal, D. *Exploring Materials through Patent Information*. Cambridge: Royal Society of Chemistry, 2014.

natural rubber

Natural rubber is derived from latex gum from the tree *Hevea brasiliensis* and has the chemical composition of polyisoprene based on the monomer cis-isoprene [$CH_2=C(CH_3)$ $CH=CH_2$], in contrast to the *trans* form known as gutta percha.[1] The latter is hard and inelastic compared to the elastic properties of natural rubber, which is an elastomer. Natural rubber is formally *cis*-1,4-polyisoprene and its synthetic equivalent is prepared by polymerisation of 2-methyl-1,3-butadiene, $CH_2=C(CH_3)-CH=CH_2$. Natural rubber has been referred to as *Hevea* rubber.

See: *elastomers; gutta percha*

1. Mark, J., B. Erman and M. Roland (eds). *The Science and Technology of Rubber*. San Diego, CA: Academic Press, 2011.

nematic liquid crystals

In nematic liquid crystals, molecules or polymers are aligned parallel to each other in the same direction but in a random way.[1,2] Poly (p-phenylene terephthalamide), which corresponds to the chemical composition of Kevlar, forms the nematic liquid crystalline phase in solution and when spun into fibres. Cyanobiphenyls form nematic phases at or near room temperatures and thus are useful materials for displays.

See: *cholesteric phases; Kevlar; liquid crystals; smectic phases*

1. Solymar, L., D. Walsh and R. R. A. Syms. *Electrical Properties of Materials*, 9th edn, p. 254. Oxford: Oxford University Press, 2014.
2. Kawamoto, H. The history of liquid crystal displays. *Proceedings of the IEEE* (Apr. 2002) 90, 460–500.

neodymium-based magnets

Neodymium is a lanthanide element (also referred to as a rare-earth element) and alloys of this metal with iron (Fe) and boron (B), for example $Nd_2Fe_{14}B$, form very powerful compact magnets, although their strength falls off as the temperature is raised from ambient to 200°C. Neodymium can be substituted by other lanthanides, for example dysprosium (Dy) or terbium (Tb). Applications for these magnets include permanent magnetic motors in automobiles and industrial equipment, hard disk drives (e.g. digital video disk, DVD), wind power generators, transducers and loudspeakers, electric bicycles and scooters, and magnetic separations as examples.[1] Modern vehicles can have up to 100 magnets, mainly in motors (e.g. windscreen wipers, fluid-level sensors and actuators).

See: *samarium-based magnets*

1. Lucas, J., P. Lucas, T. Le Mercier, A. Rollat and W. Davenport. *Rare Earths: Science, Technology, Production and Use*. London: Elsevier, 2015.

neoprene

Neoprene, a synthetic rubber, was first prepared in 1931 by polymerisation of acetylene over a suspension of cuprous chloride at around 50°C and can be considered to be a derivative of isoprene.[1] Neoprene is a polychloroprene that is made nowadays by the free-radical polymerisation of 2-chloro-1,3-butadiene, $CH_2=C(Cl)-CH=CH_2$.[2] Initial work by Julius Niewland identified divinyl acetylene ($CH_2=CH-CC-CH=CH_2$), a low boiling point hydrocarbon and monovinyl acetylene from the reaction of acetylene with cuprous chloride. Divinyl acetylene was then shown to react with sulphur dichloride to form a rubber-like material that did not retain its elasticity.[3] A team at Du Pont Chemical Company led by Wallace Carrothers (who invented nylon) continued the research and showed that monovinyl acetylene reacted with hydrogen chloride to produce a clear liquid, chloroprene, which could be polymerised to an elastic material similar to vulcanised rubber. The polymerised material was called neoprene.

See: *elastomers*

1. Nieuwland, J. A. Vinyl derivatives of acetylene and method of preparing the same. United States Patent 1,811,959, granted on 30 June 1931.
2. Gratzer, W. *Giant Molecules: From Nylon to Nanotubes*, p. 113. Oxford: Oxford University Press, 2013.
3. Frost, S., and A Cooper. The invention of neoprene. *Materials World* (2016) 24(4), 26–7.

nickel-metal hydride batteries

Rechargeable nickel-metal hydride (Ni-MH) batteries[1] are used as a power source in hybrid vehicles, for example in the Toyota Prius. These batteries contain an alloy of nickel and lanthanide elements as one electrode such as $La_{10.5}Ce_{4.3}Pr_{0.5}Nd_{1.4}Ni_{60.0}\ Co_{12.7}Mn_{5.9}Al_{4.7}$. The other electrode is nickel hydroxide and both electrodes are immersed in an electrolyte of a mixture of potassium and lithium hydroxide solutions. During charging of the battery nickel oxyhydroxide is formed at one electrode and a metal hydride is formed at the nickel alloy electrode. The battery is discharged during use in which stored hydrogen reacts with hydroxyl ions in the electrolyte to form water while at the other electrode the oxyhydroxide is converted back to nickel hydroxide. The battery is recharged from the hybrid vehicle's petrol engine.

1. Lucas, J., P. Lucas, T. Le Mercier, A. Rollat and W. Davenport. *Rare Earths: Science, Technology, Production and Use*, p. 167. London: Elsevier, 2015.

nitinol

Nitinol, an alloy of nickel and titanium (approximately 50% nickel), is a shape memory alloy and its applications include implantable coronary stents.[1] Nitinol is an acronym for Nickel Titanium Naval Ordinance Laboratories. The use of nitinol relies on a transformation between the low-temperature martensitic phase and the high-temperature austenitic phase. The low-temperature phase is malleable but reverts to the original shape of the component in the austenitic phase. The martensitic phase 'remembers' the shape of the component in the austenitic phase.

See: *shape memory alloys*

1. Lecce, L., and A. Concilio (eds). *Shape Memory Alloy Engineering for Aerospace, Structural and Biomedical Applications*, p. 12. Oxford: Butterworth-Heinemann, 2015.

Nomex

Nomex, poly (m-phenylene isophthalamide), is an aromatic polyamide or aramid fibre. It is prepared by reaction of a meta-substituted aromatic diacid dichloride and a diamine, in particular by reaction of iso-phthaloyl chloride and m-phenylenediamine.[1] Whereas Kevlar is a linear aramid, Nomex is a branched polymer used in flame-resistant protective clothing and gloves.[2]

See: *Kevlar*

1. H. B. Hockessin. United States Patent US 3,767,756, 1973.
2. J. R. Fried. *Polymer Science & Technology*, 3rd edn, pp. 399–401. Upper Saddle River, NJ: Prentice Hall, 2014.

Noryl

Noryl is a polyphenylene oxide but is a brittle solid. It is miscible with polystyrene and the blend is much easier to process than the individual polymers.[1] Polymer blends are used as they have beneficial mechanical properties compared to the individual polymer components. Polyphenylene oxide (PPO) has a glass transition temperature (T_g) of 208°C. Addition of polystyrene produces a miscible mixture with a lower value of T_g, intermediate between the values of T_g for the two polymers. The value of T_g determines the upper temperature of use for amorphous polymers such as PPO. Below T_g, the polymer is in a hard, glassy state but becomes soft above T_g. As the temperature increases to and beyond T_g, vibrational motion, bending and stretching of polymer molecules increases so that molecules are no longer frozen in one region but develop longer range motion so that the material softens.

See: *polymer blends*

1. Tyrell, J. A. *Fundamentals of Industrial Chemistry: Pharmaceuticals, Polymers and Business*, p. 141. Chichester: Wiley, 2014.

n-type semiconductors

The semiconductor silicon has a valency of four. Silicon can be doped with a Group V element such as arsenic, phosphorus or antimony and these elements are known as *n*-type dopants or donors.[1] The latter can be introduced by techniques such as ion implantation, or diffusion of dopant. The doped semiconductor has excess free electrons introduced into the crystalline structure by incorporation of the donor atom.

See: *light-emitting diodes; p-type semiconductor; p–n junction diodes*

1. Solymar, L., D. Walsh and R. R. A. Syms. *Electrical Properties of Materials*, 9th edn. Oxford: Oxford University Press, 2014.

nuclear fuel

Fuel that is used in thermal nuclear reactors is based on ^{235}U that undergoes fission on bombardment by thermal neutrons.[1] Uranium ores contain about 99% of the isotope ^{238}U with smaller amounts of fissionable ^{235}U; ^{234}U is also present in the ore. The fuel contains uranium dioxide pellets enriched with about 3 wt% of ^{235}U, which can be obtained by converting the ore to uranium hexafluoride and separating out the lighter $^{235}UF_6$ by high-speed centrifugation. Neutron bombardment of fissile material produces a range of fission products such as ^{92}Kr and ^{141}Ba and the difference in masses between the fission products and ^{235}U is converted to energy (E) according to Einstein's famous equation $E = mc^2$ from the special theory of relativity, where m is the mass difference and c is the speed of light. The fission products are highly radioactive and there are issues to be solved in the 21st century on their safe storage and disposal.

See: *sphene; Zircaloy*

1. Stanek, C. R., R. W. Grimes, C. Unal, S. A. Maloy and S. C. Scott. Nuclear energy: current and future schemes. In D. S. Ginley and D. Cahen (eds), *Fundamentals of Materials for Energy and Environmental Sustainability*, pp.148–50. Cambridge: Cambridge University Press, 2012.

nutraceuticals

The word nutraceutical is a combination of nutrition and pharmaceutical and may be defined as health-promoting compounds that are added to foods.[1] Nutraceuticals are also described as functional foods but the boundary of what constitutes a food or medicine is a legal question. Foods are currently fortified with extra nutrients; for example vitamins are added to cereals.

1. Hughes, E. Food with a function. *Chemistry World* (Oct. 2012) 9, 50–3.

nylon

Wallace Carothers at DuPont Chemical Company invented nylon, a crystalline synthetic polymer, that is, a plastic in 1935.[1,2] In early work an intermediate salt was produced by the interfacial condensation reaction at around room temperature of a diamine with a dicarboxylic acid, for example by reaction of hexamethylenediamine with adipic acid to produce hexamethylene diammonium adipate. The intermediate salt was processed further to a fibre-forming polyamide, namely nylon that could be melt-spun. The plastic has corrosion resistance, electrical insulation and toughness. An initial application was for toothbrush bristles and nylon achieved spectacular commercial success when used in women's stockings as a substitute for silk stockings. Nylon is a versatile manmade fabric with applications including lightweight fabrics, upholstery, ropes, nets, seatbelts, conveyer belts, parachutes, uniforms, tent canvas and tarpaulins. Other syntheses involve a ring-opening polymerisation of a lactam ring such as caprolactam that contains both the amine and carboxylic acid group. Wallace Carothers did not live to witness the success of his invention and committed suicide in 1937.

See: *Kevlar*

1. Ploszajski, A. Nylon. *Materials World* (Mar. 2014) 22, 58–9.
2. Carothers, W. H. Diammine-dicarboxylic acid salts and process of preparing same. United States Patent 2,130,947, granted on 20 September 1938.

odour-free adhesives

Adhesives can give rise to odours due to the high vapour pressure of their constituent liquids and this is undesirable in confined spaces such as in spacecraft if the gases are toxic. Ionic liquids have low vapour pressures and have been fabricated for use as odourless cross-linked epoxy resins for use in harsh environments.[1]

1. Paley, M. S., R. S. Libb, R. N. Grugel and R. E. Boothe. Ionic liquid epoxy resins. United States Patent 8,193,280, 2012.

oil shale

Oil shale is sedimentary rock that lies near the earth's surface and contains organic matter.[1,2] It can be mined and heated to produce oil. In contrast, shale gas involves deep wells to extract gas by fracking.

See: *fracking; shale gas*

1. Hester, R. E., and R. M. Harrison (eds). *Fracking, Issues in Environmental Science and Technology*, vol. 39. Cambridge: Royal Society of Chemistry, 2015.
2. Bunger, J. W. Oil shale and tar sands. In D. S. Ginley and D. Cahen (eds), *Fundamentals of Materials for Energy and Environmental Sustainability*, pp. 127–36. Cambridge: Cambridge University Press, 2012.

olefin-based polymers

Olefin-based polymers including synthetic fibres or polyolefins are derived from alkenes (olefins) such as ethene (ethylene) and propene (propylene) and include polyethylene and polypropylene.[1]

See: *polyethylene; polypropylene*

1. Fried, J. R. *Polymer Science and Technology*, 3rd edn. Upper Saddle River, NJ: Prentice Hall, 2014

oligonucleotides

Oligonucleotides are single-stranded fragments of DNA containing up to 40 nucleotide bases in length and can hybridise to a denatured sample of DNA in a process known as annealing.[1] Oligonucleotides are the primers used in the polymerase chain reaction. Oligonucleotides are short polymers.

See: *polymerase chain reaction; polymers*

1. Daintith, J. (ed.). *A Dictionary of Chemistry*, 6th edn. Oxford: Oxford University Press, 2008.

oligopeptides

Oligopeptides contain up to ten amino acids and include the hormone oxytocin that is an octapeptide.[1] Oligopeptides are thus short polymers.

See: *peptides; polymers*

1. Daintith, J. (ed.). *A Dictionary of Chemistry*, 6th edn. Oxford: Oxford University Press, 2008.

oligosaccharides

Oligosaccharides are produced as intermediates during the digestion of polysaccharides such as cellulose and starch.[1] They are carbohydrates containing a chain of up to twenty monosaccharide units. Oligosaccharides are thus short polymers.

See: *carbohydrates; cellulose; polymers*

1. Barber, J., and C. Rostron (eds). *Pharmaceutical Chemistry*, p. 258. Oxford: Oxford University Press, 2013.

opalescent materials

The mineral opal is used as a gemstone and consists of three-dimensional arrays of silica microspheres.[1,2] Opal can have a milky-white appearance known as opalescence and exhibit a range of colours due to the presence of impurities. Optical interference from this microstructure of close-packed silica particles contributes to the colouring of this mineral.

See: *photonic materials*

1. Ozin, G. A., A. C. Arsenault and L. Cademartiri. *Nanochemistry: A Chemical Approach to Nanomaterials*. Cambridge: Royal Society of Chemistry, 2009.
2. Tilley, R. J. D. *Understanding Solids: The Science of Materials*. Chichester: Wiley, 2013.

optical amplifiers

Optical fibres are widely used to transmit broadband and public telecommunications around the world.[1] Because of optical losses, the transmitted signals are attenuated and the signal must be regenerated or amplified about every 80 km. Amplification can be achieved with erbium-doped silica-fibre amplifiers that emit radiation with a wavelength around 1.55 nm where the erbium concentration is approximately 0.1 wt%. The fibres are optically pumped, producing a population inversion and stimulated emission. Hence the transmitted signal is amplified. Erbium-doped fibres can amplify multiple channels so that multiple sources of information can be transmitted in the technique of wavelength division multiplexing without interference.

See: *optical fibres*

1. Lucas, J., P. Lucas, T. Le Mercier, A. Rollat and W. Davenport. *Rare Earths: Science, Technology, Production and Use.* London: Elsevier, 2015.

optical fibres

Broadband communications and public telecommunications networks using optical fibres are increasingly used in the 21st century;[1] in optical communication information is encoded on light waves that are propagated along the fibres. Light is guided by total internal reflection in the core of an optical fibre. The silica-based core is surrounded by a silica-based cladding of lower refractive index (Fig. G.18). The core and cladding are drawn from a preform containing dopants that ensure the core has a higher refractive index than the cladding so that light is trapped inside the core. The cladding is surrounded by a protective barrier layer, usually a plastic coating.

1. Ploszajski, A. Fibre optics. *Materials World* (July 2014) 22, 62–4.
2. Tilley, R. J. D. *Understanding Solids: The Science of Materials*, p. 480. Chichester: Wiley, 2013.

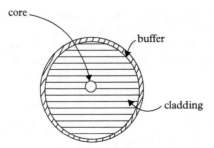

Fig. G.18 *Schematic diagram for the cross-section of an optical fibre (not to scale).* *(W.H. Culver. United States Patent 5,923,694, 1999.)*

optical tweezers

There is much interest in the 21st century in nanotechnology in which small particles are prepared and characterised. A technique known as optical tweezers is used to manipulate small particles (0.1–10 μm), including living cells, DNA, bacteria and metallic particles for example.[1] The technique is non-invasive and involves holding a microscopic particle near the focus of a laser beam. Measurement of elasticity, force, torsion and position can be measured with this technique with forces in the range 1–10 piconewtons.

See: *nanotechnology*

1. Novotny, L., and B. Hecht. *Principles of Nano-optics*, 2nd edn. Cambridge: Cambridge University Press, 2006.

optical waveguides

The silica-based core in an optical fibre is surrounded by a silica-based cladding of lower refractive index so that light is guided by total internal reflection in the core of an optical fibre. The fibre acts as an optical waveguide but optical waveguides do not have to be in the form of a fibre.[1] If a layer of gallium arsenide (GaAs) is grown on top of a substrate of aluminium gallium arsenide (AlGaAs) which has a lower refractive index than GaAs then the structure acts as a waveguide in which light is confined to the layer of higher refractive index. Optical waveguides are important for the operation of lasers and the optical properties of photonic materials.

See: *optical fibres*

1. Solymar, L., D. Walsh and R. R. A. Syms. *Electrical Properties of Materials*, 9th edn, p. 354. Oxford: Oxford University Press, 2014.

optoelectronic materials

Optoelectronic materials utilise the electronic properties of a material to produce an optical effect.[1] For example, the semiconducting properties of gallium arsenide (GaAs) enable visible light in the red region of the electromagnetic spectrum to be emitted by electroluminescence in light-emitting diodes. Blue light-emitting lasers are produced using gallium nitride (GaN).

See: *electroluminescence*

1. Solymar, L., D. Walsh and R. R. A. Syms. *Electrical Properties of Materials*, 9th edn. Oxford: Oxford University Press, 2014.

organic aerogels

Aerogels based on silica, alumina or other metal oxides are described as inorganic. Organic aerogels are derived from organic compounds with linking groups, for example resorcinol-formaldehyde and melamine-formaldehyde aerogels.[1] Potential applications for organic aerogels include insulation in refrigerators as a replacement for polyurethane-based foams. Unlike the latter, organic aerogels do not require blowing agents such as chloro-fluorohydrocarbons as there is concern over the effect of such hydrocarbons on depletion of the ozone layer in the atmosphere.

See: *aerogels*

1. Reade, L. Full of air. *Chemistry and Industry* (Feb. 2013) 77, 32–5.

organic light-emitting diodes

Organic light-emitting diodes (OLEDs) are a type of light-emitting diode in which emission of light takes place from organic molecules by the process of electroluminescence, namely by application of an electrical voltage.[1,2] Applications are for low-power-consuming light sources, flat panel displays and particularly displays in flexible electronic devices. OLEDs are used as a thin semiconducting layer about 50 nm in thickness. An advantage of OLEDs over LEDs is that they can be deposited from solution rather than from the gas phase in order to produce the layer. Poly-(p-phenylenevinylene) is an example of an OLED. In an inorganic semiconductor, electroluminescence is explained in terms of valence and conduction bands. In an OLED electrons and holes can recombine in the organic layer and emit light with a wavelength depending on the properties of the organic material. It is customary to use the terms HOMO (highest occupied molecular orbital level) and LUMO (lowest unoccupied molecular orbital level) so that electrons from a LUMO and holes from a HUMO combine, resulting in light emission.

See: *AMOLED; poly-(p-phenylenevinylene)*

1. Solymar, L., D. Walsh and R. R. A. Syms. *Electrical Properties of Materials*, 9th edn, p. 427. Oxford: Oxford University Press, 2014.
2. Osram GmbH. Introduction to OLED technology, available at http://www.osram.com/oled.

oxide–oxide composites

Oxide–oxide composites are more environmentally stable in oxidising atmospheres than non-oxide-based composites such as SiC/SiC but have inferior mechanical properties.[1] An example of an oxide–oxide composite consists of continuous Nextel fibre (63% Al_2O_3; 25% SiO_2; 12% B_2O_3) in an aluminosilicate or mullite-alumina matrix. Fibre-reinforced metals have been prepared by melt infiltration of liquid metal into a fibre mat, for example infiltration by aluminium. Applications of oxide–oxide composites include combustor liners in gas turbine engines. Addition of fibres to a matrix increases the toughness of the component that is the resistance to fracture.

1. Bansal, N. P., and J. Lamon (eds). *Ceramic Matrix Composites: Materials, Modeling and Technology*. Chichester: Wiley, 2015.

Parkesine

In 1856 the British chemist Alexander Parkes mixed nitrocellulose with an alcohol, ethanol and a plasticiser, camphor, to produce a tough material that could be moulded when hot, coloured and polished.[1] This material was called Parkesine. Camphor is a cyclic ketone that can now be synthesised but in the 19th century it was obtained from laurel trees. Use of different additives led to a manufacturing process by John Hyatt Jr in 1869 and the material was referred to as celluloid.

See: *celluloid; cellulose nitrate*

1. Gratzer, W. *Giant Molecules: From Nylon to Nanotubes*. Oxford: Oxford University Press, 2013.

partially crystalline materials

Crystalline solids such as sodium chloride have an ordered arrangement of atoms, whereas non-crystalline solids such as glasses are disordered solids. Glass ceramics contain small crystallites dispersed in an amorphous matrix. Some materials, especially synthetic polymers, can contain regions of crystallinity.[1] For example, polyethylene molecules can contain 10,000 or more monomer units in a long chain. A melt, if cooled quickly, contains the long chains but if cooled slowly then some chains fold up into crystalline regions around 20 nm thick. Polythene that has been cooled slowly looks slightly opaque as the crystalline regions have a higher refractive index compared with the non-crystalline regions.

See: *polyethylene*

1. Askeland, D. K., and W. J. Wright. *The Science and Engineering of Materials*, 7th edn, p. 332. Boston, MA: Cengage Learning, 2011.

patents

A patent represents a legal relationship between an inventor and a nation state.[1,2] In exchange for disclosing in the public domain how the invention works, the patent owner (who may not be the inventor) has a right for a limited time to prevent others from exploiting the invention, namely manufacturing, importing or selling the invention without permission in the country where the patent is in force, usually 20 years from the filing date. The concept of a patent and its associated right dates back many years. In the 14th century 'letter patents' were assigned to Flemish weavers by Edward II to teach English residents weaving techniques and gain an income. Also, Henry VI invited Flemish artisans to train others to produce stained glass technology in England.

1. McManus, J. *Intellectual Property: From Creation to Commercialisation*. Oxford: Oak Tree Press, 2012.
2. Jackson Knight, H. *Patent Strategy for Researchers and Research Managers*. New York: Wiley, 2012.

patent trolls

The development of materials is associated with patents taken out by their inventors to give legal protection to their inventions. The first two decades of the 21st century has seen the rise of non-practicing entities (NPEs) sometimes referred to as 'patent trolls', although when used in this way the phrase 'patent troll' has derogatory overtones.[1] NPEs are individuals or firms who own patents but do not directly use their patented technology to produce goods or services. NPEs assert their patents in the courts against companies that do produce goods or services. These infringement cases have been reported in the popular press particularly in the mobile phone arena.

See: *patents*

1. Bessen, J., and M. J. Leurer. The direct costs from NPE disputes. Boston University School of Law Working Party number 12-34 (25 June), revised 28 June 2012.

PEEK (polyether ether ketone)

PEEK, i.e. polyether ether ketone, is a thermoplastic that has high chemical resistance, for example to solvents, good abrasion and wear resistance and it is a tough material which is resistant to fracture. Polyether ether ketone belongs to the class of polyaryl ether ketones $-(OC_6H_4-O-C_6H_4-CO-C_6H_4)-$ and it has been manufactured by Victrex in the United Kingdom.[1] PEEK can be melt-processed and the polymer is used in ABS braking systems, lightweight gears, high-temperature seals, medical implants to replace damaged spinal discs, lightweight pipes as a replacement for heavier metal piping and casings for consumer electronic goods; it has potential application as the matrix in graphite composites.

1. Walton, D., and P. Lorimer. *Polymers*, p. 109. Oxford: Oxford University Press, 2005.

peptides

Peptides are organic compounds in which two or more amino acids are linked by a peptide bond formed between adjacent carboxyl groups ($-COOH$) and amino groups ($-NH_2$) with elimination of water.[1] Polypeptides contain more than ten amino acids and usually 100–300. Dipeptides contain two amino acids while tripeptides contain three amino acids, as examples. An example of a tripeptide is glutathione.

See: *amino acids*

1. Barber, J., and C. Rostron (eds). *Pharmaceutical Chemistry*, p. 252. Oxford: Oxford University Press, 2013.

perovskite-based solar cells

The mineral perovskite $CaTiO_3$ gives its name to a class of materials that have the same chemical structure often represented as ABX_3 (here A = Ca, B = Ti and X = O). Although traditionally this class of materials is occupied by inorganic materials, namely oxides, hybrid organic–inorganic perovskite structures have been identified. For example, methylammonium lead triiodide, $CH_3NH_3PbI_3$ is made by mixing methylammonium iodide (CH_3NH_3I) and lead iodide (PbI_2). This perovskite is a semiconductor and has been used as a light-absorbing perovskite when absorbed onto anatase particles in a solar cell;[1] the anatase forms the anode in the solar cell. An advantage of using this perovskite is that it can be deposited by spin-coating, a relative easy process, and in addition there is evidence of high efficiency compared to existing materials, namely approaching 20% for conversion of incident light.

See: *dye-sensitised materials*

1. Extance, A. The power of perovskites. *Chemistry World* (Sept. 2014) 11, 46–9.

personalised medicines

There has been interest in the early part of the 21st century in 'personalised medicine' in improving the health of populations. In this approach the genome of an individual would be sequenced with the intention that this approach will lead to specific drugs that can be used for targeting diseases.[1,2] The human genome contains 3165×10^6 base pairs, that is hydrogen-bonded linkages between four bases, (i) adenine (A) linked to thymine (T) and (ii) cytosine (C) linked to guanine (G) in a characteristic sequence. Each A–T or C–G linkage represents a base pair. Adenine is said to be complementary to thymine and cytosine complementary to guanine. The 100,000 Genomes Project in the United Kingdom aims to sequence the genome for 100,000 people and combine data with information from health records.

See: *DNA; nanoparticle-based medicines*

1. DNA mapping. *The Times*, 3 January 2015.
2. Could a DNA cream get rid of wrinkles? *Daily Mail*, p.5, 1 December 2014.

phase-change chalcogonides

Chalcogenides are alloys formed between a metal and group 16 elements in the periodic table, for example with sulphur, selenium, tellurium and oxygen.[1] An example is GST, a chalcogenide glass consisting of germanium, antimony and tellurium (GeSbTe). The stable amorphous phase in the glass can be reversibly converted into a stable crystalline phase by heat. The amorphous phase can be recovered by applying a pulse of heat to crystalline material to raise the temperature above the melting point followed by rapid cooling. Heat can be applied electrically or from a solid-state laser. Phase-change chalcogenides have the potential for acting as switches and hence for use as phase-change computer memories.

1. Solymar, L., D. Walsh and R. R. A. Syms. Electrical Properties of Materials, 9th edn, p. 459. Oxford: Oxford University Press, 2014.

phase-change materials

Phase-change materials can act as a latent heat storage material for regulating the temperature within buildings.[1] For example, a proprietary paraffin mixture can absorb heat during the day, causing the paraffin to melt, while at night this heat is released as the material recrystallises. Another application for a phase-change material, in this case a concentrated salt solution, is in a self-warming baby bottle that has a trade name of yoomi.[2]

1. Reade, L. Building a future. *Chemistry and Industry* (Apr. 2014) 78, 24–7.
2. Shaikh, J. Self-heating fluid connector and self-heating fluid container. International application WO 2006/109098, published 19 October 2006.

phase-separated glasses

Borosilicate glasses are glasses that contain boron and silicon. High-silica content glasses such as vitreous silica that consists of 100 wt% SiO_2 are difficult to process because silica sublimes at around 1900°C and becomes very viscous on heating to high temperatures. High-silica-containing glasses can be prepared by heating a borosilicate glass, after which boron-rich phases can be leached out, leaving a silica-based matrix.[1] Heat treatment of the remaining matrix reduces voids (pores) and increases the density of the component. Vycor is an example of a high-silica-containing glass with a typical composition 96% SiO_2; 3% B_2O_3; 1% Na_2O and has a low coefficient of thermal expansion.[2] It has applications as windows in space vehicles.

See: *Pyrex*

1. Burggraaf, A. J., and K. Keizer. Ceramic membranes. In R. J. Brook (ed.), *Concise Encyclopaedia of Advanced Ceramic Materials*, pp. 62–7. Oxford: Pergamon Press, 1991.
2. Bolton, W., and R. A. Higgins. *Materials for Engineers and Technicians*, 6th edn. London: Routledge, 2015.

phase-separated polymers (LYCRA)

The properties of polyurethane elastomers result from phase separation of the polymer into hard segments (urethane) and soft segments (polyether or polyester), known as domains in the polymer backbone.[1] The material LYCRA, produced by DuPont and also known as spandex and elastane, is used in textiles (e.g. in socks to add elasticity to the textile) and contains short polymeric chains of a polyglycol about 40 repeat units in length that is soft and rubbery and hard rigid segments based on urethane linkages. The hard and soft domains represent two different phases and LYCRA can be considered to be a block copolymer.[2]

See: *polymers; polyurethanes*

1. Hudgins, R. G., and J. M. Criss Jr. Phase separated branched copolymer hydrogel. United States Patent 7,919,542, granted on 5 April 2011.

2. Stewart, C., and J. P. Lomont. *The Handy Chemistry Answer Book*, p. 176. Canton, MI: Visible Ink Press, 2014.

phenol-formaldehyde resin (bakelite)

Leo Baekeland invented Bakelite, one of the first commercially successful synthetic plastics through the condensation reaction between phenol and formaldehyde.[1] The condensation product was a viscous cross-linked resin that set to a brittle solid on heating, although the strength and toughness of Bakelite was markedly improved by incorporation of fibres or powder (e.g. mica) into the resin. Bakelite is a thermosetting electrically insulating plastic and found many uses in everyday life, replacing natural materials such as shellac (obtained from excretions of the kerria lacca insect) for gramophone records and ivory billiard balls, hence for buttons, knitting needles, ashtrays, telephones and fountain pens as examples.[2] Phenolic-based resins had the disadvantage of being available only in dark colours. However, urea-formaldehyde and melamine-formaldehyde resins as well as resorcinol-formaldehyde resins could be obtained in a range of colours, white, cream, pale or bright and with a porcelain-type texture. These resins found applications in lacquers and laminates such as table tops.

1. Baekeland, L. H. Method of making insoluble products of phenol and formaldehyde. United States Patent 942,699, granted on 7 December 1909.
2. Gratzer, W. *Giant Molecules: From Nylon to Nanotubes*. Oxford: Oxford University Press, 2013.

phenolic resins

Thermosetting phenol-formaldehyde resins are known as phenolic resins.[1] When the resins are prepared under acidic conditions (formaldehyde:phenol ratio less than 1) the resin is known as a novolac but preparation under basic conditions (formaldehyde:phenol ratio greater than 1) produces a resole. Phenolic resins have applications as adhesives, particularly for wood composites such as plywood, and have good flame resistance. The latter property has found application as brake linings and clutch facings. An additional application is as an abrasive in waterproof sandpaper.

See: *phenol-formaldehyde resin; plywood*

1. Tyrell, J. A. *Fundamentals of Industrial Chemistry: Pharmaceuticals, Polymers and Business*, pp. 133–4. New York: Wiley, 2014.

PHOLED

PHOLED is an acronym for a phosphorescent organic light-emitting diode.[1] A PHOLED is an electroluminescent material in which light is emitted from an organic phosphorescent material. Phosphorescent OLEDs are potentially more energy-efficient than non-phosphorescent emitters as emission takes place from triplet rather than singlet electronic states.

See: *electroluminescence; phosphors*

1. Hack, M., J. J. Brown, P. Levermore and M. S. Weaver. Organic light emitting device lighting panel. United States Patent Application US 2011/0284899, published 24 November 2011.

phosphors

Phosphors are materials that emit radiation when supplied with a source of energy.[1] This emission is known as luminescence and if the energy source is electromagnetic radiation the emission is known as photoluminescence. If photoluminescence is immediate then fluorescence occurs, whereas if the conversion of the exciting radiation is slow then the photoluminescence is known as phosphorescence (Fig. G.19). Fluorescent radiation has a longer wavelength than the exciting radiation. For example, Eu-doped Y_2O_3 absorbs ultraviolet radiation and emits visible light and this conversion is used in fluorescent lighting. $CaWO_4$ emits visible light when bombarded with X-rays and Tb-doped $Y_3Al_5O_{12}$ emits visible light under electron bombardment. The latter process is known as cathodoluminescence. Examples of other compositions for phosphors include silver-doped zinc sulphide (blue light), thulium-doped lanthanum oxybromide (blue light), europium-doped yttrium oxide (red light) and manganese-doped zinc silicate (green light). Phosphors have been used in radar screens, television monitors, oscilloscopes, emergency lighting in buildings and security markings. Semiconductor quantum dots behave as phosphors. Other sources of excitation include an electric field (electroluminescence), chemical reactions (chemiluminescence) and biological processes (bioluminescence).

See: *after-glow pigments; quantum dots*

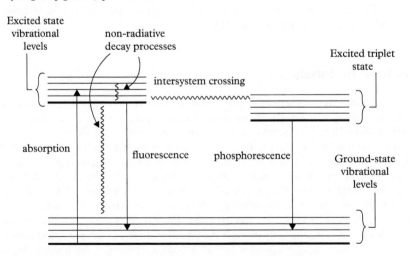

Fig. G.19 *Schematic representation of fluorescence and phosphorescence.*
(R. M. Christie. Colour Chemistry, *2nd edn, p. 33. Cambridge: Royal Society of Chemistry, 2015.)*

1. Lucas, J., P. Lucas, T. Le Mercier, A. Rollat and W. Davenport. *Rare Earths: Science, Technology, Production and Use.* London: Elsevier, 2015.

phosphorus-based flame retardants

Phosphorus compounds are used as flame retardants in textiles and plastics (e.g. polyurethane foams).[1] For example, organic phosphonates, phosphate esters and phosphinates. These compounds act by forming a carbon-based crust or char from degradation products of the substrate that acts as a thermal barrier shielding the polymer from the gas phase.

1. Segal, D. *Exploring Materials through Patent Information.* Cambridge: Royal Society of Chemistry, 2014.

photochromic materials

Photochromic materials change colour when exposed to light of varying intensity.[1] They are particularly associated with spectacle lenses and sunglasses. Spirooxazines and naphthopyrans are photochromic materials. Coatings of photochromic materials on spectacle lenses react quickly with ultraviolet light in sunlight and change colour from a colourless state to give a strong colour that reversibly fades to the colourless state when the source of ultraviolet light becomes less intense. Absorption of ultraviolet light produces a different chemical species in a reversible reaction.

See: *smart materials*

1. Reduwan Billah, S. M. Photo-responsive dyed textiles for advanced applications. In J. Fu (ed.), *Dyeing: Processes, Techniques and Applications*, pp. 19–38. Hauppauge, NY: Nova Science Publishers, 2014.

photonic materials

Photonic materials are those in which optical diffraction occurs with visible light from periodic structures within the material in which the periodicity is on the order of the wavelength of light.[1,2] Photonic materials are also known as photonic crystals. The latter are associated with a photonic gap, analogous to the band gap in a semiconductor. The dielectric band gap in a photonic crystal is analogous to the valence band in a semiconductor and the air band is analogous to the conduction band. An example of a photonic crystal involves a close-packed structure of silica spheres (855 nm diameter) on a silicon wafer in which interstitial spaces are filled with silicon, after which the silica spheres have been removed by wet etching.[2] Photons that have energies within the photonic gap cannot propagate through the crystal and are confined to defect regions so that a line of defects in a photonic crystal acts as a waveguide for transporting light around corners as light is repelled from the bulk crystal. Interference takes place between diffracted components of visible light. Defects can be introduced using, for example, a microfabricated hexagonal array of air holes in InGaAsP. Colours in the wings of butterflies and peacocks and the colours in opal are characteristic of photonic materials.

See: *band gaps; iridescent organisms; opalescent materials*

1. Ozin, G. A., A. C. Arsenault and L. Cademartiri. *Nanochemistry: A Chemical Approach to Nanomaterials.* Cambridge: Royal Society of Chemistry, 2009.
2. Novotny, L., and B. Hecht. *Principles of Nano-optics,* 2nd edn, p. 338. Cambridge: Cambridge University Press, 2006.

photopolymers

A photopolymer is a liquid which on exposure to visible or ultraviolet light is cured by a polymerisation reaction, forming a solid body.[1] Photopolymers have an important role as photoresists that are used with photolithography to fabricate electronic circuits containing millions of transistors on integrated circuits. In photolithography, a surface is coated with a layer of photoresist, after which selected surface regions are masked. The unmasked regions are exposed to a light source, for example an ArF excimer laser producing wavelengths of 193 nm, and the resist in these areas undergoes polymerisation (i.e. curing) and solidifies. Non-cured regions of resist can be removed with a solvent known as a developer so that a pattern of lines on the exposed silicon surface of the wafer is produced. Photopolymers have been used in stereolithography to build up components layer by layer and an example of their use in everyday life is as a curable adhesive for artificial nails.

See: *artificial nails; stereolithography*

1. Tiwari, A., and A. Polykarpov (eds). *Photocured Materials.* Cambridge: Royal Society of Chemistry, 2014.

photoresists

Semiconductor devices such as transistors on integrated circuits are built up on a silicon wafer by using photolithography and etching combined with curable organic resins known as photoresists, also referred to as resists.[1,2] Regions of a layer of photoresist on the wafer that are exposed to ultraviolet radiation (e.g. 193 nm wavelength from an argon fluoride excimer laser) that passes through a photolithographic mask with a specified pattern of holes in it are polymerized, resulting in an insoluble solid region. The unexposed areas in these positive resists is washed away by a solvent known as a developer, leaving the required pattern on the surface of the wafer. In negative photoresists it is the unexposed regions that become insoluble and after removing the soluble resist a negative pattern for the resist is produced on the wafer. Photoresists are photopolymers.

See: *photopolymers*

1. Thackray, A., D. C. Brock and R. Jones. *Moore's Law.* New York: Basic Books, 2015.
2. Tiwari, A., and A. Polykarpov (eds). *Photocured Materials.* Royal Society of Chemistry, 2014.

piezoelectric materials

Piezoelectric materials are materials that produce a voltage when a mechanical stress is applied; conversely, a stress is produced when a voltage is applied.[1] The electric field or mechanical stress generates surface charges in the crystal. Examples of materials that exhibit the piezoelectric effect include lead zirconate titanate-based ceramics, barium titanate ($BaTiO_3$), ammonium dihydrogen phosphate ($NH_4H_2PO_4$) and Rochelle salt (potassium sodium tartrate tetrahydrate, $NaKC_4H_4O_6.4H_2O$). Applications for piezoelectric materials include microphones when an electrical signal is produced by variations in pressure produced by sound waves, spark ignition systems for ignition systems and loud speakers. The natural resonance frequency of a slice of single crystal of a piezoelectric material can act as a frequency standard. An applied varying voltage causes the crystal to vibrate but the vibrations will be at a maximum when the frequency of the varying voltage has the same frequency as the mechanical vibrations of the crystal, thus at resonance frequencies. Quartz crystal resonators provide accurate time-keeping in clocks and watches. For example, if a quartz crystal in the shape of a tuning fork can generate a frequency of 32,768 Hz (i.e. 2^{15} Hz) then this frequency value can be repeatedly halved to give an interval of one second that can be displayed.

See: *poly (vinylidene fluoride); smart materials*

1. Tilley, R. J. D. *Understanding Solids: The Science of Materials*, p. 333. Chichester: Wiley, 2013.

piezoelectric polymers

Crystalline polymers, for example poly (vinylidene fluoride), can be fabricated into flexible piezoelectric plastic sheets for use as sensors and transducers in microphones, keyboards and flat panel speakers.[1] The polymer must be heated above its glass transition temperature and annealed in an electric field, a process known as poling.

See: *poly (vinylidene fluoride)*

1. Walton, D., and P. Lorimer. *Polymers*, p. 15. Oxford: Oxford University Press, 2005.

PlantBottle

Polyesters are used for plastic bottles as containers for bottled water and soft drinks and polyethylene terephthalate, the most widely used polyester, is usually derived from petrochemical sources, one of which is ethylene glycol. The PlantBottle (Coca Cola Company) uses a bio-based ethylene glycol derived from bioethanol for some of the bottles.[1,2] Ethanol is dehydrated to ethylene, which is then converted to ethylene oxide and then ethylene glycol.

See: *polyester*

1. O'Driscoll, C. Next generation polyester beats PET. *Chemistry and Industry* (Dec. 2012) 76, 15.

2. Tyrell, J. A. *Fundamentals of Industrial Chemistry: Pharmaceuticals, Polymers and Business*, p. 119. New York: Wiley, 2014.

plant dyes

The synthetic dye industry dates from the work of Sir William Perkin. Natural textile dyes can be obtained from plants, for example indigo. Yellow dye was obtained from the leaves of *Eucalyptus macrorhyncha* as early as the 1880s, Turkey red dye from dried madder root (*Rubia tinctorum*), yellow dye from the wild chamomile (*Anthemis chia*) and black dye from the bark of the knobbly oak (*Quercus macrolepis*).[1] Turkey red dye has been used to dye wool for use in traditional Turkish carpets. It has been estimated that one ton of textile fabric requires about 270 tons of water in the dyeing process. There is increasing interest in environmentally friendly industrial processes and whether natural dyes are used in large quantities remains to be seen.

See: *indigo; mauveine*

1. Carmichael, H. Natural reds. *Chemistry and Industry* (Jan. 2014) 78, 36–9.

plant oils

Plant oils (e.g. vegetable oils) are triglycerides, namely esters of glycerol, a trihydric alcohol, and esterification can take place at one, two or all three hydroxyl groups. Thus, these esters are formed by reaction of fatty acids with alcohols, although the biochemical pathway in plants for their synthesis is a complex process. Plant oils are renewable resources and have been used for the preparation of polymers including epoxies, polyurethanes and nylons.[1]

See: *biodiesel; fats*

1. Sarin, A. *Biodiesel: Production and Properties*. Cambridge: Royal Society of Chemistry, 2012.

plasmonics

The study of the interaction of electromagnetic radiation, in particular light with metals, and the response of the metal is known as plasmonics or nanoplasmonics.[1,2] This area is associated with oscillating values of surface charge density at a metal–dielectric interface and these oscillations are known as surface plasmons. The interaction can give rise to enhanced optical fields at the surface of the metal. Colours of metallic films and the optical properties of precious metal particles are associated with plasmonics.

See: *metallic nanoparticles*

1. Novotny, L., and B. Hecht. *Principles of Nano-optics*, 2nd edn, p. 369. Cambridge: Cambridge University Press, 2006.
2. Extance, A. Plasmons with a purpose. *Chemistry World* (Sept. 2012) 56–9.

plastics

Plastics may be defined as usually derived from synthetic polymers that are shaped by application of heat or pressure.[1] There are two types of plastic, thermoplastics and thermosetting materials. The former can be repeatedly shaped by cycles of heating and cooling, whereas thermosetting plastics or thermosets are resins that can be set to the desired shape by heating and forming. This heat treatment causes cross-linked polymer chains in an irreversible process so that re-shaping is not possible. The word plastic is derived from the Greek word *plastikos* that means 'capable of being shaped and moulded'. The global production of plastics in 2011 was 280 million tonnes, almost all derived from oil, corresponding to about 5% of global oil production. The global production capacity in 2011 for bioplastics was just over 1 million tonnes.

See: *recycling plastics; thermoplastics; thermosetting plastics*

1. Askeland, D. K., and W. J. Wright. *The Science and Engineering of Materials*, 7th edn, p. 572. Boston, MA: Cengage Learning, 2011.

PLEDs

PLED is an acronym for a polymer light-emitting diode that emits light when subjected to electroluminescence.[1]

See: *electroluminescence; organic light-emitting diodes; poly-(p-phenylenevinylene)*

1. Segal, D. *Exploring Materials through Patent Information*. Cambridge: Royal Society of Chemistry, 2014.

plenum cables

Ducts in a building that from part of a heating, ventilation and air conditioning (HVAC) system form an open space known as a plenum that is often hidden from the view of occupants and circulates air throughout the building.[1] Cables often occupy the length of these ducts but the absence of fire blocks in the plenum combined with air flow throughout the ducts is a potential fire hazard. It is important for the safety of staff in the buildings that cables in the plenum do not burn easily. Fluoroplastics have very good fire resistance and fluorinated ethylene propylene resins are widely used as protective sheaths for cables. Data cables require stringent dielectric properties of the polymer and these are met by use of fluoroplastics.

1. Smith, D. W. Jr, S. T. Iaccono and S. S. Iyer (eds). *Handbook of Fluoropolymer Science and Technology*, pp. 603–6. New York: Wiley, 2014.

plywood

Plywood consists of thin layers of wood known as plies that are cut from logs and stacked together so that grains in adjacent plies are oriented at 90° to each other.[1-3] This arrangement reduces the anisotropic structure of wood, increases its toughness and lowers the tendency of warping in use. The plies are bonded together with a thermosetting resin (e.g. a phenolic resin) which is deposited between the plies, after which the plies are pressed together when hot that results in polymerisation of the resin. Plywood can be used as a facing or visible material for a lower-quality wood product.

See: *cross-laminated timber; laminated safety glass*

1. Askeland, D. K., and W. J. Wright. *The Science and Engineering of Materials*, 7th edn, p. 666. Boston, MA: Cengage Learning, 2011.
2. Coulson, J. *Wood in Construction*. Oxford: Wiley-Blackwell, 2012.
3. Miodownik, M. Wood 2.0 reaches for the skies as digital techniques give it a new lease of life. *Observer Tech Monthly*, p. 38, 14 June 2015.

PMOLED

PMOLED refers to passive matrix organic light-emitting diode and there are similarities between PMOLEDs and AMOLEDs.[1] However, in a PMOLED each pair of conductors only supplies current to a pixel but PMOLEDs are less responsive than AMOLEDs and can show 'smearing effects' when images change quickly.

See: *AMOLED*

1. Platt, C., and F. Jansson. *Encyclopaedia of Electronic Components*, vol. 2. San Francisco, CA: Maker Media, 2014.

p–n junction diodes

If a layer of *n*-type semiconductor is produced by exposure of a semiconductor to a source of donor dopant and a layer of *p*-type semiconductor is deposited on top of the *n*-type semiconductor using an acceptor dopant so that half of the semiconductor is *p*-type and the other half represents *n*-type, then the electronic properties at the *p–n* interface constitute a *p–n* junction diode, usually referred to as a diode.[1] Electrons diffuse from the *n*-type region into the *p*-type region and holes move from the *p*-type region into the *n*-type region. Holes can combine with electrons with the result that a potential barrier is set up at the junction or interface, preventing further movement of electrons from the *n*-type region and holes from the *p*-type region (Fig. G.20). The region at the junction is known as a depletion layer. If an applied voltage is applied across the depletion layer as a forward bias in which the *p*-type region is biased with a positive potential, then the potential barrier is lowered and current flows across the junction. If the *p*-type region is connected to a negative potential that is negatively or reversed bias then the height of the potential barrier is increased and there is

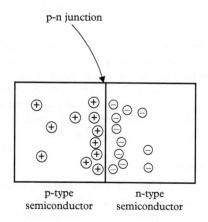

Fig. G.20 *Schematic diagram for electric charge distribution in a* p–n *semiconductor diode.*
(M. Nixon. Digital Electronics: A Primer, *p. 83. London: Imperial College Press, 2015.)*

very limited current flow. Hence the *p–n* junction acts as a diode controlling the direction of flow of current.

See: *light-emitting diodes*

1. Solymar, L., D. Walsh and R. R. A. Syms. *Electrical Properties of Materials*, 9th edn. Oxford: Oxford University Press, 2014.

polyacetals

Polyacetals (also known as acetal resin) are a crystalline form of polymerised formaldehyde and these polymers are also referred to as polyoxymethylene. Polyacetals are polyethers. They are produced by polymerisation of formaldehyde or its cyclic trimer trioxane in acidic or basic solutions.[1,2] The repeating unit in the polymer backbone is $-O-CH_2-O-CH_2-$. Polyoxymethylene can be shaped by injection moulding or extrusion and has been used as bearings, car instrument panels and handles, as examples. Polyacetals have low moisture absorption, high chemical resistance, low wear and friction and good creep resistance and are good electrical insulators.

1. Fried, J. R. *Polymer Science and Technology*, 3rd edn, pp. 405–6. Upper Saddle River, NJ: Prentice Hall, 2014.
2. Walton, D., and P. Lorimer. *Polymers*, p. 108. Oxford: Oxford University Press, 2005.

polyacetylene

Pioneering work by Heeger and co-workers showed that doped films of polyacetylene could have electrical conductivities characteristic of semiconductors or metals.[1] Acetylene monomer was polymerised in the presence of titanium butoxide and triethyl aluminium catalyst. Polyacetylene is a conjugated molecule.[2]

See: *poly-(p-phenylenevinylene)*

1. Heeger, J., A. G. MacDiarmid, C. K. Chiang and H. Shirakawa. P-type electrically conducting doped polyacetylene film and method of preparing the same. United States Patent 4,222,903, 1980.
2. Geoghegan, M., and G. Hadziioannou. *Polymer Electronics*. Oxford: Oxford University Press, 2013.

polycarbonate

Polycarbonate was invented in 1953 at around the same time at GE Plastics in the USA and at Bayer in Germany and has been sold under the trade names of Lexan (GE Plastics) and Makrolon (Bayer). Polycarbonate is a thermoplastic polymer and is prepared by condensation of bisphenol-A [$HOC_6H_4CH_3CCH_3C_6H_4OH$, that is 2, 2 bis (4-hydroxyphenyl) propane] with phosgene ($COCl_2$) in an interfacial reaction;[1] the repeating unit in polycarbonate is $-(OC_6H_4CH_3CCH_3C_6H_4COO)-$. It can also be made in a melt process when diphenyl carbonate ($C_6H_5-O-CO-O-C_6H_5$) is used instead of phosgene. Polycarbonate is a tough amorphous polymer and has a high light transmittance and high impact strength. Applications include a substitute for window glass where impact resistance is required, for example on buses, and is the material for compact discs.

1. Fried, J. R. *Polymer Science and Technology*, 3rd edn. Upper Saddle River, NJ: Prentice Hall, 2014.

polycarbosilanes

Polysilanes can be converted to polycarbosilane polymers with repeating Si–C linkages in the polymer backbone by a thermal treatment.[1] Polycarbosilanes are soluble in solvents and can be spun to fibres that can be converted to silicon carbide fibres by a heat treatment. Conversion to silicon carbide fibres is an important industrial application for polycarbosilanes. The thermal treatment causes cleavage of the polysilane molecules at the Si–Si linkages followed by a free radical rearrangement. Polycarbosilanes are ceramic precursors.

See: *ceramic precursors; polysilanes; silicon carbide fibres*

1. Segal, D. *Chemical Synthesis of Advanced Ceramic Materials*. Cambridge: Cambridge University Press, 1989.

polyester

Polyesters are polymers formed by the reaction of a polyhydric alcohol with a polybasic acid.[1] For example, an important industrial polyester is polyethylene terephthalate (PET) that is prepared by first reacting ethylene glycol with dimethyl terephthalic acid at around $200°C$, after which the product of this transesterification reaction is heated at approximately $280°C$ to yield PET with ethylene glycol as a by-product;[2] the repeating unit in polyethylene terephthalate is $-(COOC_6H_4COOCH_2CH_2)-$. PET can be melt-spun into fibres. Polyesters are thermoplastics and electrical insulators and have mechanical stability at elevated temperatures. They can be injection-molded and applications include textiles, clothing, sheets, blankets, soft-drink bottles and packaging for consumer products, as examples. Polyethylene terephthalate has the trade name of Terylene while Dacron is another polyester. Dacron tubing has application in vascular grafts in which blood vessels are joined together by the tubing in a surgical procedure.

See: *furan-2; 5-dicarboxylic acid*

1. Daintith, J. (ed.). *A Dictionary of Physics*, 6th edn. Oxford: Oxford University Press, 2009.
2. Ploszajski, A. Polyester. *Materials World* (Nov. 2014) 22, 60–2.

polyetherimide

Polyetherimide belongs to the class of thermoplastic polyimides that are high-temperature solvent resistant polymers.[1] Polyimides can be derived from polycondensation between aromatic diamines and a dianhydride to give an intermediate that is dehydrated on heating to yield the polyimide. Polyimides have applications as matrix materials for carbon composites. Imides contain the imido group, -CONHCO-, here derived from the dianhydride. The transparent polyimide Kapton (DuPont) has been evaluated as a substrate for flexible photovoltaic cells based on cadmium telluride with a reported efficiency of about 14%.[2]

See: *silicon-based solar cells*

1. Fried, J. R. *Polymer Science and Technology*, 3rd edn. Upper Saddle River, NJ: Prentice Hall, 2014.
2. Burke, M. Solar future looks bright. *Chemistry and Industry* (Sept. 2011) 19–21.

polyethylene

Polyethylene, or polythene, was first prepared accidentally in 1933 by Reginald Gibson and Eric Fawcett of Imperial Chemical Industries (ICI, United Kingdom), who heated benzaldehyde and ethylene to $170°C$ at high pressure, 1900 atm, to yield a waxy solid rather than the intended product ethyl phenyl ketone. Polythene had an important role in World War Two as an insulator for cables carrying high-frequency signals in radar equipment installed in RAF aircraft. Since that time, a number of methods that have widespread applications including transparent food packaging, crates, electrical insulation for wires and injection-molded products have been used to prepare polyethylene. For example, polymerisation of ethylene takes place over a supported chromium oxide catalyst on alumina in a fixed bed column at temperatures up to $260°C$ and pressures of around 30 atm.[1] A stirred reactor can be used where the catalyst is suspended in a carrier liquid such as 2,2,4-trimethylpentane.

Other catalysts include Ziegler catalysts, complexes of aluminium trialkyls with titanium.[2] Low-density polyethylene (LDPE) and high-density polyethylene (HDPE) differ in terms of their crystallinity, namely their density.

See: *olefin-based polymers; partially crystalline materials; Ziegler–Natta catalysts*

1. Hogan, J. P., and R. L. Banks. Polymers and production thereof. United States Patent 2,825,721, granted on 4 March 1958.
2. Fried, J. R. *Polymer Science and Technology*, 3rd edn. Upper Saddle River, NJ: Prentice Hall, 2014.

polyhydroxyalkanoates

Some bacteria produce plastics[1] directly as a result of their metabolism and polyhydroxyalkanoates (PHA) represent a class of linear polymers produced by bacteria during fermentation as storage compounds. *Ralstonia eutropha* is a wild-type strain that has been used for the industrial production of polyhydroxybutyrate (PHB).

See: *commercial bioplastics*

1. Northen, T. R. Biofuels and biomaterials from microbes. In D. S. Ginley and D. Cahen (eds), *Fundamentals of Materials for Energy and Environmental Sustainability*, p. 326. Cambridge: Cambridge University Press, 2012.

poly-3-hydroxybutyrate

Poly-3-hydroxybutyrate is a polyester and biopolymer produced by some bacterial and plant species.[1] It is synthesised by the organism in the presence of sugar and can be microbially digested into carbon dioxide and water. This biopolymer has properties similar to those of polypropylene derived from petrochemical sources; namely, it can be injection molded, extruded and electrospun into fibres and can be manufactured in principle from renewable sources such as plant biomass. Genetically engineered *Escherichia coli* have been used to prepare poly-3-hydroxybutyrate by digestion of plant biomass and while the biopolymer decomposes to 3-hydroxybutyrate, this process is slow and the polymer cannot be recycled by melting.

See: *biopolymers; green chemistry*

1. Pilcher, J. Plastic fantastic. *Materials World* (Dec. 2014) 22, 3235.

polylactic acid

Lactic acid ($CH_3CH(OH)COOH$), an α-hydroxycarboxylic acid, can undergo polymerisation to yield polylactic acid (PLA), a biodegradable ester; the repeating unit in the ester is ($-COOCH_3CH-$). Lactic acid can be produced from conventional petroleum resources, namely by hydrocyanation of acetaldehyde or by fermentation of corn starch by *Lactobacillus acidophilus* bacteria. It can be converted to PLA by dehydration over a tin oxide catalyst, although the polymer can also be obtained directly from fermentation of food wastes using bacteria and spun into fibres for potential use in clothing.[1] PLA degrades to lactic acid and as the latter is produced from pyruvic acid in muscle tissue, PLA has biomedical applications, for example in drug delivery due to its biocompatibility. Lactic acid is an intermediate for production of commodity chemicals such as acrylic acid and pyruvic acid ($CH_3COCOOH$) and has been prepared from glycerol, a by-product from the production of biodiesel in which glycerol is enzymatically oxidised to dihydroxyacetone that was isomerised to lactic acid over a tin-based zeolite catalyst.[2] Other applications for PLA include packaging (NatureWorks) and disposable cutlery and containers (Cereplast).

See: *biodiesel*

1. Walton, D., and P. Lorimer. *Polymers*, p. 145. Oxford: Oxford University Press, 2005.
2. Kitley, G. Glycerol cascades towards plastic. *Chemistry World* (Feb. 2015) 12, 21.

polymerase chain reaction

The polymerase chain reaction (PCR) was pioneered by Kary Mullis and amplifies (i.e. copies) a fragment of DNA known as the template DNA in a mixture of starting material; that is, it results in the generation of new DNA molecules with the same sequence as existing ones.[1,2] There are three steps in carrying out the PCR and these steps constitute one cycle. In the first step the double-stranded DNA is denatured, that is, converted to single strands by heating the starting mixture in an aqueous medium at around 90°C. In the second step, known as annealing, forward and reverse oligonucleotide primers bind to their complementary sites that flank the target DNA using a reaction temperature around 50°C. This binding is referred to as hybridisation. In the third step, known as extension and carried out at around 70°C, a thermostable DNA polymerase such as *TAQ* DNA polymerase extends the target sequences to create the polymer. After about 40 cycles, millions of identical copies of the target DNA are produced.

See: *DNA; oligonucleotides*

1. Mullis, K. B. Process for amplifying nucleic acid sequences. United States Patent 4,683,202, granted on 28 July 1987.
2. Rapley, R., and D. Whitehouse (eds). *Molecular Biology and Biotechnology*. Cambridge: Royal Society of Chemistry, 2015.

polymer blends

In use, many polymers are blended with other polymers to produce materials with superior properties than either of the individual polymers possesses.[1] An example of a blended polymer is based on Noryl, a polyphenylene oxide. Other polymers are difficult to shape, for example the glassy polymer polyvinyl chloride (PVC), and the addition of plasticiser to PVC enables its shaping into protective film or coatings for wire. Lubricants, fillers and flame retardants are also added to polymer blends to help their ease of handling.

See: *Noryl; polyphenylene oxide; polyvinylchloride*

1. Fried, J. R. *Polymer Science and Technology*, 3rd edn. Upper Saddle River, NJ: Prentice Hall, 2014.

polymer foams

Soft, flexible, elastic and porous foams[1] can be produced from many polymers and are widely used in linings for clothes, mattresses and cushions and in furniture as examples. Rigid foams are used as insulation materials, for example in refrigerators. Polyurethane foams are used in packaging, noise reduction and insulation, rigid epoxy foams are used in the manufacture of rigid composite materials, while polyethylene foam has applications in thermal insulation and in damping vibrations. Foaming agents are added to the foam precursors in order to generate the foams.

1. Liu, P. S., and G. F. Chen. *Porous Materials*. London: Elsevier, 2014.

polymers

Polymers are macromolecules and include bio-polymers that are naturally occurring as well as synthetic polymers derived from petrochemical sources in which monomers are chemically joined together by polymerisation to form larger polymer molecules.[1] In homopolymers only one type of monomer is used; for example, polyethylene is derived from ethylene monomer. Heteropolymers or copolymers are made from two or more different monomers. Nylon and styrene-acrylonitrile-butadiene are examples of heteropolymers. DNA (deoxyribonucleic acid) can be considered to be a heteropolymer. Block copolymers consist of two polymer chains or blocks that are covalently bonded to each other and which are chemically different: for example, polystyrene–polymethylmethacrylate, which is derived from styrene and methylmethacrylate monomers and contains polystyrene and polymethylmethacrylate polymers.

See: *amyloid fibrils; DNA; nylon; polyethylene; hydrophilic polymers*

1. Askeland, D. K., and W. J. Wright. *The Science and Engineering of Materials*, 7th edn, p. 571. Boston, MA: Cengage Learning, 2011.

polymethyl methacrylate

Esters of methacrylic acid can be polymerised in suspension by free-radical initiators such as peroxides. For example, polymethyl methacrylate (PMMA) is produced by polymerisation of methyl methacrylate monomer in the presence of benzoyl peroxide.[1] The polymer is obtained as a fine powder or as sheets. PMMA is a component of Perspex, a trade name in the United Kingdom, and is known as Plexiglass in the USA. It has widespread applications, including bathroom fixtures, light fittings for domestic use, building panels, advertising displays, windows and the automotive market such as light covers.

1. Rohm, O., and E. Trommadorff. Process for the polymerization of methyl methacrylate. United States Patent 2,171,765, granted on 5 September 1939.

polypeptides

Polypeptides consist of ten or more amino acids.[1] Polypeptides that are proteins contain usually 100–300 amino acids. The type and sequence of amino acids determine the properties of the polypeptide.

See: *peptides; proteins*

1. Daintith, J. (ed.). *A Dictionary of Chemistry*, 6th edn. Oxford: Oxford University Press, 2008.

polyphenylene oxide

Polyphenylene oxide (PPO) is a polyether made by oxidatively coupling 2, 6 xylenol.[1–4] It is difficult to process but is miscible with polystyrene to form a polymer blend that has a glass transition temperature lower than that for PPO (208°C) but higher than that for polystyrene; the blends with varying amounts of each polymer have applications as automotive bumpers for impact resistance and where resistance to solvents (e.g. petrol) is required. The blends have been marketed under the trade name Noryl.

See: *glass transition temperature; polymer blends*

1. Hay, S. United States Patent 3,306,874 1967.
2. Hay, S. United States patent 3,306,875, 1967.
3. Hay, S. United States Patent 4,028,341 1975.
4. Tyrell, J. A. *Fundamentals of Industrial Chemistry: Pharmaceuticals, Polymers and Business*, p. 141. New York: Wiley, 2014.

poly-(p-phenylene sulphide)

Poly-(p-phenylene sulphide) is a crystalline thermoplastic suitable for use in demanding environments where chemical resistance to solvents, stability to high temperatures (melting point 285°C), good flame resistance and low coefficient to friction is required. It has been prepared by reaction of sodium sulphide with p-dichlorobenzene in a polar organic solvent and has trade names of Ryton and Fortron.[1,2] It is a brittle thermoplastic and can be reinforced with glass or carbon fibres or mineral fillers such as talc. Applications include bearings and pump parts for corrosive environments, for example where resistance to automotive fluids is required. Coatings are obtained by spraying a slurry of the polymer.

1. Edmonds, J. T., and H. W. Hill. United States Patent 3,354,129, 1967.
2. Tyrell, J. A. *Fundamentals of Industrial Chemistry: Pharmaceuticals, Polymers and Business*, pp. 128–9. New York: Wiley, 2014.

poly-(p-phenylenevinylene)

Poly-(p-phenylenevinylene) (PPV) is a conjugated semiconducting polymer discovered by Friend and co-workers.[1] The inventors describe a conjugated polymer as 'a polymer which possesses a delocalised Π-electron system along the polymer backbone; the delocalised Π-electron system confers semiconducting properties to the polymer and gives it the ability to support positive and negative charge carriers with high mobilities along the polymer chain'.[2] It has potential application as an organic light-emitting diode on undergoing electroluminescence. As it is a polymer the acronym PLED, for polymer light-emitting diode, is used to describe it.

See: *organic light-emitting diodes*

1. Friend, R. H., J. H. Burroughes and D. D. Bradley. Electroluminescent devices. United States Patent 5,247,190, 1993.
2. Geoghegan, M., and G. Hadziioannou. *Polymer Electronics*. Oxford: Oxford University Press, 2013.

polyphosphazenes

Polyphosphazenes or polyorganophosphazenes are polymers with a phosphorus–nitrogen repeat unit of alternating phosphorus and nitrogen atoms, namely, $-N=P-$ units in the backbone.[1] Each phosphorus atom is attached to two side-groups that can be the same or different. Examples of polyphosphazenes are poly [bis (trifluoroethoxy) phosphazene] and poly(dichlorophosphazene). The chlorine side-group can be substituted by other side-groups, particularly by fluorine-based groups that form polyfluorophosphazenes. The latter have potential applications as hydrophobic thermoplastics, hydrophobic elastomers, non-linear optical materials and fibre-forming materials and uses as superhydrophobic coatings.

See: *superhydrophobic materials*

1. Smith, D. W. Jr, S. T. Iaccono and S. S. Iyer (eds). *Handbook of Fluoropolymer Science and Technology*. New York: Wiley, 2014.

polypropylene

Polypropylene, an olefin-based polymer, is a thermoplastic used in the form of pipes, films, textile fibres and containers including bottles made by blow molding in which a jet of air is used to shape a sheet of polypropylene in a mold.[1,2] High-density (HDPP) and low-density polypropylene (LDPP) are commercially available and are made by polymerisation of propylene (CH_3CHCH_2). It is an isotactic polymer whereby substituent groups, here methyl (CH_3) lie in the same side of the plane formed by the polymer chain. HDPP is made by passing propylene into an inert solvent containing a heterogeneous catalyst based on a titanium compound such as titanium (III) chloride and an aluminium compound, for example a trialkyl aluminium. Density is affected by the degree of branching in the polymer. Developments in the synthesis of polypropylene include homogeneous polymerisation of propylene using metallocene catalysts, for example those based on cyclopentadienyl ligands. Metallocene compounds are also known as 'sandwich compounds' and include ferrocene.

See: *olefin-based polymer*

1. Walton, W., and P. Lorimer. *Polymers*, p. 78. Oxford: Oxford University Press, 2005.
2. Fried, J. R. *Polymer Science and Technology*, 3rd edn. Upper Saddle River, NJ: Prentice Hall, 2014.

polysaccharides

Polysaccharides are carbohydrates with a larger molecular weight and complexity than simple sugars (monosaccharide) and consist of long chains of monosaccharides.[1] Examples of polysaccharides include starch, cellulose and chitin and molecular weights of these polymers that can be linear or branched can be as large as several million. Humans are unable to digest cellulose as they do not possess suitable enzymes that can break down the close-packed structure. However, they do possess enzymes that can break down carbohydrates such as starches into simple sugars, for example glucose that is stored in the body as glycogen (see Chapter 2).

See: *carbohydrates; hemicellulose; cellulose; hyaluronic acid; monosaccharides*

1. Gratzer, W. *Giant Molecules: From Nylon to Nanotubes*, p. 73. Oxford: Oxford University Press, 2013.

polysilanes

Polysilanes are inorganic polymers comprising a backbone of Si–Si linkages. Early work on polysilanes was carried out in the 1920s and a general preparative method involves the reaction of molten sodium with a dichlorosilane, for example dimethyldichlorosilane. Interest in polysilanes was initially limited because of their low solubility in solvents. Nowadays they have applications as ceramic precursors and photoresists.[1,2] Polysilanes are ceramic precursors.

See: *polycarbosilanes*

1. Segal, D. *Chemical Synthesis of Advanced Ceramic Materials*. Cambridge: Cambridge University Press, 1989.
2. Mark, J., D. Schaefer and G. Lin. *The Polysiloxanes*. Oxford: Oxford University Press, 2015.

polysilastyrene

Polysilastyrene is a polysilane prepared by condensation followed by polymerisation between dimethyldichlorosilane and phenylmethyldichlorosilane and it belongs to the group of soluble phenylmethylpolysilanes.[1] This polymer can be molded and cured, that is cross-linked to a solid by exposure to ultraviolet radiation, cast into films from solution and drawn into filaments that can be converted on heating to silicon carbide whiskers. Polysilastyrene is a ceramic precursor.

See: *polysilanes*

1. West, R. C., L. D. David, P. I. Djurovich, H. Yu and R. Sinclair. Polysilastyrene: phenylmethylsilane-dimethylsilane copolymer as precursors to silicon carbide. *American Ceramic Society Bulletin* (1983) 62, 899–903.

polysilazanes

Silazanes are compounds with $Si-N-Si$ bonds such as hexamethyldisilazane and hexaphenylcyclotrisilazane.[1] Polymers of silazanes are known as polysilazanes for example perhydropolysilazane containing the repeating units $-(SiH_2NH)_n-$ with a weight average molecular weight in the range 4500–7000.[2] A film of polysilazane on a substrate such as a semiconductor silicon wafer is hydrolysed to a layer of electrically insulating silica when heated at around 900°C in an atmosphere containing water vapour. This property allows polysilazanes to be used in the semiconductor industry to produce insulating layers on a substrate when the surface topography is irregular and uneven due to the presence of interconnected surface devices. The planarised layers of insulator prevent electrical interference between layers of metal wiring on the wafer. The polysilazane layer acts as a spin-on-glass and its deposition onto the substrate is referred to as trench isolation.

See: *ceramic precursors; spin-on-glass*

1. Segal, D. *Chemical Synthesis of Advanced Ceramic Materials*. Cambridge: Cambridge University Press, 1989.
2. Lee, J.-H., J.-H. Cho, J.-S. Choi and D.-J. Lee. Spin-on-glass composition and method of forming silicon oxide layer in semiconductor manufacturing process using the same. United States Patent 7,270,886, granted on 18 September 2007.

polystyrene

The thermosetting plastic polystyrene is prepared by emulsion polymerisation of styrene monomer ($C_6H_5CHCH_2$) by using, for example, peroxide initiators.[1] It is used widely for packaging and insulation in the form of foam and bead. Styrene monomer is a precursor for the widely used ABS copolymer (styrene-acrylonitrile-butadiene).

See: *latex; styrene-acrylonitrile-butadiene; thermoplastics*

1. Tyrell, J. A. *Fundamentals of Industrial Chemistry: Pharmaceuticals, Polymers and Business*, pp. 85–91. New York: Wiley, 2014.

polysulphones

Polysulphones are thermoplastics with good chemical resistance to mineral acids and alkali.[1] They can be sterilised and are used as membranes for reverse osmosis. They have been prepared by electrophilic substitution in a Friedel–Crafts reaction of sulphonyl chlorides by using Lewis acid catalysts such as ferric chloride, for example in the polymerisation of polyethersulphone.

See: *reverse osmosis*

1. Fried, J. R. *Polymer Science and Technology*, 3rd edn. Upper Saddle River, NJ: Prentice Hall, 2014.

polyurethanes

Polyurethanes were developed by Otto Bayer and co-workers at I. G. Farbenindustrie in 1937.[1] These elastomers are prepared by condensation of diisocyanates (O=C=N–R–N=C=O) with diols (HOR'OH) where the repeating unit in the polymer is ($-OC-NH-R-NH-CO-O-$ $-R'-O-$) and R and R' are alkyl or aryl groups; the polymer contains the urethane group –NHCOO– (the carbamide group). Isocyanates are esters of cyanic acid, HNC=O, and include hexamethylene diisocyanate, $O=C=N-(CH_2)_6-N=C=O$. An example of a polyurethane involves the product from reaction of hexamethylene diisocyanate with ethylene glycol. Diols include hydroxyl-terminated polyesters and polyethers. Polyurethane elastomers can be shaped by injection molding and these polymers have applications as solid foams for furniture, mattresses and insulation and also as soles in shoes.

See: *phase-separated polymers*

1. Tyrell, J. A. *Fundamentals of Industrial Chemistry: Pharmaceuticals, Polymers and Business*, pp. 130–3. New York: Wiley, 2014.

polyvinylchloride

Polyvinylchloride (PVC) is manufactured by the free-radical polymerisation of vinyl chloride in an exothermic reaction.[1-3] The polymer has high chemical-resistance, for example, to acids and alkalis and is weather resistant and while many plastics are flammable, the presence of chlorine in the structure makes PVC flame retardant. Early work on PVC was carried out by Waldo Semon who showed that while as-prepared polyvinylchloride is difficult to shape it could be blended with a plasticiser and then easily shaped at room temperature so that physical properties such as elasticity and flexibility can be adjusted.

1. Jones, R. G. PVC. *Materials World* (May 2014) 22, 50.
2. Semon, W. L. Synthetic rubber-like composition and method of making the same. United States Patent 1,929,453, granted on 10 October 1933.
3. Semon, W. L. Method of preparing polyvinyl halide products. United States Patent 2,188,396, granted on 30 January 1940.

poly (vinylidene chloride)

Cling film has been widely used for wrapping sandwiches and other foods. Cling film is poly (vinylidene chloride) that was first prepared in 1933 by Ralph Wiley (Dow Chemical Company) and it has also been called Saran Wrap.[1] Nowadays the film used to wrap sandwiches and other food tends to be low-density polyethylene.

> 1. Stewart, C., and Lomont, J. P. *The Handy Chemistry Answer Book*. Canton, MI: Visible Ink Press, 2014.

poly (vinylidene fluoride)

Poly (vinylidene fluoride) (PVdF) is manufactured by a radical-initiated emulsion polymerisation of CH_2CF_2 monomer and is used as a porous separator in lithium-ion batteries to prevent electrical shorting of the electrodes.[1] This polymer can also be processed into a piezoelectric strip that can act as a microphone or loud speaker. This processing involves casting a crystalline film from a melt and mechanically stressing the film by stretching or rolling to about five times the original length to produce an oriented film. The piezoelectric strip can be used to detect the irregular breathing of a sleeping baby or when combined with a microchip and incorporated into a greetings card can play the tune 'Happy Birthday' to an unsuspecting recipient of the card.

See: *lithium-ion batteries; piezoelectric materials*

> 1. Smith, D. W. Jr, S. T. Iaccono and S. S. Iyer (eds). *Handbook of Fluoropolymer Science and Technology*, pp.583–4. New York: Wiley, 2014.

polytetrafluoroethylene

Polytetrafluoroethylene (PTFE), often referred to by its trade name of Teflon, was accidentally invented by Roy Plunkett who in 1938, while researching new refrigerants, identified a white waxy powder in a gas cylinder containing tetrafluoroethylene (C_2F_4) at high pressure at ambient temperatures.[1,2] The polymerisation of tetrafluoroethylene monomer was enhanced in the presence of a catalyst, for example, silver nitrate and a solvent such as methyl alcohol. Nowadays, PTFE is produced by the emulsion free-radical polymerisation of tetrafluoroethylene under pressure, around 50 atm.[3] Polytetrafluoroethylene is a fluoropolymer, has high chemical-resistance and decomposes at around 300°C and the polymer is an electrical insulator. It has a low coefficient of friction, that is it is slippery, and it is hydrophobic so that water does not spread across a Teflon surface. The initial application for polytetrafluoroethylene was in the Manhattan Project in the USA for the development of the atomic bomb in which its chemical resistance to the highly corrosive uranium hexafluoride was ideal for use in isotope separation. Applications now extend to gaskets, sealing tape, pipe liners, bearings, gears, slide plates, medical prosthetics, insulation for wiring and cables and Teflon is particularly associated with its non-stick properties for cooking utensils. Polytetrafluoroethylene cannot be melt-processed and products are fabricated by mechanical working of a pressed and sintered preform. Additions of low-molecular-weight PTFE powders with a particle size 3–4 μm to printing inks protects the ink pigments from abrasion and prevents printed pages from sticking to one another, a process known as blocking.[4]

See: *fluoropolymers; Gore-Tex*

1. Felice, M. Polytetrafluoroethylene. *Materials World* (Dec. 2012) 20, 54.
2. Plunkett, R. J. Tetrafluoroethylene polymers. United States Patent 2,230,654, granted on 4 February 1941.
3. Fried, J. R. *Polymer Science and Technology*, 3rd edn. Upper Saddle River, NJ: Prentice Hall, 2014.
4. Carlick, D. J., R. W. Bassemir, R. Krishnan and R. R. Durand. Low rub printing ink. United States Patent 5,158,606, granted on 27 October 1992.

porcelain

While many materials are used for their functional properties, others are admired for their aesthetic properties. In 1709 Johann Böttger created porcelain at the town of Meissen in Germany and this earthenware is known as Meissen porcelain characterised by its translucent white appearance.[1-3] Porcelain is produced by firing raw materials including clays and silica sand in a reducing atmosphere and is non-porous so that an external glaze is not required. Meissen porcelain is a hard-paste porcelain with a higher kaolin content than earlier soft Chinese porcelains from the Ming period (1368–1644). Meissen porcelain is also known as European porcelain.

See: *Chinese porcelain*

1. Hennicke, H. W., and A. Hesse. Traditional ceramics. In *Concise Encyclopaedia of Advanced Ceramic Materials*, p. 489. Oxford: Pergamon Press, 1991.
2. Miller, J. (Gen. Ed.). *Miller's Antiques Encyclopedia*. London: Octopus Publishing, 2013.
3. Ploszajski, A. Porcelain. *Materials World* (June 2016) 24, 60–2.

portland cement

Portland cement was introduced in the 19th century by Joseph Aspdin.[1] The raw materials are limestone (calcium carbonate) and clay that are pulverised to form an aqueous slurry that is heated in a kiln at around 1500°C. The result of the heat process that initially drives off the water and decomposes the limestone to calcium oxide (quicklime) is a lumpy clinker that is ground to a fine powder. A small amount of gypsum (2–5 wt%) that is hydrated calcium sulphate is added to produce Portland cement powder that consists of five main ingredients. These are tricalcium silicate (Ca_3SiO_5, known as alite and referred to in shorthand as C_3S), dicalcium silicate (Ca_2SiO_4, known as belite and in shorthand C_2S), tricalcium aluminate ($Ca_3Al_2O_6$, in shorthand C_3A), tetracalcium aluminoferrite ($Ca_4Al_2Fe_2O_{10}$, in shorthand C_4AF) and gypsum, calcium sulphate dihydrate ($CaSO_4.2H_2O$, in shorthand CSH_2). Portland cement sets or hardens on addition of water and the hardening process involves hydration between water and the silicates, aluminates and gypsum. Hydration produces a hard, porous and rigid crystalline mass.

See: *macro-defect-free cement*

1. Bolton,W., and R.A. Higgins. Materials for Engineers and Technicians, 6th edn, p. 332. London: Routledge, 2015.

post-it notes

Adhesives are widely used in everyday life, for example the weak adhesive at the back of Post-it notes.[1] This reusable pressure-sensitive tacky adhesive consists of elastomeric copolymer microspheres approximately 1–250 μm in diameter with the majority in the range 10–150 μm in diameter and is made by an emulsion polymerisation process. The reactants involve an alkyl acrylate ester such as iso-octyl acrylate and an ionic monomer, for example trimethylamine methacrylimide, and polymerisation takes place in the presence of an anionic emulsifier. The emulsifier, for example ammonium lauryl sulphate or an alkyl aryrlpolyethylene oxide sodium sulphonate, is used above its critical micelle concentration. The resulting tacky microspheres can be dispersed in solvents and act as pressure-sensitive adhesives for adhering objects to surfaces. The copolymers can be easily removed from surfaces and the objects, for example paper, can be repositioned. The adhesive was first produced by Spencer Silver in 1968 by chance while working on the development of a strong adhesive.[2]

See: *bioadhesives; epoxy adhesives; superglue*

1. Silver, S. F. Acrylate copolymer microspheres. United States Patent 3,691,140, granted 12 September 1972.
2. Donald, G.. *The Accidental Scientist: The Role of Chance and Luck in Scientific Discovery*, pp. 98–102. London: Michael O'Mara Books, 2013.

prepregs

There is much interest in the use of lightweight composite components as structural materials in aircraft and vehicles in order to save fuel and to reduce operating costs. A prepreg consists of a felt or fabric or mat of woven or non-woven fibres (e.g. glass fibres) that is impregnated with an unpolymerised resin.[1] Layers of the impregnated fibres are stacked and pressed together or inserted in a laminar structure and undergo a heat treatment that polymerises the resin.

See: *glass epoxy composites; microlaminates*

1. Askeland, D. K., and W. J. Wright. *The Science and Engineering of Materials*, 7th edn, p. 638. Boston, MA: Cengage Learning, 2011.

proppants

Proppants are used in the technique of hydraulic fracturing or fracking for the extraction of shale gas.[1] Proppants are hard material in granular form, for example silica sand, resin-coated sand and other ceramic materials that are injected into wells in order to hold open fissures in the rock to release the shale gas.

See: *shale gas*

1. Hester, R. E., and R. M. Harrison (eds). *Fracking, Issues in Environmental Science and Technology*, vol. 39, p.8. Cambridge: Royal Society of Chemistry, 2015.

proteins

Proteins are organic compounds expressed by DNA that encodes them in the cells of organisms.[1] Proteins have molecular weights from around 6000 to several million and are polymers made up of amino acids linked together in a characteristic sequence; polypeptides are proteins. The order of base pairs in DNA determines the type of amino acid that is expressed. In particular, a sequence of three bases out of adenine, cytosine, thymine and guanine known as a codon encodes a specific amino acid.

See: *amino acids; DNA; polypeptides*

1. Gratzer, W. *Giant Molecules: From Nylon to Nanotubes*, pp. 26–77. Oxford: Oxford University Press, 2013.

p-type semiconductors

The semiconductor silicon has a valency of four. Silicon can be doped with a Group III element such as indium, aluminium, gallium or boron and these elements are known as *p*-type dopants or acceptors.[1] The latter can be introduced by techniques such as ion implantation, or diffusion of dopant. The doped semiconductor has excess of holes or electron deficit introduced into the crystalline structure by incorporation of the acceptor atom.

See: n-*type semiconductor; light-emitting diodes;* p–n *junction diodes*

1. Solymar, L., D. Walsh and R. R. A. Syms. *Electrical Properties of Materials*, 9th edn. Oxford: Oxford University Press, 2014.

pure materials

The exploitation of materials in the 21st century is very dependent on the ability to prepare them with exceptional purity. This requirement for purity is particularly important for silicon-based semiconductors for the preparation of pure silicon and the introduction of dopant atoms by, for example, ion implantation and thermal diffusion.[1] Vast financial investments of hundreds of millions of US dollars are required to construct industrial facilities to manufacture semiconductors. Telecommunication companies that sell mobile phones as well as producers of other consumer products such as washing machines and televisions will often design their equipment but source electronic components such as integrated circuits from third parties.

1. Bolton, L., and R. A. Higgins. *Materials for Engineers and Technicians*, 6th edn, p. 334. London: Routledge, 2015.

Pyrex

Pyrex is a borosilicate glass with an approximate composition of 81% SiO_2; 13% B_2O_3; 4% Na_2O; 2% Al_2O_3. It has a low coefficient of thermal expansion and a softening temperature around 830°C. It has applications in laboratory glassware and kitchenware. The Pyrex brand of laboratory glassware was introduced in 1915.[1]

See: *phase-separated glasses*

1. L. Morrey. A glass act to follow. *Chemistry World* (Sept. 2015) 12, 46–7.

pyroelectric materials

Pyroelectric crystals develop opposite electrical charges on faces of the crystal when heated and all pyroelectric materials are piezoelectric but not all piezoelectric materials exhibit pyroelectric behaviour.[1] Examples of pyroelectric materials are tourmaline (a boron-containing silicate mineral) and crystals with the perovskite structure as highlighted by barium titanate, $BaTiO_3$. Cadmium sulphide and lithium niobate are also pyroelectric materials. Pyroelectric crystals have a spontaneous polarisation in the absence of an electric field and have application in thermal imaging cameras.

See: *ferroelectric materials*

1. Tilley, R. J. D. *Understanding Solids: The Science of Materials*. London: Wiley, 2013.

quantum cascade lasers

There has been much development on laser systems since the ruby laser was invented by Theodore Maiman in 1960. The quantum cascade laser is a solid-state laser.[1,2] Whereas in a light-emitting diode electron transitions take place from the conduction to the valence band, resulting in light emission, in a cascade laser electrons in a quantum will undergo transitions within interbands in the conduction band. These transitions result in emission of laser radiation. Terahertz radiation with wavelengths between the infrared and microwave part of the electromagnetic spectrum can be obtained from a quantum cascade laser. As with other lasers a population inversion and stimulated emission from the material of the laser are required.

See: *ruby lasers*

1. Solymar, L., D. Walsh and R. R. A. Syms. *Electrical Properties of Materials*, 9th edn, pp. 318–19. Oxford: Oxford University Press, 2014.
2. Faist, J. *Quantum Cascade Lasers*. Oxford: Oxford University Press, 2013.

quantum dots

Quantum dots (or boxes) are semiconductors subject to quantum confinement where the electronic wave function is restricted to small regions of space in a material. Atoms have discrete energy levels, whereas a macroscopic piece of a semiconductor has electronic bands. However, as the size of the piece of semiconductor decreases to around 10 nm a quantum dot (or box) is produced that behaves as if it has discrete energy levels. Quantum dots are regarded as zero-dimensional with regards to electronic properties and have a three-dimensional confinement. A quantum dot may be considered to be a small section of a quantum wire.[1] Electronic and optical properties of semiconductors differ markedly from bulk materials as the size decreases to the nanometre range (Fig. G.21). Quantum dots are semiconductor phosphors and exhibit fluorescence. As the particle size decreases the wavelength of emitted light decreases, corresponding to an increase in band gap with decreasing particle size, thus a shift from red to blue light with decreasing particle size.

See: *band gap; phosphors; semiconductors*

1. Tilley, R. J. D. *Understanding Solids: The Science of Materials*. London: Wiley, 2013.

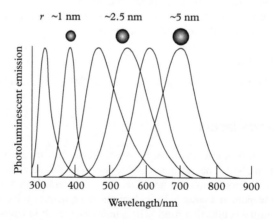

Fig. G.21 *Photoluminescent colours emitted by CdS quantum dots with radius* r *nm.*
(R. J. D. Tilley. Understanding Solids, *2nd edn, p. 488. Chichester: Wiley, 2013.)*

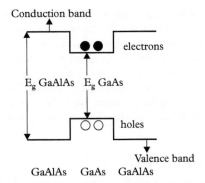

Fig. G.22 *Schematic illustration of a quantum well formed by sandwiching a thin layer of gallium arsenide (GaAs) between layers of gallium aluminium arsenide (GaAlAs) where the band gap (E$_g$) for GaAs is less than that for GaAlAs.*
(R. J. D. Tilley. Understanding Solids, *2nd edn, p. 424. Chichester: Wiley, 2013.)*

quantum wells

A quantum well consists of a thin layer (around 10 nm) of a semiconductor sandwiched between two layers of a different semiconductor that has a higher band gap than the well layer.[1] This structure is known as a heterostructure: for example, a layer of GaAs (gallium arsenide) between layers of GaAlAs (gallium aluminium arsenide) (Fig. G.22). Quantum wells are composite materials and they can be deposited by molecular beam epitaxy or metalorganic chemical vapour deposition. In a quantum well, electrons are confined to the plane of the layer and the well is considered to be two-dimensional with respect to its electronic properties with one-dimensional confinement. Quantum wells are thus subjected to quantum confinement. In a quantum well, electrons in, for example, the GaAs layer are trapped in a well in the conduction band of the composite while holes are trapped on 'hills' in the valence band. The band gap of the well layer increases with decreasing thickness. Quantum wells have applications as quantum well lasers where the frequency and wavelength of emitted length depends on the structure and composition of the semiconductors.

See: *band gap; heterostructures; quantum cascade lasers; semiconductors*

1. Tilley, R. J. D. *Understanding Solids: The Science of Materials.* Chichester: Wiley, 2013.

quantum wires

A quantum wire is a strip of semiconductor material in which electrons are confined in two dimensions so that the wire is a one-dimensional conductor.[1] Quantum wires are subject to quantum confinement and are one-dimensional materials with regard to electronic properties and have a two-dimensional confinement. A quantum wire is a linear electrical conductor where electrons have quantised energy levels for motion perpendicular to the axis.[2] These discrete energy bands are in contrast with continuous energy bands for electron states in bulk crystalline solids, thus valence and conduction bands. Possible future applications for quantum wires are in quantum devices to operate as faster semiconductor switches than are currently available. There are limitations to the miniaturisation of field-effect transistors as current leakage reduces the effectiveness of their switching properties and electron scattering can increase due to lattice defects and impurities in the semiconductor. Quantum devices incorporating quantum wires may offer a solution to these limitations.

See: *band gap; quantum dot; semiconductors*

1. Tilley, R. J. D. *Understanding Solids: The Science of Materials.* Chichester: Wiley, 2013.
2. Daintith, J. (ed.). *Oxford Dictionary of Physics*, 6th edn. Oxford: Oxford University Press, 2009.

quasicrystals

Quasicrystals are solids that have regions with a structure equal to or greater than a fivefold rotational symmetry. Concepts in conventional crystallography indicate that unit cells with this type of rotational symmetry, that is fivefold or greater, cannot exist as the cells are not able to be packed together to form a crystal. Although quasicrystals have a rotational symmetry they do not have translational order and can be considered to have regions of structural units separated by disordered regions. Quasicrystals were first identified by Professor Daniel Schectman and co-workers in the early 1980s in alloys of aluminium and manganese, in particular in Al_6Mn.[1] Other alloy compositions such as $Al_{88}Mn_{12}$ were shown to have a tenfold rotational symmetry and long-range translational order.[2,3]

1. Janot, C. *Quasicrystals: A Primer*, 2nd edn. Oxford: Oxford University Press, 2012.
2. Schectman, D., et al. Metallic phase with long-range orientation order and no translational symmetry. *Physical Review Letters* (1984) 53, 1951–3.
3. Tilley, R. J. D. *Understanding Solids: The Science of Materials.* Chichester: Wiley, 2013.

rapeseed oil

Fields of yellow oil seed rape are familiar to many observers. Conventional esters are made by reaction of alcohols with acids in the presence of a catalyst in a solvent derived from petrochemical sources and have applications as solvents. Rapeseed oil from crops has been converted directly to esters for use as biosolvents by reaction with a mixture of potassium hydroxide in methyl alcohol.[1]

1. Sarin, A. *Biodiesel: Production and Properties.* Cambridge: Royal Society of Chemistry, 2012.

rayon

Rayon is a textile fibre derived from cellulose (e.g. wood pulp).[1] Cellulose is dissolved in a mixture of carbon disulphide and sodium hydroxide, after which the resulting aqueous cellulose xanthate solution is converted into continuous fibres of regenerated cellulose (i.e. rayon) by passage through fine nozzles into sulphuric acid. The fibres are known as viscose rayon and can be woven into fabrics. Acetate rayon is derived from cellulose acetate.

See: *cellulose; cellulose acetate; lyocell*

1. Emsley, J. *Molecules at an Exhibition*, p. 119. Oxford: Oxford University Press, 1998.

reactive textile dyes

Reactive textile dyes react with molecules in fibres to form permanent covalent bonds between the dye and fibre.[1] The dyes can be used for cellulose-based fibres (cellulosics) where the dye reacts with hydroxyl groups and for protein-based fibres such as wool where reaction occurs with amine groups in the fibre. Reactive dyes consist of four components: a water-solubilising group (e.g. COONa) to make the dye water-soluble, a chromophore to produce the colour (e.g. azo or anthroquinone derivative), a reactive chemical group to react with hydroxyl groups on cellulose or amine (NH_2) groups on protein fibres and a bridging group to link the chromophore to the reactive group. Examples of reactive dyes include vinylsulphones and chlorotriazines that are derivatives of cyanuric chloride (2, 4, 6–trichloro-1,3, 5-triazine) and bis-monofluorotriazines. Trade names include Procion (ICI), Avitera SE (Huntsman) and Cibacron (Ciba). Reactive dyes are soluble in cold water and dissolve to produce anionic species that bind to the fibre.

1. Patra, A. K., and R. Paul. Reactive dyeing of textiles: practices and developments. In J. Fu (ed.). *Dyeing: Processes, Techniques and Applications*, pp. 1–18. Hauppauge, NY: Nova Science Publishers, 2014.

recombinant 6-aminocaproic acid

6-Aminocaproic acid is a precursor for nylon and worldwide demand for this material in 2005 was 2.7 million metric tonnes.[1,2] This feedstock was obtained from petroleum sources. 6-Aminocaproic acid has been prepared from the amino acid lysine and a recombinant micro-organism. The latter consisted of a recombinant nucleic acid encoded with a peptide that catalyses a substrate-to-product conversion of 6-amino-3-hydroxyhexanoyl-CoA to 6-aminohex-2-enoyl-CoA that underwent fermentation, yielding 6-aminocaproic acid. This example shows how a genetically engineered micro-organism can be used to produce a useful product, here a peptide, that can be used to catalytically convert a reactant into an important chemical, 6-aminocaproic acid, in a fermentation process.[2]

See: *nylon*

1. DeRosa, T. F. *Engineering Green Chemical Processes: Renewable and Sustainable Design*. New York: McGraw-Hill, 2015.

2. Baynes, B. M., et al. Biological synthesis of 6-aminocaproic acid from carbohydrate feedstocks. United States Patent 8,404,465, granted on 26 March 2013.

recombinant *cis*-isoprene

The monomer isoprene is derived from petrochemical sources by thermal cracking of petroleum or naphtha feedstocks and is obtained in the *cis*- and *trans*- isomer forms.[1,2] Isoprene obtained in this way is not a renewable resource. An application of recombinant DNA technology for the preparation of *cis*-isoprene uses glucose as a starting material and a yeast extract as a growth medium and modified *Escherichia coli* cells containing the pTrcKudzu + yIDI + DXS plasmid in a bioreactor. The *cis*-isoprene could be extracted after the fermentation process and converted by a neodymium-based catalyst at 30°C to *cis*-1,4-polybutadiene that has a structure similar to that of natural rubber. This example illustrates how advances in recombinant DNA technology since its inception in the 1970s allow the production of useful industrial chemicals. In this fermentation process, a starting reactant and growth medium are required, here glucose and a yeast extract, respectively. The plasmid DNA in the bacterium *E. coli* is carefully modified to ensure that glucose is enzymatically converted to *cis*-isoprene.

See: *recombinant DNA technology; synthetic rubber*

1. Feher, F. J., et al. Polymers of isoprene from renewable resources. United States Patent 8,420,759, granted on 16 April 2013.
2. DeRosa, T. F. *Engineering Green Chemical Processes: Renewable and Sustainable Design*. New York: McGraw-Hill, 2015.

recombinant DNA technology

Recombinant DNA technology, which was pioneered by Stanley Cohen and Herbert Boyer at Stanford University in the 1970s, underpins the worldwide biotechnology industry; the technology is associated with genetic engineering techniques.[1,2] In the original technique, cells from bacteria, viruses, animals or plants are broken up and their DNA is extracted. Restriction enzymes are used to chemically cut the DNA at specific points and a DNA fragment of interest is isolated. The nucleotide sequence in the isolated gene can express a particular compound such as a protein with a particular sequence of amino acids. Circular rings of plasmid DNA are then extracted from *Escherichia coli*, the rings are opened with a restriction enzyme and the fragment of DNA from the other cell is inserted into the plasmid. The ring is then closed (i.e. recombined) and put back into a host bacterium. The new bacteria are referred to as 'DNA chimera'. Combination of the two sources of DNA is sometimes referred to as gene-splicing. The new gene sequence in the hybrid DNA is passed onto new generations of bacteria on cell division so that the colony contains genetic clones. The inserted fragment can be obtained through synthesis rather than from an organism.

See: *bio-based chemicals; yeasts; ZMapp*

1. Cohen, S. N., and H. W. Boyer. Biological functional molecular chimeras. United States Patent 4,740,470, granted on 26 April 1988.
2. Cohen, S. N., and H. W. Boyer. Process for producing biologically functional molecular chimeras. United States Patent 4,237,224, granted on 2 December 1980.

recombinant face cream

Propanediol distearate has applications in cosmetic formulations including face creams, skin cleaning formulations and skin care.[1] This material is traditionally prepared by chemical synthesis. However, in a fermentation process, an *Escherichia coli* strain, ECL707, containing the *Klebsiella pneumoniae* dha regulon cosmids pKP1 or pKP2, the *K. pneumoniae* pdu operon pKP4 or the Supercos vector alone was grown in a fermenter and the batch was then used to produce 1,3-propanediol from D-glucose in a growth medium. Propanediol distearate was then obtained through reaction of the recombinant 1,3-propanediol, stearic acid and *p*-toluene sulphonic acid. This example is given to illustrate the range of materials that can be obtained by using recombinant DNA technology. The technology is characterised by use of fermentation processes, a reactant, growth medium and a modified micro-organism to express the desired product.

See: *recombinant DNA technology*

1. Fenyvesi, G., et al. Personal care and cosmetics compositions comprising biologically based mono- and di-esters. United States Patent 8,309,116, granted on 13 November 2012.

recycling ceramics

The question of recycling ceramics and end-of-life products will be an important issue in the 21st century if populations continue to grow and resources are finite. Annual worldwide production for some ceramics is as follows with amounts in brackets:[1] concrete (about 4×10^{10} tonnes), cement (about 10^9 tonnes), glass (about 7×10^7 tonnes), brick (about 6×10^6 tonnes) and carbon fibre (about 5×10^4 tonnes).

1. Ashby, M. F., D. F. Balas and J. S. Coral. *Materials and Sustainable Development.* Oxford: Butterworth, 2016.

recycling metals

The question of recycling metals and end-of-life products will be an important issue in the 21st century if populations continue to grow and resources are finite. Annual worldwide production for some metals is as follows with amounts in brackets:[1] lanthanides (about 10^5 tonnes), gold (about 10^3 tonnes), silver (about 10^4 tonnes), titanium alloys (about 10^5 tonnes), nickel alloys (about 10^6 tonnes), copper alloys (about 10^7 tonnes) and steel (about 10^9 tonnes). About 75% of the total U.S. steel production is from recycled steel.[2] Some metals are critical materials, for example indium that is used as transparent indium tin oxide electrically conducting coatings in portable devices such as smartphones.

See: *critical materials; steel*

1. Ashby, M. F., D. F. Balas and J. S. Coral. *Materials and Sustainable Development.* Oxford: Butterworth, 2016.
2. Hosford, W. F. *Iron and Steel.* Cambridge: Cambridge University Press, 2012.

recycling plastics

The question of recycling plastics and polymers will be an important issue in the 21st century if populations continue to grow and resources are finite. Annual worldwide production (1) for some plastics is as follows with amounts in brackets: polyethylene (about 5×10^7 tonnes), polystyrene (about 2×10^7 tonnes) and polyester (about 10^7 tonnes). Natural fibres (e.g. wool) correspond to around 3×10^7 tonnes per year. Biodegradable plastics can be recycled although there are still significant amounts of non-biodegradable polymers.

See: *plastics*

1. Ashby, M. F., D. F. Balas and J. S. Coral. *Materials and Sustainable Development.* Oxford: Butterworth, 2016.

refractory materials

Refractory materials are traditional ceramics that are incorporated into equipment used for the manufacture of metals and glasses as well as linings for furnaces.[1] The compositions need to survive high temperatures without being corroded or weakened by corrosive gaseous environments in the equipment. Examples of refractories are magnesite, chromite-magnesia and olivine (a solid solution of magnesium silicate, Mg_2SiO_4, and iron silicate, Fe_2SiO_4). Carbon acts as a refractory material in the absence of oxygen atmospheres.

See: *ceramic materials*

1. Askeland, D. R., and W. J. Wright. *The Science and Engineering of Materials*, 7th edn. Boston, MA: Cengage Learning, 2011.

reverse osmosis

When a porous membrane has pure water on one side and salt water on the other side then pure water is expected to diffuse through the membrane from a region of low salt concentration to a region of high salt concentration. In reverse osmosis an applied pressure is used to force water from a salty solution through pores in a membrane so that the membrane prevents salt from passing through it.[1] The process is thermodynamically driven and not kinetically driven. Typical pore diameters in reverse osmosis membranes are around 1 nm, sufficiently small to hold back sodium ions with a diameter (hydrated) of around 0.4 nm. Polysulphones have been used as reverse osmosis membranes.

See: *polysulphones*

1. Fried, J. R. *Polymer Science and Technology*, 3rd edn, pp. 497–8. Upper Saddle River, NJ: Prentice Hall, 2014.

RNA

RNA (ribonucleic acid) is a nucleic acid found in living cells and is responsible for protein synthesis. It is a heteropolymer and consists of a single polymer chain of nucleotides in which the polymer backbone consists of molecules of the sugar ribose linked though phosphate groups (Fig. G.23).[1] One of four bases, adenine, cytosine, guanine and uracil designated by the letters A, C, G and U, is linked to the ribose molecule. In protein synthesis, a stretch of DNA (a gene) is transcribed into messenger RNA (mRNA);[2] that is, mRNA is synthesised *in situ*, utilising the property that complementary base pairing takes place between the RNA and DNA nucleotides. The sequence of bases in DNA codes for a protein and the code is retained in mRNA because of the complementary base pairing. The synthesis of mRNA is known as transcription. After its synthesis, mRNA is transferred to a large macromolecular complex outside the cell nucleus known as a ribosome, which is composed of proteins and RNA molecules. In the ribosome the message contained in RNA (i.e. the order of bases) is translated from nucleotides into amino acids. Transfer RNA (tRNA) in the ribosome reads the sequence of codons in mRNA and can base pair with mRNA and transfer-specific amino acids to form a growing protein.

See: *DNA; heteropolymers*

1. Barber, J., and C. Rostron (eds). *Pharmaceutical Chemistry*, pp. 242–8. Oxford: Oxford University Press, 2013.
2. Baldwin, G., T. Bayer, R. Dickinson, T. Ellis, P.S. Freemont, R.I. Kitney, K. Polizzi and G-B Stan. *Synthetic Biology: A Primer*. London: Imperial College Press, 2012.

Roundup

Roundup is a herbicide for controlling weeds and is associated with genetically modified seeds that are resistant to this herbicide and with genetically engineered crops.[1] The biosynthesis of aromatic amino acids such as tryptophan and phenylalanine takes place in plants and micro-organisms by the shikimate pathway. An enzyme in this pathway is 5-enolpyruvyl shikimate 3-phosphase synthase (EPSPS). Glyphosate (Roundup) is water soluble and can be sprayed onto plant leaves when it is absorbed and inhibits EPSPS. Roundup Ready plants carry the gene coding for a glyphosate-insensitive form of this enzyme obtained from *Agrobacterium* sp. strain CP4. When incorporated into the plant

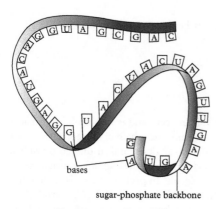

Detail of molecular structure of the sugar-phosphate backbone. The carbon atoms of ribose are numbered 1′ (1 'prime') to 5′. Each ribose unit is attached via its 5′ carbon atom to a phosphate group, and via its 1′ carbon atom to a base.

Single-stranded structure of RNA

The four bases of RNA

Fig. G.23 *Molecular structure of RNA.*

(R. S. Hine (ed.). Oxford Dictionary of Biology, 7th edn, p. 521. Oxford: Oxford University Press, 2015.)

genome, the gene product CP4 EPSP synthase confers crop resistance to glyphosate. Understanding the terminology in the biochemical area can be confusing for anyone with no direct experience of this subject. The shikimate pathway is the major metabolic pathway in plants to produce aromatic amino acids.[2] The pathway involves condensation of erythrose-4-phosphate and phosphoenolpyruvate and the product undergoes cyclisation followed by reduction to form the intermediate shikimate that has a phenolic ring structure characteristic of aromatic amino acids. The intermediate combines with phosphoenolpyruvate to yield chorismate that has a pathway to tryptophan and phenylalanine. Glyphosate blocks the action of EPSPS and the production of aromatic amino acids in plants. These are essential amino acids and the weeds will die without them. Roundup Ready plants contain a gene that expresses an enzyme that is unaffected by glyphosate.

1. Tyrell, J. A. *Fundamentals of Industrial Chemistry: Pharmaceuticals, Polymers and Business*. New York: Wiley, 2014.
2. Hine, R. S. (ed). *Oxford Dictionary of Biology*, 7th edn. Oxford: Oxford University Press, 2015.

ruby laser

The ruby laser was invented in 1960 by Theodore Maiman.[1,2] Rubies are alumina crystals containing about 0.5 wt% of chromium (III) ions. A single-crystal rod of ruby was placed inside a helical coil of a flash lamp and two partially reflecting mirrors were positioned at the end of the rod (Fig. G.24). The rod is optically pumped with light from the flash lamp. Light produced by stimulated emission is trapped between the two mirrors and is reflected back and forth. Each time the light passes through the ruby further stimulated emission is induced. This arrangement produces a population inversion and amplification of the emitted light. Red laser light with a wavelength of 694 nm emerges from the partially reflecting mirror.

See: *laser-based materials*

1. Maiman, T. United States Patent 3,353,115, 1960.
2. Tilley, R. J. D. *Understanding Solids: The Science of Materials*, p. 456. Chichester: Wiley, 2013.

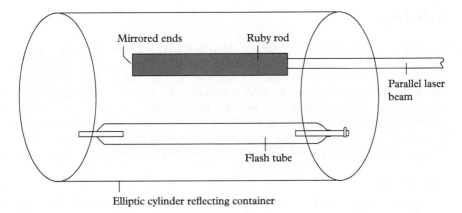

Fig. G.24 *General arrangement of a ruby laser.*

(L. Solymar, et. al. Electrical Properties of Materials, *9th edn, p. 303. Oxford: Oxford University Press, 2014.)*

saccharification

Saccharification refers to the process in which cellulose is hydrolysed to produce monosaccharides, disaccharides and oligosaccharide products and glucose is a major product. Cellulose hydrolysis is catalysed by mineral acids or enzymes, in particular by cellulases. Biomass is a source of cellulose. The sugar xylose obtained from saccharification is an important chemical intermediate and can be chemically converted to furfural by, for example, heating at elevated temperature and pressure.[1] Other furfural-based products include furfural alcohol, tetrahydrofuran and polyethers and these products have applications as solvents for mineral oils and resins.

See: *biomass; cellulose*

1. Medoff, M., T. C. Masterman, C. Cooper and J. Khan. Conversion of biomass. United States Patent Application 2014/0011248, published on 9 January 2014.

samarium-based magnets

Samarium is a lanthanide element (also referred to as a rare-earth element) and alloys of this metal with cobalt (Co), in particular $SmCo_5$ and Sm_2Co_{17}, are used as compact permanent magnets;[1] $SmCo_5$ has a hexagonal crystal structure. Samarium-based magnets are not as powerful as neodymium-based magnets at ambient temperature but are stronger at higher temperatures, around $200°C$.

See: *neodymium-based magnets*

1. Lucas, J., P. Lucas, T. Le Mercier, A. Rollat and W. Davenport. *Rare Earths: Science, Technology, Production and Use.* London: Elsevier, 2015.

scaffolds

Scaffolds are semi-solid substrates used in tissue engineering in which immobilised cells such as stem cells can grow *in vitro* or *in vivo* to form new tissue.[1] Examples of scaffolds include collagen–hydroxyapatite composite hydrogels for bone regeneration and polysaccharides such as dextran, hyaluronic acid, chitosan and heparin. By tissue engineering is meant synthetic tissue that can be used instead of human tissue in surgery or for research.[2] The word tissue refers to a collection of cells that are organised to carry out one or more specific functions. Thus, nervous tissue in animals can detect and transmit stimuli. However, an organ such as a lung or kidney can also be referred to as 'tissue'.

See: *bioprinting*

1. Fried, J. R. *Polymer Science and Technology*, 3rd edn, p. 532. Upper Saddle River, NJ: Prentice Hall, 2014.
2. Hine, R. S. (ed.). *Osford Dictionary of Biology*, 7th edn. Oxford: Oxford University Press, 2015.

security markers

There is much concern in governments relating to the counterfeiting[1] of currencies and important documents, such as passports, credit cards and driving licenses. Phosphors based on lanthanide elements which have a wide range of optical emissions in the visible regions are incorporated into sensitive documents, for example incorporation into banknotes that will emit light under specific excitation.

1. Lucas, J., P. Lucas, T. Le Mercier, A. Rollat and W. Davenport. *Rare Earths: Science, Technology, Production and Use*, pp. 313–14. London: Elsevier, 2015.

self-cleaning glass

The lotus plant grows in marshes and lagoons, in particular in Asia and the Middle East. The leaves are superhydrophobic so that water on them rolls up into droplets and falls off, removing dirt in the process. The surface of the leaves has a hierarchical structure consisting of raised asperities of epidermal cells known as papillae, forming an uneven surface, and each papillae has a nanostructure of protruding tubular asperities of waxes. The combination of microscale and nanoscale features make up the hierarchic structure that prevents contact and adhesion between water droplets and the surface of the leaf. The surface has a series of 'mountains and valleys'. This property of the lotus plant has been referred to as the 'lotus effect'. An approach to making self-cleaning glass is to fabricate a patterned surface that contains a hierarchical structure on the microscale and nanoscale.[1]

See: *superhydrophobic materials; self-cleaning titania-coated glass*

1. Daoud, W. A. *Self-Cleaning Materials and Surfaces: A Nanotechnology Approach.* New York: Wiley, 2013.

self-cleaning paint

Lotusan (Sto Corporation, Germany) is a self-cleaning paint[1] that applies the 'lotus effect' to produce a superhydrophobic self-cleaning coating on the exterior of buildings. The exterior coating is known as StoCoat. The paint combines the ability to produce surface roughness using sol-gel latex particles with a fluorinated hydrophobic coating. The combination of microscale and nanoscale features makes up the hierarchic structure that prevents contact and adhesion between water droplets and the coating.

See: *lotus leaves; self-cleaning glass*

1. Daoud, W. A. *Self-Cleaning Materials and Surfaces: A Nanotechnology Approach.* New York: Wiley, 2013.

self-cleaning titania-coated glass

Window glass that has been coated with a thin transparent layer of crystalline anatase titanium dioxide is self-cleaning. The coating absorbs ultraviolet light that produces a photocatalytic reaction, breaking down dirt on the glass surface, after which residues are swept away by rainwater. An example of this type of coated glass is Pilkington Activ.[1] An overview of self-cleaning materials is given in reference (2).

See: *self-cleaning glass*

1. van Dulken, S. *Inventing the 21st Century*. London: British Library, 2014.
2. Daoud, W. A. *Self-Cleaning Materials and Surfaces: A Nanotechnology Approach*. New York: Wiley, 2013.

self-healing materials

Structural materials fail due to the propagation of cracks and existing methods for increasing the toughness of materials and hence the resistance to failure include the use of fibre-reinforced composites. Self-healing materials have the capability to recover from damage without external intervention. Two approaches have been used. In the first method micro-capsules containing a chemical healing agent are dispersed throughout a composite matrix that contains a catalyst.[1] As a crack spreads and ruptures a capsule, the healing agent leaks out from the capsule and flows into the region of the crack where it reacts with the catalyst, causing polymerisation that seals the crack. In the second approach, hollow glass fibres have been used to reinforce concrete and polymers.[2] In the case of reinforced polymers, some of the fibres contained a resin that was released when the advancing crack ruptured the fibres. The resin reacted with a hardener in the matrix, causing polymerisation and sealing the crack.

See: *composite materials*

1. White, S. R., N. R. Sottos, P. H. Geubelle, J. S. Moore, M. R. Kessler, S. R. Sriram, E. N. Brown and S. Viswanathan. Autonomic healing of polymer composites. *Nature* (2001) 409, 794–7.
2. Ghosh, S. K. Self-healing materials: fundamentals, design strategies and applications. In Self-Healing Materials: Fundamentals, Design Strategies and Applications, pp. 1–28. Berlin: WWiley-VCH, 2009.

semiconducting-grade silicon

The early decades of the 21st century are an age of digital electronics that rely on the production of integrated circuits. Manufacture of the latter requires chemically pure silicon wafers. The Czochralski technique is used to produce single crystals of silicon.[1] In the technique, a seed crystal is rotated, inserted into a bath of molten silicon and slowly withdrawn. Silicon atoms deposit onto the rotating crystal to form the desired orientation. Silicon wafers are then cut from the silicon ingot.

See: *integrated circuits*

1. Bolton, W., and R. A. Higgins. *Materials for Engineers and Technicians*, 6th edn, p. 334. London: Routledge, 2015.

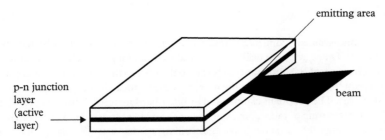

Fig. G.25 *Schematic diagram of a semiconductor laser. The beam is emitted from a thin* p–n *junction active layer.*

(R. J. D. Tilley. Understanding Solids, *2nd edn, p. 460. Chichester: Wiley, 2013.)*

semiconductor lasers

Solid-state semiconductor lasers consist of a *p—n* junction under forward bias and are analogous to light-emitting diodes (LEDs); the *n*-type region is heavily doped (Fig. G.25).[1] Electrons and holes recombine as in a LED but under an increasing current, a population inversion occurs so that the number of electrons in the conduction band (*n*-type region) is greater than in the valence band. Stimulated emission takes place when the population inversion is achieved. Emission comes from the material in the junction not from a dopant.

See: *lasers;* p–n *junction diodes*

1. Tilley, R. J. D. *Understanding Solids: The Science of Materials.* Chichester: Wiley, 2013.

semiconductors

Semiconductors have an electrical conductivity of around 10^5–10^{-7} siemens per metre (S m^{-1}), intermediate between that of a conductor (up to10^9 S m^{-1}) and an insulator (as low as 10^{-16} S m^{-1}).[1,2] For example copper, a conductor, has a conductivity of approximately 10^8 S m^{-1}; polystyrene, an insulator, has conductivity in the region of 10^{-14} S m^{-1}; while silicon, a semiconductor, has a conductivity of approximately 10^{-3} S m^{-1}. While individual atoms have discrete energy levels, the proximity of atoms in a crystalline solid allows orbitals of their electrons to overlap and spread out to occupy energy bands. Such a band model applies to crystalline semiconductors such as silicon and germanium. The electrical conductivity of semiconductors increases with temperature as the number of electrons promoted across the band gap to the conduction band from the valence band increases.

See: *band gaps; intrinsic semiconductors; extrinsic semiconductors*

1. Daintith, J. (ed.). *A Dictionary of Physics,* 6th edn. Oxford: Oxford University Press, 2009.
2. Tilley, R. J. D. *Understanding Solids: The Science of Materials.* Chichester: Wiley, 2013.

shale gas

Shales are fine-grained sedimentary rocks that can contain oil reserves or natural gas (methane).[1,2] They have been drilled by vertical wells and injecting chemicals in solution and a propping agent (i.e. proppant) to break up the rock so that entrapped gas is released. The proppant 'props up' fissures in the rock, preventing their closure and allowing release of the shale gas. The proppant can be sand or other hard material. This process is known as hydraulic fracturing or more commonly it is referred to as 'fracking'. Vertical shafts can be turned through 90° to allow horizontal drilling so that a large area of land can be explored from a single well. The technique has been widely used in the USA and there is interest in using fracking in the United Kingdom. However, the technique has become controversial and environmental concerns have been raised by opponents relating to the large volumes of water that are injected into the wells and the possible effects of this water on the quality of drinking water and on possible contamination of water supplies by methane.

See: *fracking; proppants*

1. Hester, R. E., and R. M. Harrison (eds). *Fracking, Issues in Environmental Science and Technology*, vol. 39. Cambridge: Royal Society of Chemistry, 2015.
2. Bunger, J. W. Oil shale and tar sands. In D. S. Ginley and D. Cahen (eds), *Fundamentals of Materials for Energy and Environmental Sustainability*, pp. 127–36. Cambridge: Cambridge University Press, 2012.

shape memory alloys

Shape memory alloys exhibit a reversible martensitic–austenitic phase transformation and remain malleable in the low-temperature martensitic form but revert to their original stiff shape in the austenitic form above the transformation temperature.[1,2] They can exhibit superelasticity when subjected to a stress-induced martensitic transformation. Nitinol, an alloy of nickel and titanium (approximately 50% nickel), has a transformation temperature in the range 10–20°C and was identified by Buehler and Wiley;[3] the alloy is made by metallurgical processes of melting, hot forging and rolling. Nitinol has good erosion, corrosion and wear resistance, is biocompatible and has applications as the material for implantable coronary stents. Other applications include connectors for pipes in aircraft hydraulic systems producing a tight fit between components, valves, seals, actuators and jewellery. When used in coronary stents the transformation temperature is below the body temperature where the martensitic phase is present and the stent is inserted in a compressed state, after which it expands to its final shape that has the austenitic phase. Forces produced due to volume changes in the martensite–austenite transformation can be utilised in switches in, for example, electric kettles. Other alloys that exhibit the shape memory effect include those based on copper–zinc–aluminium, coppe–aluminium–nickel, iron–manganese-silicon, gold–cadmium and iron-manganese-aluminium-nickel compositions.[4]

See: *biodegradable stents; nitinol; shape memory polymers; superelastic materials*

1. Noebe, R. D., S. L. Draper, M. V. Nathal and E. A. Crombie. Precipitation hardenable high temperature shape memory alloy. United States Patent 7,749,341, granted on 6 July 2010.
2. Yamauchi, K., K. Mori and S. Yamashita. Balloon expandable superelastic stent. United States Patent 7,658,761, granted on 9 February 2010.
3. Buehler, W. J., and R. C. Wiley. Nickel-based alloys. United States Patent 3,174,851, granted in March 1965.
4. Lecce, L., and A. Concilio (eds). *Shape Memory Alloy Engineering for Aerospace, Structural and Biomedical Applications*. Oxford: Butterworth-Heinemann, 2015.

shape memory hydrogels

Hydrogels can absorb 10 times their dry weight as fluid and have applications in personal sanitary towels. Smart memory polymers have been used in mattresses and cushions. Shape memory hydrogels that contain features of hydrogels and shape memory polymers have been used in sanitary pads such as those marketed under the trade name of 'SmartFoam always infinity'.[1,2]

See: *hydrogels; shape memory polymers*

1. Segal, D. *Exploring Materials through Patent Information*. Cambridge: Royal Society of Chemistry, 2014.
2. *You Magazine*, p. 68, 12 April 2015, available at http://You.co.uk.

shape memory polymers

Shape memory polymers are analogous to shape memory alloys and have mechanical properties that occupy the region between the elastic zone and plastic deformation in the stress–strain curve.[1] Applications include energy absorbing sportswear and self-healing materials. A shape memory polymer based on poly (e-caprolactone) has potential application in facial reconstruction surgery.[2] Tempur is a viscoelastic shape memory foam developed in the 1970s to improve g-force protection and comfort in seating in spacecraft.[3] It has wider commercial applications including cushions and mattresses.

See: *shape memory alloys; Tempur*

1. Hu, J. Shape Memory Polymers: Fundamentals, Advance and Applications. Shrewsbury: Rapra, 2014.
2. O'Driscoll. C. Bone filler for faces. *Chemistry and Industry* (Sept. 2014) 78, 13.
3. *Chemistry World* (2015) 12 (4), 34.

shark skin

The skin of sharks has small surface projections known as denticles about 0.2 mm in length that are aligned in the same direction and have surface grooves in the direction parallel to the flow of water across the skin.[1] This surface texture seems to prevent the adhesion of bacteria perhaps due to the surface curvature. Whether the denticles improve the swimming capability of sharks has not yet been evaluated. A material Sharklet AF that mimics the surface texture of shark skin with potential applications in preventing the build-up of pathogens or algae on surfaces has been developed.

See: *biofilms*

1. Lee, M. (ed.). *Remarkable, Natural Material Surfaces and Their Engineering Potential,* p. 21. London: Springer, 2014.

sialons

Replacement of some silicon by aluminium in the crystal lattice of β-silicon nitride (β-Si_3N_4), together with substitution of some nitrogen by oxygen in the crystal lattice to maintain charge neutrality, produces solid solutions in the Si-Al-O-N system known as β'-sialons with structures identical to that of with β-Si_3N_4.[1,2] Their mechanical behaviour is similar to that of silicon nitride with features of aluminium oxide but sialons have a stronger Al-O bond strength than the oxide. β'-Sialons retain their hardness at higher temperature than does alumina and can be densified by pressureless sintering. They are tough with high-temperature and thermal-shock resistance and applications include cutting tool materials, dies for drawing wires and tubes, thermocouple sheaths, crucibles and other areas where temperatures are as high as around 1250°C.

See: *silicon nitride*

1. Jack, K. H. Sialons. In R. J. Brook (ed.), *Concise Encyclopaedia of Advanced Ceramic Materials*, pp. 411–16. Oxford: Pergamon Press, 1991.
2. Somiya, S. (ed.). *Handbook of Advanced Ceramics*. San Diego, CA: Academic Press, 2013.

silicon-based solar cells

In a solar or photovoltaic cell, photons are absorbed by a semiconductor, a process that excites electrons from the semiconductor's valency band across the band gap into the conduction band.[1] Electron–hole pairs are separated in the solar cell to generate a photocurrent and photovoltaic effect (Fig. G.26). Single-crystal silicon has been used as a rigid substrate for photovoltaic panels with efficiencies up to 20%: the efficiency is the percentage of sunlight striking the cell that generates electricity. Amorphous silicon (a-Si) can be deposited from gas phase reactants onto a low-cost flexible backing material such as plastic or stainless steel. Silicon is an indirect band gap semiconductor, which means it is not an efficient absorber of light so that a thick layer of single crystal is required for its use in solar cells. Efficiencies for amorphous silicon thin-film solar cells are around 10% but they are cheaper to produce than cells based on single-crystal silicon.

See: *amorphous silicon*

1. Eisberg, N. Solar bonanza. *Chemistry & Industry* (Nov. 2014) 78, 38–41, 2014.

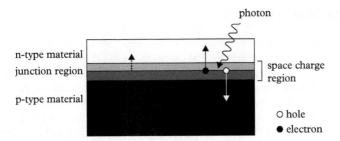

Fig. G.26 *Schematic diagram for a solar cell. Sunlight falling on the cell generates electron–hole pairs. Electrons move from the* p *to the* n *region and holes from the* n *to the* p *region in a drift process. A photovoltage and photocurrent is produced as the* p *region becomes more positive and the* n *region more negative.*

(R. J. D. Tilley. Understanding Solids, *2nd edn, p. 484. Chichester: Wiley, 2013.)*

silicon carbide fibre composites

Silicon carbide fibre is produced as a continuous length or as a woven fabric. Fibre-reinforced silicon carbide composites can be fabricated in which the matrix is silicon carbide by infiltrating a SiC woven mat from, for example, the gas phase by chemical vapour infiltration (CVi). Lightweight SiC/SiC composites have potential applications in gas turbines as replacements for superalloys as they have higher operating temperatures.[1] The addition of fibres to a matrix increases the toughness of the component, which is the resistance to fracture.

See: *silicon carbide fibres*

1. Bansal, N. P., and J. Lamon (eds). *Ceramic Matrix Composites: Materials, Modeling and Technology.* New York: Wiley, 2015.

silicon carbide fibres

Continuous polycarbosilane fibres can be drawn from solutions or by melt-spinning.[1,2] Nicalon (Nippon Carbon) is a β-silicon carbide fibre manufactured as a continuous fibre and continuous yarn. This fibre is one of the strongest materials known with a tensile strength of around 3.0 GPa and has applications in reinforced ceramic or metallic matrices for use at elevated temperatures where high strength and fracture toughness are required along with low thermal expansion and density. The fibre has a maximum operating temperature of approximately 1450°C. A similar continuous fibre is Tyranno (Ube Industries) that is melt-spun from a polytitanocarbosilane with a maximum operating temperature approaching 1900°C.

See: *polycarbosilanes*

1. Bansal, N. P., and J. Lamon (eds). *Ceramic Matrix Composites: Materials, Modeling and Technology.* New York: Wiley, 2015.
2. Somiya, S. (ed.). *Handbook of Advanced Ceramics.* San Diego, CA: Academic Press, 2013.

silicon carbide LEDs

Silicon carbide (SiC) does not occur in nature and was first synthesised in the early part of the 19th century. Its commercial production from the reaction of sand with coke at around 2700°C was developed by Acheson in 1892 and the material is widely used as an abrasive grit in cutting and grinding wheels while as an electrical conductor it is used as a heating element. Silicon carbide is an indirect band gap semiconductor and electroluminescence was first observed in this material. It emits blue light but is not as bright as blue light emissions from direct band gap semiconductors such as gallium nitride.[1]

See: *electroluminescence; indirect band gap semiconductors*

1. Johnstone, B. *Brilliant!: Updated edition,* p. 136. Prometheus Books, 2015.

silicon chips

A silicon chip is a piece of single crystal of semiconducting silicon that has been doped with impurity atoms and contains electronic circuits for carrying out logical operations in an algorithm. Silicon chips can contain millions of transistors, resistors and capacitors that are produced by photolithographic techniques and form parts of an integrated circuit that form the core of all electronic devices (Fig. G.27). The smallest feature that can be obtained in current lithographic processes using argon fluoride excimer gas lasers (wavelength 193 nm) is approximately 32 nm where this dimension refers to the width of deposited lines on the wafer. This linewidth is approximately equal to the spacing between electrodes on the chip. Continual reduction of the minimum feature size (here 32 nm) allows more transistors to be packed onto the chip, which increases the computing power of the device.[1,2] Reducing the minimum feature size allows the size of the chip to be reduced with the same number of transistors. Assuming that the minimum feature size corresponds to the length of one transistor, then about 2 billion (2×10^9) transistors can be packed onto a silicon wafer of the size of 1 cm^2.

See: *amorphous silicon; integrated circuits; Moore's law; photoresists;* n-*type semiconductors;* p-*type semiconductors*

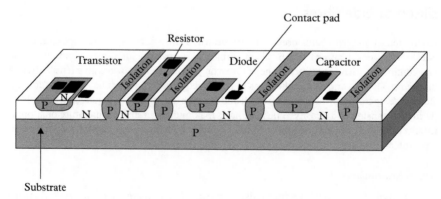

Fig. G.27 *Basic components of integrated circuits where the dark areas are contact pads.* (R. E. Hummel. Electronic Properties of Materials, *4th edn, p. 164. Berlin: Springer, 2013.)*

1. Boysen, E., and N. Boysen. *Nanotechnology for Dummies*, 2nd edn. New York: Wiley, 2011.
2. Askeland, D. K., and W. J. Wright. *The Science and Engineering of Materials*, 7th edn. Boston, MA: Cengage Learning, 2011.

silicones

Silicones are polymers that contain alternating silicon and oxygen atoms $(-Si-O-)$ atoms in the polymer backbone.[1,2] An example of a silicone is polydimethylsiloxane. Silicones can be water-soluble or water-insoluble. Applications for silicones include foaming agents for toothpastes and hair shampoos, sealants for bathroom tiles and aquaria and potting agents for securing electrical wires and cables. Low-molecular-weight silicones are liquids known as silicone oils. Fluorosilicones include poly [methyl (3,3,3-trifluoropropyl) siloxane], which is a fluorosilicone elastomer; fluorosilanes are not formally fluoropolymers. Fluorosilicone elastomers have applications as seals, gaskets and valves in the auto and aerospace industries as they have good resistance to oil and fuel as well as good thermal stability.

See: *fluoropolymers*

1. Smith, D. W. Jr, S. T. Iaccono and S. S. Iyer (eds). *Handbook of Fluoropolymer Science and Technology*. Chichester: Wiley, 2014.
2. Fried, J. R. *Polymer Science and Technology*, 3rd edn. Upper Saddle River, NJ: Prentice Hall, 2014.

silicon nitride

Silicon nitride, Si_3N_4, occurs in two phases, the α and β forms.[1] It can be sintered to dense components by addition of basic additives such as alumina or magnesium oxide that give rise to liquid phases which form a glassy network that cements grains of silicon nitride together. Both silicon nitride and sialons are structural or engineering ceramics and applications for the former include engine components and cutting tools.

See: *sialons*

1. Somiya, S. (ed.). *Handbook of Advanced Ceramics*. San Diego, CA: Academic Press, 2013.

silver nanoparticles

Silver has been used as a microbial agent in particular silver ions as silver nitrate and sulfadiazine for treatment of burns and bacterial infections.[1,2] Silver nanoparticles can be produced by chemical reduction of silver salts in aqueous and organic solution. Examples of chemical reducing agents include sodium citrate, sodium borohydride, hydroxylamine hydrochloride, ascorbic acid and ethylenediaminetetracetic in the presence of a stabilising agent to prevent aggregation of silver nanoparticles. Stabilising agents include polyacrylic acid and polygalacturonic acid. Silver nanoparticles have a diameter less than 50 nm.

See: *colloidal systems; nanoparticles*

1. Zhang, J., and C. Shachaf. Methods of making silver nanoparticles and their applications. International Patent Applications WO 2014/052973, published on 3 April 2014.
2. Tani, T. *Silver Nanoparticles*. Oxford: Oxford University Press, 2015.

small molecule OLEDs

Semiconducting conjugated organic polymers that exhibit electroluminescence have potential for use in displays such as in televisions and smartphones. Polymers are particularly suited for flexible displays. Small organic molecules can exhibit electroluminescence and can be used in principle for organic light-emitting diodes (OLEDs).[1,2] For example, aluminium quinolate can be evaporated under vacuum to a conducting thin layer around 0.2 µm thick on an electron–hole layer that is deposited by vacuum evaporation onto an indium tin oxide electrode on a glass substrate. An electrode that is deposited on top of the aluminium quinolate layer completes the construction of the light-emitting diode.

See: *conducting polymers; poly-(p-phenylenevinylene)*

1. Geoghegan, M., and G. Hadziioannou. *Polymer Electronics*. Oxford: Oxford University Press, 2013.
2. Solymar, L., D. Walsh and R. R. A. Syms. *Electrical Properties of Materials*, 9th edn, p. 427. Oxford: Oxford University Press, 2014.

smart materials

Smart materials, also referred to as intelligent materials, are materials that can sense and respond to their environment and change their properties to respond to the environmental conditions.[1,2] Smart materials include piezoelectric, electrochromic, thermochromic and photochromic materials.

1. Bolton, W., and R. A. Higgins. *Materials for Engineers and Technicians*, 6th edn, p. 427. London: Routledge, 2015.
2. Aguilar, M. R., and J. S. Roman (eds). *Smart Polymers and their Application*. Cambridge: Woodhead, 2014.

smartphones

Smartphones are ubiquitous in everyday life and while users will be aware of the plastic cases and glass screens, materials inside the smartphone are hidden from view.[1,2] These materials include indium tin oxide used as a transparent electrically conducting coating on the inside of the screen, lithium-ion batteries and lanthanide elements, for example europium, terbium and praseodymium that are used as phosphors to generate colours on the screens. Some materials used in smartphones are critical materials, for example gallium, indium, platinum, gold, iridium, palladium, neodymium, gadolinium, erbium, terbium and lithium. Smartphones contain accelerometers and magnetometers that are fabricated by methods used for the manufacture of integrated circuits.

See: *critical materials*

1. Ashby, M. F., D. F. Balas and J. S. Coral. *Materials and Sustainable Development.* Oxford: Butterworth, 2016.
2. Lawler, R. What's in your laptop? *Materials World* (June 2013) 21, 31.

smart textiles

Smart textiles, also referred to as smart clothing, have a function other than the traditional role of textiles.[1] For example, a thin layer of titanium dioxide on the surface of textile fibres can photocatalytically oxidise stains and dirt attached to clothing to carbon dioxide and water. Absorption of ultraviolet light by titanium dioxide produces an excited electronic state in the oxide that interacts with oxygen in the air to produce reactive free radicals that are responsible for oxidation of the stains and dirt on clothing.

1. Norman, N. Clothing gets smart. *Chemistry World* (Oct. 2012) 9, 58–61.

smectic phases

Liquid crystals can form smectic phases,[1] in which molecules are arranged side by side in layers and the molecules that are elongated or have a cigar-like shape are perpendicular to the plane of the layers (Fig. G.28).

See: *cholesteric phases; liquid crystals; nematic liquid crystals*

1. Kawamoto, H. The history of liquid crystal displays. *Proceedings of the IEEE* (Apr. 2002) 90, 460–500.

SMOLED

SMOLED is an acronym for a small molecule organic light-emitting diode that emits light by electroluminescence and which has lighting applications, for example in displays for smartphones and in television screens.[1] Examples of SMOLED are anthracene, naphthalene and polyaromatic amines. A feature of these molecules is that they are conjugated.

See: *electroluminescence; poly-(p-phenylenevinylene)*

1. Hack, M., J. J. Brown, P. Levermore and M. S. Weaver. Organic light emitting device lighting panel. United States Patent Application US 2011/0284899, published 24 November 2011.

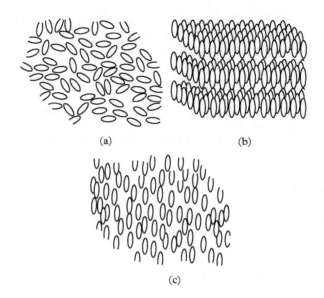

Fig. G.28 *Schematic representation of molecular alignments in liquid and liquid crystals: (a) liquid, (b) smectic and (c) nematic.*

(H. Kawamoto. The history of liquid crystal displays. Proceedings of the IEEE *(2002) 90, 460–500.)*

soft contact lenses

Soft contact lenses[1] are mainly hydrogels made out of cross-linked hydrophilic monomers or polymers and have been made by casting a pre-polymer mixture into a mould using two mould halves, pressing the two halves together and then curing the reactants by, for example, exposure to ultraviolet light that promotes polymerisation.

See: *hydrogels*

1. Segal, D. *Exploring Materials through Patent Information.* Cambridge: Royal Society of Chemistry, 2014.

sol-gel processes

Sol-gel processes are a set of techniques for preparing ceramic materials, particularly advanced ceramics.[1] The key feature of the methods is that they involve use of wet chemical techniques rather than the conventional preparation of ceramics that mix and mill dry powders. Ceramics can be made in the form of powders, coatings, fibres and monoliths by sol-gel processing, often with compositions that cannot be prepared by conventional methods. The techniques often involve handling of colloidal dispersions or alcoholic solutions of metal alkoxides.

See: *colloidal systems; insulating fibres*

1. Segal, D. *Chemical Synthesis of Advanced Ceramic Materials*. Cambridge: Cambridge University Press, 1989.

solid-state lighting

About 20% of the total electric energy production in the USA is consumed for lighting applications.[1] Energy-efficient solid-state lighting based on light-emitting diodes is rapidly replacing other lighting sources such as incandescent bulbs and fluorescent tubes in domestic, industrial and environmental applications such as street lamps. This replacement is anticipated to continue in the 21st century. In addition, the proliferation of portable devices such as smartphones and tablet computers helps to drive innovation in lighting technology, whether in light-emitting diodes, quantum dots, liquid crystals or organic light-emitting diodes (OLEDs).

1. Hack, M., M.-Hao, M. Lu and M. S. Weaver. Organic light-emitting devices for illumination. United States Patent Application 2011/0284899, 2011.

sphene

The development of nuclear power in the 20th century has resulted in the generation of high-level radioactive wastes that are currently stored as solutions. One way of immobilisation of these wastes is to incorporate radionuclides into crystalline phases from which they will not leach out. An example[1] of such a phase is sphene ($CaTiSiO_5$). Other crystalline phases have been suggested for immobilisation including zircon and Synroc, a combination of zirconolite ($CaZrTi_2O_7$), perovskite ($CaTiO_3$) and hollandite ($BaAl_2Ti_6O_{16}$).

See: *glass ceramics*

1. Reeve, K. D. Nuclear waste storage material. In R. J. Brook (ed.), *Concise Encyclopaedia of Advanced Ceramic Materials*, pp. 331–5. Oxford: Pergamon Press, 1991.

spider silk

The 20th century saw considerable research and development on synthetic fibres, such as nylon and Kevlar. Spider silk proteins[1,2] used by spiders for constructing spider webs form strong and elastic fibres about 1 μm (micron) in diameter that is one-thousandth of a millimetre as can be observed by how insects are trapped in the webs; the proteins are oriented in the fibre, effectively forming liquid crystals. Glycine, an amino acid, is a major component of spider silk. In the 21st century attention has turned to the use of recombinant DNA technology to express proteins, for example from transgenic tobacco.

See: *recombinant DNA technology; ZMapp*

1. Lee, M. Spider silk: a sticky situation. In M. Lee (ed.), *Remarkable Natural Material Surfaces and Their Engineering Potential*, pp. 145–57. New York: Springer, 2014.

2. Ozin, G. A., A. C. Arsenault and L. Cademartiri. *Nanochemistry: A Chemical Approach to Nanomaterials*, p. 680. Cambridge: Royal Society of Chemistry, 2009.

spin-on-glass

The manufacture of integrated circuits on a semiconductor wafer involves building up layers of interconnected surface devices using metal wiring.[1] The surface topography of each layer can be irregular and uneven. Electrically insulating layers of glass are deposited onto the layers, for example by spin coating to produce even surfaces that electrically isolate neighbouring layers of devices. Materials used to produce the insulating layers are known as spin-on-glasses.

See: *polysilazanes*

1. Lee, J.-H., J.-H. Cho, J.-S. Choi and D.-J. Lee. Spin-on glass composition and method of forming silicon oxide layer in semiconductor manufacturing process using the same. United States Patent 7,270,886, granted on 18 September 2007.

spintronic materials

Computers based on the use of silicon chips and integrated circuits rely on the transfer of electric charge (through charging and discharging of capacitors) to carry out logic operations using binary notation and the use of transistors as switches that recognise 0 and 1 in the circuits. Spintronic materials utilise the quantum mechanical property of electron spin for potential applications in computers for carrying out logic operations.[1]

1. Solymar, L., D. Walsh and R. R. A. Syms. *Electrical Properties of Materials*, 9th edn, p. 287. Oxford University Press, 2014.

sporting equipment

Materials used in sporting equipment have the potential to give the user a competitive advantage. Fibre-reinforced polymer composites have been used in tennis rackets so that the racket is lightweight with high strength. Incorporation of graphene flakes into a polymer blend used for a fibre-reinforced composite can potentially increase the compressive strength of the composite when it is moulded into the shape of a tennis racket. Fibre-reinforced composites have been used in the construction of skis and surfboard fins.[1,2]

1. Felice, M. A slippery slope. *Materials World* (Jan. 2013) 21, 54–5.
2. Perkins, J. Materials in sport: surfing. *Materials World* (Aug. 2015) 23, 24–5.

spray-drying

Many materials in particular ceramics are used in the form of powders. For example, in thermal spraying spherical powders have good flow properties for passing through a high-temperature plasma source onto the substrate. In spray-drying,[1] a slurry or colloidal dispersion is thrown off a rotating disc that is kept at an elevated temperature around 200°C. Water evaporates from the liquid droplets, producing dry spherical powders.

See: *thermal barrier coatings*

1. Buzak, S., and D. Rende. *Colloid and Surface Chemistry: A Laboratory Guide for Exploration of the Nano World*, pp. 74–5. Boca Raton, FL: CRC Press, 2014.

stainless steel

Stainless steel was first developed at the Krupp steelworks in 1912.[1] A minimum content of around 12 wt% chromium in steel together with a small carbon content conferred rust and stain resistance. Stainless steel cutlery was first produced in 1913 at the Portland Works in Sheffield. The 18/8 alloy containing 18 wt% chromium, 8 wt% nickel, 0.08 wt% carbon is widely used in chemical plant and domestic equipment such as refrigerators.

See: *steel*

1. Tindell, H. *Engineering Materials*, p. 15. Marlborough, UK: Crowood Press, 2014.

stain-resistant clothing

The hierarchical surface structure of leaves in the lotus plant has been used as a model for fabrication of stain-resistant clothing by incorporation of microscopic hydrophobic whiskers onto the surface of textile fibres.[1] The whiskers attach to the fibres in the fabric and prevent contact between the fabric so that water droplets then roll off the fabric. The fabric is sold under the name Aquapel and manufactured by Nano-Tex.[2]

See: *superhydrophobic materials*

1. Lee, M. Lotus leaves: humble beauties. In M. Lee (ed.), *Remarkable Natural Material Surfaces and Their Engineering Potential*, p. 62. New York: Springer 2014.
2. Nano-tex. Available at http://www.nanotex.com

starch

Starches are polysaccharides found in plants, straw, fruit and shells as examples and are often used as animal feeds.[1–3] Starches are energy sources for animals and yield glucose on digestion. Hence starches can be viewed as a form of biomass and a potential source of biofuels. Starch consists of a mixture of amylose and amylopectin; amylose is soluble in hot water, whereas amylopectin remains insoluble. Amylose is a linear polysaccharide containing between 100 and 1000 linked glucose molecules while amylopectin is a polysaccharide containing branched chains of glucose molecules. The group of amylase enzymes cleaves the polysaccharide chains producing glucose and the sugar maltose that consists of two linked glucose molecules. Maltose occurs in barley seeds and is critical for the manufacture of beer and malt whisky. Mesoporous carbonised starches (pores with diameter 2–50 nm), referred to as Starbons, have potential applications[4] in liquid-phase chromatographic separations, catalyst supports for catalytic hydrogenation of succinic acid to 1,4-butanediol and butyrolactone as well as adsorbents, including removal of dyes from wastewaters.

See: *biomass; carbohydrates; monosaccharides; polysaccharides; starch nanoparticles*

1. Daintith, J. (ed.). *A Dictionary of Chemistry*, 6th edn. Oxford: Oxford University Press, 2008.
2. Atkins, P. *Atkins' Molecules*, 2nd edn, pp. 115–17. Cambridge: Cambridge University Press, 2003.
3. Goldstein, D. (ed.). *The Oxford Companion to Sugar and Sweets*, pp. 653–4. Oxford: Oxford University Press, 2015.
4. Clark, J. A star is born. *Chemistry & Industry* (July 2013) 77, 115–17.

starch nanoparticles

Starch is a renewable material and starch nanocrystals (i.e. crystalline nanoparticles) can be extracted, for example by treatment of waxy corn starch with sulphuric acid; the nanoparticles had a size around 50 nm and an aggregate size of approximately 1–5 µm. Biodegradable sheets and bags for potential use as food packaging have been prepared from a mixture of glycerol, starch and starch nanoparticles.[1]

See: *starch*

1. Smith, C., et al. Biodegradable, biocompatible and non-toxic material, sheets consisting of said material and the use thereof in food, pharmaceutical, cosmetic and cleaning products. United States Patent Application 2013/0034638, published on 7 February 2013.

statins

Cardiovascular disease accounts for about 30% of deaths worldwide, more than any other disease, and is associated with the build-up of cholesterol in the body.[1] Cholesterol is a steroid and its biosynthesis *in vivo* involves the conversion of mevalonate to squalene that is then converted to cholesterol. Mevalonate is produced *in vivo* by the catalytic action of the enzyme hydroxymethylglutaryl coenzyme A reductase (abbreviated to HMGR) and the formation of mevalonate is the rate-determining step in cholesterol biosynthesis. Human HMGR consists of a polypeptide chain of 888 amino acids. Statins are small molecules that occupy the binding sites of HMGR and limits the production of mevalonate.[2] Examples of statins are Crestor (rosuvastatin calcium) and Lipitor(atorvastatin calcium).

1. Tyrell, J. A. *Fundamentals of Industrial Chemistry: Pharmaceuticals, Polymers and Business.* New York: Wiley, 2014.
2. Cordes, E. K. *Hallelujah Moments: Tales of Drug Discovery*, p. 123. Oxford: Oxford University Press, 2014.

stealth materials

Stealth materials are radar-absorbing materials and have the ability to minimise the radar profile of military equipment.[1] Hexagonal ferrites, or hexaferrites, such as those based on barium hexaferrite ($BaFe_{12}O_{19}$) have high-frequency properties in the microwave region of the electromagnetic spectrum that give the materials potential as stealth materials.

1. Pullar, R. A magnetic future. *Materials World* (May 2013) 21, 10.

steel

Iron and steel are essential for modern societies and annual worldwide production of steel is around 10^9 tonnes. Low-carbon steels (less than 0.06% carbon) are used for automobile bodies, cans and appliances while higher carbon contents provide greater strength. Hence steel tools contain around 1.2% carbon but some steels contain very little carbon. By definition steels are iron-based alloys.[1] Introduction of the Bessemer process in 1856 allowed low-carbon steel to be produced quickly but nowadays this technique has been replaced by the basic oxygen process.

See: *recycling metals; stainless steel*

1. Hosford, W. F. *Iron and Steel.* Cambridge: Cambridge University Press, 2012.

stereolithography

Stereolithography is an automated process for fabrication of three-dimensional objects by exposure of layers of a photopolymer liquid to visible or ultraviolet light that causes layers to solidify and adhere to each other.[1] Two-dimensional layer data representing cross-sections of the object are stored on a computer and the movement of the light source (e.g. a laser) is determined by the data sets so that the object is built up layer by layer. Stereolithography is a three-dimensional printing technique.[2] Pioneering work on stereolithography was carried out by Charles Hull, who proposed this word.[3]

See: *photopolymers*

1. Partanan, J. P., and D. R. Smalley. Apparatus and method for controlling exposure of a solidifiable medium using a pulsed radiation source in building a three-dimensional object using stereolithography. United States Patent 6,215,095, granted on 10 April 2001.

2. Hausman, K., and R. Horne. *3D Printing for Dummies*. New York: Wiley, 2014.
3. Hull, C. W. Apparatus for production of three-dimensional objects by stereo-lithography. United States Patent 4,575,330, 1986.

styrene-acrylonitrile-butadiene

The thermoplastic styrene-acrylonitrile-butadiene (ABS) copolymer resin is prepared from polymerisation of styrene, acrylonitrile and butadiene, for example by using emulsion polymerisation in suspensions of reactants. Acrylonitrile imparts strength, butadiene imparts impact resistance and styrene aids processing of the polymer.[1] Injection-molded ABS resins have many applications: the body shells of office equipment such as computers, home appliances, automotive applications, ducts to contain electrical wiring and toys as examples. The mechanical properties of ABS can be varied by changing the relative amounts of the components.

See: *polystyrene; thermoplastics*

1. Lee, W. K., B. D. Kim, C. D. Jung and S. H. Jin. Flame resistant thermoplastic resin composition. United States Patent Application 2012/0172502, published on 5 July 2012.

styrene-butadiene rubber

Styrene-butadiene thermoplastic elastomers are the major polymer used as artificial rubber in tyres for cars and other vehicles.[1] In a styrene-butadiene block copolymer, elastic behaviour does not depend on cross-linking of butadiene chains. Spherical domains of glassy styrene at the end of butadiene repeat units between the styrene domains are rigid and hold the chains together. The butadiene polymer regions are above the glass transition temperature at ambient temperatures and induce soft rubbery properties to the elastomer.

See: *living polymers; tyre compositions*

1. Askeland, D. K., and W. J. Wright. *The Science and Engineering of Materials*, 7th edn, p. 602. Boston, MA: Cengage Learning, 2011.

subtractive colour mixing

Colours are an integral part of everyday life, whether in the printed medium of newspapers and magazines, in textiles or in electronic displays such as in mobile phones or in television and computer screens. Visible light contains a narrow range of wavelengths within the electromagnetic spectrum from around 360 to 780 nm. Violet has a wavelength in the region of 400 nm while red has a longer wavelength around 750 nm. Red, green and blue are known as primary additive colours as they cannot be obtained by mixing light of other colours (i.e. wavelength). Magenta, cyan and yellow are complementary colours that arise when one component of visible light is absorbed and the other components are transmitted or reflected. Thus, if blue light is absorbed, the reflected light contains red and green components that appear as yellow. Magenta, cyan and yellow are the subtractive primary colours[1] and are used for coloured printed material and in inkjet printers. Subtractive colour mixing is associated with dyes and pigments that selectively absorb primary colours. When magenta, cyan and yellow are simultaneously used on an object, then the object will appear black as all visible wavelengths are absorbed.

See: *additive colour mixing; laser printing inks*

1. Christie, R. M. *Colour Chemistry*, 2nd edn. Cambridge: Royal Society of Chemistry, 2015.

subtractive machining

The conventional method used in the production of components involves a finishing step of machining by means of milling, drilling and polishing so that parts meet requirements on shape, size and quality. Machining removes material from the component and this step is known as subtractive machining.[1] Hence waste material is produced and tools need to be replaced due to wear.

See: *additive manufacturing*

1. Andersson, C. Innovation in rapid prototyping and additive manufacturing. *European Medical Device Technology*, 2012 (January/February).

sugary foods

There is much debate in countries with public health services on the rising levels of obesity and the associated costs to those countries and the role of different types of food in contributing to obesity. If foods are considered to be materials then their components such as sugars[1,2] and carbohydrates can be defined in the same way as components in renewable resources such as wood and plants. Carbohydrates are compounds with the general formula $C_6(H_2O)_y$ and include the simple sugars (monosaccharides) that cannot be broken down into smaller units. Examples of monosaccharides are glucose ($C_6H_{12}O_6$) and fructose ($C_6H_{12}O_6$). Sucrose is a carbohydrate consisting of a molecule of glucose chemically linked to a molecule of fructose. Polysaccharides are carbohydrates with a larger molecular weight and complexity than simple sugars and consist of long chains of mono-saccharides, such as glucose. Examples of polysaccharides include starch and cellulose. Formally starches are not sugars even though they can contain chains of glucose molecules. About 210 g of granulated sugar can be dissolved in 100 g of water at room temperature and potential high concentrations of sugar in beverages may not be appreciated by the general public.

See: *carbohydrates; polysaccharides*

1. Goldstein, D. (ed.). *The Oxford Companion to Sugar and Sweets*, pp. 662–9. Oxford: Oxford University Press, 2015.
2. Atkins, P. *Atkins' Molecules*, 2nd edn, pp.108–13. Cambridge: Cambridge University Press, 2003.

sunscreens

Nanoparticles of titanium dioxide (TiO_2) are used in sunscreens[1,2] to protect the skin from ultraviolet (UV) light and to both reflect and, in particular, scatter UV light as it falls onto the skin. Nanoparticles of zinc oxide (ZnO) have also been used in sunscreens. Both types of nanoparticles can leave a clear film on the skin rather than a white residue.

See: *nanoparticles*

1. Boysen, E., and N. Boysen. *Nanotechnology for Dummies*, pp.156–7. Chichester: Wiley, 2011.
2. Ngo, C., and M. van de Voorde. *Nanotechnology in a Nutshell: From Simple to Complex Systems*. Amsterdam: Atlantis Press, 2014.

superabsorbent polymers

Superabsorbent polymers form hydrogels capable of absorbing aqueous fluids.[1] The hydrogels can be used in the form of particles around 250 μm in diameter. Monomers that form superabsorbent polymers include acrylic acid, methacrylic acid and vinylsulphonic acid.

See: *hydrogels; disposable nappies*

1. Hermeling, D., U. Stuven and U. Hoss. Super-absorbing hydrogel with specific particle size distribution. United States Patent Application 2004/0265387, published on 30 December 2004.

superalloys

Nickel-based superalloys are used as turbine blades in gas turbines in aircraft and power generation plants.[1] Oxidation and corrosion resistance, toughness, ductility and low rate of creep are desirable properties at operating temperatures around 1500°C. Blades are cast from single crystals eliminating grain boundaries, resulting in low creep. Intricate cooling channels are built into the blades. Three-dimensional printing is being explored for the fabrication of turbine blades.

See: *thermal barrier coatings; three-dimensional printing*

1. Felice, M. Materials for aeroplane engines. *Materials World* (May 2013), 52–3.

supercapacitors

Conventional capacitors store electric charge on their surfaces. Supercapacitors, also known as ultracapacitors or electrochemical capacitors, have properties intermediate between conventional capacitors and batteries in how they store electric charge.[1] The latter is stored in the double layer around the surface. Graphene has been considered as a component of a supercapacitor electrode.

1. Whittingham, M. S. Electrochemical energy storage: batteries and capacitors. In D. S. Ginley and D. Cahen (eds), *Fundamentals of Materials for Energy and Environmental Sustainability*, p. 618. Cambridge: Cambridge University Press, 2012.

superconducting metals and alloys

Superconductors are materials that have no electrical resistance. The phenomenon was discovered in 1911 when a superconducting transition or critical temperature (often referred to as T_c) of 4.2 K was observed for mercury in liquid helium. The value of T_c was raised to 23.2 K for alloys of niobium and germanium (Nb_3Ge) in 1973. Certain alloys, for example a niobium-titanium alloy in the form of a coil, when immersed in liquid helium, is used to generate intense magnetic fields (up to 2 T) used in magnetic resonance imaging (MRI). The explanation of this conventional superconductivity is explained by the BCS theory (after J. Bardeen, L. N. Cooper and R. Schrieffer) in which current in the superconductor is carried by bound pairs of electrons known as Cooper pairs.[1,2]

See: *density functional theory*

1. Daintith, J. (ed.). *A Dictionary of Physics*, 6th edn. Oxford: Oxford University Press, 2009.
2. Kresin, V., H. Morawitz and S. Wolf. *Superconducting State*. Oxford: Oxford University Press, 2013.

superconducting oxide ceramics

High-temperature oxide or ceramic superconductors that exhibit superconductivity were identified by Georg Bednorz and Karl Muller.[1] Transition temperatures for ceramic superconductors are higher than those for superconducting alloys, around 100 K compared to 22.3 K for niobium-germanium alloys, and exhibit loss of electrical resistance when immersed in liquid nitrogen at 77 K.[2] Examples of superconducting oxide compositions are based on yttrium barium copper oxide ($YBa_2Cu_3O_{6.95}$) and bismuth strontium calcium copper oxide ($Bi_2Sr_2Ca_2Cu_3O_{10}$) with values of T_c around 90 and 125 K, respectively. $Hg_{0.8}Tl_{0.2}Ba_2Ca_2Cu_3O_{8.33}$ has a T_c of 135 K. A recent application for ceramic superconductors is as a power cable to replace copper cabling in a power network in Essen, Germany.[3] Challenges for exploiting high-temperature superconductors are described in reference (4).

1. Bednorz, J. G., and K. A. Muller. Possible high-T_c superconductivity in the Ba-La-Cu-O system. *Zeitschrift fur Physik B* (1986) 64, 189–93.
2. Qiu, X. G. (ed.). *High-Temperature Superconductors*. Cambridge: Woodhead, 2011.

3. Frost, S. *Materials World* (Nov. 2014), 22, 16–17.
4. Melham, Z. (ed.). *High-Temperature Superconductors (HTS) for Energy Applications.* Cambridge: Woodhead, 2012.

superelastic materials

Superelastic materials are associated with the shape memory effect in alloys.[1] Here if a material such as nitinol is deformed at low temperature its original shape can be recovered by heating above a characteristic temperature. However, if the alloy is deformed at high temperature in the austenite phase, then its original shape can be recovered by removing the applied load. This property, that is the deformation, is known as superelasticity or pseudoelasticity. Nitinol can deform to nearly 8% strain, whereas most metals can deform to less than 1% strain.[1] The applied load induces the phase transformation from the austenite to martensite phase. Superelastic properties of shape memory alloys are exploited in dental braces and spectacle frames.

See: *nitinol; shape memory alloys*

1. Lecce, L., and A. Concilio (eds). *Shape Memory Alloy Engineering for Aerospace, Structural and Biomedical Applications.* Oxford: Butterworth-Heinemann, 2015.

superglue

Superglues, which are widely available to the general public, are based on cyanoacrylates, $CH_2=C(CN)COOR$, where R is an alkyl group.[1,2] The methyl and ethyl esters corresponding to $R=CH_3$ and $R=C_2H_5$ are components of superglues, in particular in Super Glue. These adhesives polymerise quickly in air in the presence of moisture and are very strong adhesives. Polymerisation also occurs in the presence of a base such as $N,N,$-dialkyl anilines.[1] Superglues can be compared to the weak tacky adhesives used in Post-it notes and biodegradable hydrogels used for surgical adhesives.

See: *bioadhesives; biodegradable hydrogels; epoxy adhesives; hairy adhesives; Post-it notes*

1. Huver, T., C. Nicolaison and S. Camp. United States Patent 5,561,198, 1995.
2. Walton, D., and P. Lorimer. *Polymers*, p. 72. Oxford: Oxford University Press, 2005.

superhydrophobic materials

Superhydrophobicity is a property of the surface of a material and superhydrophobic materials are very difficult to wet by water; that is, water does not spread easily over the surface.[1] Superhydrophobic surfaces have a contact angle greater than 140°. Many plants have leaves that are superhydrophobic, for example the leaves of the lotus plant.

See: *contact angle; self-cleaning glass*

1. Felice, M. Superhydrophobic materials. *Materials World* (Aug. 2013) 21, 50–1.

superlattice

A superlattice consists of an array of multiple quantum wells and superlattices have applications in quantum well solid-state lasers.[1] Thus, a superlattice consists of a periodic array of wide band gap and narrow band gap materials.[2] The periodicity is artificial and has the potential for engineering new electronic properties.

See: *quantum cascade lasers*

1. Tilley, R. J. D. *Understanding Solids: The Science of Materials*. Chichester: Wiley, 2013.
2. Hummel, R. E. *Electronic Properties of Materials*, 4th edn. London: Springer Science, 2011.

super twisted nematics

Super twisted nematic liquid crystals have a faster response time and wider viewing angle than twisted nematic liquid crystals when used in displays.[1] A qualitative understanding of super twisted nematics (STN) liquid crystals can be gained by an analogy with a twisted piece of elastic.[2] The more the elastic is twisted, the more it 'twangs' on release, thus a sharper response to an applied force. In a STN liquid crystal in a display the twist is increased from a quarter turn (90°) characteristic of a nematic liquid crystal display. STN liquid crystals have a faster speed of response to the applied voltage than nematic liquid crystals. In practice, STN liquid crystals produce an improved picture quality.

See: *nematic liquid crystals*

1. Kawamoto, H. The history of liquid crystal displays. *Proceedings of the IEEE* (2002) 90, 460–500.
2. Dunmur, D., and T. Sluckin. *Soap, Science and Flat-Screen TVs: A History of Liquid Crystals*. Oxford: Oxford University Press, 2014.

synthesis gas

Synthesis gas, or syngas, is a mixture of carbon monoxide and hydrogen, often prepared by steam reforming of methane in natural gas or partial oxidation of hydrocarbons.[1] Syngas is a feedstock for conversion to synthetic diesel using the Fischer–Tropsch process and a source of hydrogen. Synthesis gas can also be produced by coal gasification and when produced in this way was known as town gas for domestic and industrial use in Europe until the 1950s when it was replaced by natural gas.

See: *coal gasification; synthetic diesel*

1. Tyrell, J. A. *Fundamentals of Industrial Chemistry: Pharmaceuticals, Polymers and Business*, p. 17. New York: Wiley, 2014.

synthetic biology

Synthetic biology is a multidisciplinary field with contributions from natural product chemistry, materials science, engineering, biotechnology and genetic engineering.[1-3] Potential applications include medical diagnostics, biofuels, regenerative medicine and tailored microbes to produce medicines. In synthetic biology, biological structures are designed and manipulated to produce useful products.

1. Ginsberg, D., J. Calvert, P. Schyfter, A. Eflick and D. Enby. *Synthetic Aesthetics: Investigating Synthetic Biology's Designs on Nature.* Cambridge, MA: MIT Press, 2014.
2. Freemont, P. S., and R. I. Kitney (eds). *Synthetic Biology: A Primer.* London: Imperial College Press, 2012.
3. Baldwin, G., T, Bayer, R. Dickinson, T. Ellis, P.S. Freemont, R.I. Kitney, K. Polizzi and G.-B. Stan. *Synthetic Biology: A Primer.* Singapore: World Scientific, 2015.

synthetic diesel

Methane (CH_4) is often found in association with oil deposits and can be converted to a mixture of carbon monoxide and hydrogen by steam reforming in which steam reacts with methane at pressures up to 20 atm, often at around 2 atm, using a catalyst such as nickel or platinum. The resulting gaseous mixture is known as synthesis gas and can be converted in the Fischer–Tropsch reaction to synthetic diesel, that is to hydrocarbon liquids at around 300°C and 20 atm in the presence of a catalyst, for example cobalt.[1,2]

See: *associated gas*

1. Bowe, M. J., D. L. Segal, C. D. Lee-Tuffnell, D. C. W. Blaikley, J. A. Maude, J. W Stairmand and I. F. Zimmerman. Catalytic reactor. United States Patent 8,118,889, granted on 21 February 2012.
2. Maitlis, M., and A. de Klerk. *Greener Fischer-Tropsch Processes for Fuels and Feedstocks.* Chichester: Wiley, 2013.

synthetic rubber

The monomer isoprene is derived from petrochemical sources by thermal cracking of petroleum or naphtha feedstocks and is obtained in the *cis-* and *trans-* isomer forms.[1] Synthetic rubber obtained by polymerisation of *cis-*isoprene obtained from petroleum or naphtha feedstocks corresponds to *cis-*1,4-polybutadiene.

See: *natural rubber*

1. DeRosa, T. F. *Engineering Green Chemical Processes: Renewable and Sustainable Design.* New York: McGraw-Hill, 2015.

tar sands

Tar sand is a sandstone located near the Earth's surface and containing bitumen, a very viscous petroleum-based material that does not flow at the temperatures used to extract oil.[1] The bitumen has formed over millions of years from prehistoric organisms such as algae. Sources of tar sands are in Canada and Venezuela. Bitumen can be recovered by injection of steam into the sands that reduces its viscosity and the molecular weight of the bitumen is reduced by hydrocracking to produce useful petroleum products.

1. Bunger, J. W. Oil shale and tar sands. In D. S. Ginley and D. Cahen (eds), *Fundamentals of Materials for Energy and Environmental Sustainability*, pp. 127–36. Cambridge: Cambridge University Press, 2012.

tempur

Tempur is a viscoelastic shape memory polyurethane foam.[1]

See: *shape memory polymers*

1. *Chemistry World* (Apr. 2015) 12, 34.

textile printing

Coloured patterns and images are often printed onto textile fabrics and a manufacturing process involves rotary screen printing. In this process an image is produced on a plastic mesh by a photopolymerisation technique which is analogous to photolithography so that the image is contained in the open areas of the mesh. The mesh is placed in contact with the fabric and the image is built up by squeezing the dye through the gaps in the mesh. Additives are used in the dye compositions to ensure the pigments adhere to the fibres. While indigo is an established vat dye, the technique of indigo printing, a direct printing processes known as Pencil Blue or China Blue for producing white on blue or blue on white patterns, was developed in the 19th century.[1] In the former technique brushes are used to apply reduced indigo pastes to the fabric while in the China Blue method indigo pastes are printed onto the fabric and subjected to a series of reductions in a bath of ferrous sulphate solution followed by air oxidation. The name China Blue arose as the final appearance of the printed fabric was similar to Chinese blue and white porcelain.

See: *indigo*

1. Ratnapandian, S., S. Fergusson L. Wang and R Padhye. Indigo colouration. In J. Fu (ed.). *Dyeing: Processes, Techniques and Applications*, pp. 65–76. Hauppauge, NY: Nova Science Publishers, 2014.

thermal barrier coatings

Thermal barrier coatings are applied to turbine blades to protect them from melting.[1] Barrier coatings are usually made from a ceramic powder, yttria-stabilised zirconia, and applied by thermal spraying, for example by plasma spraying. A metallic bondcoat layer is deposited between the surface of the superalloy and the ceramic coating and helps to prevent cracking in the barrier layer due to thermal expansion. Plasma-sprayed coatings tend to be porous.

See: *superalloys*

1. Probert, T. Blade runner: Improving gas turbine coatings. *Materials World* (Nov. 2013) 38–41.

thermochromic liquid crystals

Thermochromic materials change colour with changes in temperature and this behaviour is characteristic of some liquid crystals used as a thermometer.[1–3] If a backing strip contains layers of a liquid cholesteric crystal and is exposed to light then light that has a wavelength (i.e. colour) corresponding to the pitch is reflected and the rest of the light is adsorbed by the backing material. As the temperature rises, the pitch length decreases as the twist in each layer increases, making the helical structure tighter so that blue light is reflected (short wavelength). As the temperature falls, the helix is not wound so tightly and the pitch length increases so that longer wavelength light (red) is reflected. Regions of the strip contain liquid crystals with slightly differing compositions and pitches so that the same wavelength of light (i.e. colour) is reflected at slightly different temperatures.

See: *cholesteric phases*

1. Bolton, W., and R. A. Higgins. *Materials for Engineers and Technicians*, 6th edn, p. 427. London: Routledge, 2015.
2. Dunmur, D., and T. Sluckin. *Soap, Science, and Flat-Screen TVs: A History of Liquid Crystals*. Oxford: Oxford University Press, 2014.
3. L. C. Hallcrest Inc. Available at: http://www.hallcrest.com.

thermoplastics

Thermoplastics are polymers that can be processed either as a melt or as a softened material on heating into a variety of products (e.g. food packaging, cups, trays and car parts), for example by extrusion, molding and spinning.[1,2] Thermoplastics include polyethylene, polyesters, polycarbonate and polyvinylchloride. Thermoplastics contain additives including plasticisers for flexibility, pigments and fillers and also flame retardants.

1. Fried, J. R. *Polymer Science and Technology*, 3rd edn. Upper Saddle River, NJ: Prentice Hall, 2014.
2. Askeland, D. K., and W. J. Wright. *The Science and Engineering of Materials*, 7th edn, p. 573. Boston, MA: Cengage Learning, 2011.

thermosetting plastics

Thermosetting plastics are derived from resins that set to a hard rigid shape on heating through a polymerisation process.[1,2] These plastics cannot be processed by softening or in a melt process because polymerisation results in cross-linked polymer chains and the chains are resistant to softening and deformation. Thermosetting resins include phenol-formaldehyde polymers. Thermosetting plastics contain additives including plasticisers for flexibility, pigments and fillers such as flame retardants.

1. Fried, J. R. *Polymer Science and Technology*, 3rd edn. Upper Saddle River, NJ: Prentice Hall, 2014.
2. Askeland, D. K., and W. J. Wright. *The Science and Engineering of Materials*, 7th edn, p. 573. Boston, MA: Cengage Learning, 2011.

thin-film transistors

Thin-film transistors (TFTs) displays are frequently advertised, for example as a display in the Fiat 500 Cult automobile. A TFT display is equivalent to an active matrix display in which each pixel on the screen is backed with a thin-film transistor in a two-dimensional sheet to store its state during transitions in the energising voltage.[1] Instead of using an assembly of individual resistors, transistors and capacitors soldered together to form complex electronic circuits, TFTs are deposited as a thin layer onto the inner surface of a display such as a television screen.[2] The individual TFTs act as separate switches for the voltages applied to the pixels of the display. Pixels in the display are switched on and off by electrical pulses applied to elements of the matrix in which pixels are located at the intersection of current-carrying conductors arranged in rows and columns. This arrangement avoids pisels being switched on unintentionally.

See: *AMOLED*

1. Platt, C., and F. Jansson. *Encyclopaedia of Electronic Components*, vol. 2, p. 246. San Francisco, CA: Maker Media, 2014.
2. Dunmur, D., and T. Sluckin. *Soap, Science, and Flat-Screen TVs: A History of Liquid Crystals*. Oxford: Oxford University Press, 2014.

three-dimensional printed metals

Selective laser sintering (SLS) is a variant of three-dimensional printing in which objects are built up layer by layer from a metal powder.[1,2] In SLS metal powders are deposited layer by layer on a surface and each layer of powder particles is sintered by a laser that scans the surface before deposition of the next powder layer.

See: *three-dimensional printing*

1. Hausman, K., and R. Horne. *3D Printing for Dummies*. Chichester: Wiley, 2014.
2. Deckard, C. R. Method and apparatus for producing parts by selective sintering. United States Patent 4,863,538 1989.

three-dimensional printed plastics

Fused deposition modelling (FDM) is a variant of three-dimensional printing in which objects are built up layer by layer from a thermoplastic material.[1,2] In FDM a thermoplastic material in the form of a reel or a rod is fed to a dispensing head, melted and dispensed through an extrusion tip onto a substrate. Layers are built up from molten plastic droplets. Examples of plastics that can be used in FDM include nylon, polycarbonate, polylactic acid and styrene-acrylonitrile-butadiene (ABS).

See: *styrene-acrylonitrile-butadiene; three-dimensional printing*

1. Hausman, K., and R. Horne. *3D Printing for Dummies*. Chichester: Wiley, 2014.
2. Crump, S. S. Modeling apparatus for three-dimensional objects. United States Patent 5,340,433, 1994.

three-dimensional printing

Three-dimensional, or 3D printing, is an additive manufacturing technique in which a three-dimensional model of an object stored in a computer in a computer-aided design system is sliced into thin cross-sections and then the results are translated into two-dimensional position data.[1,2] The latter is used to control equipment for manufacture of a three-dimensional part in a layerwise manner. Materials used for printing include plastics, ceramic, metal powders and living cells.

See: *abrasives; additive manufacturing; bioprinting; stereolithography; superalloys*

1. Hausman, K., and R. Horne. *3D Printing for Dummies*. Chichester: Wiley, 2014.
2. Milmo, S. Reprint: Rebirth of an industry. *Chemistry and Industry* (Sept. 2015) 79, 44–7.

three-way automotive catalysts

Three-way automotive catalysts are incorporated into the exhaust systems of automobiles to reduce pollutants in the vehicle emissions.[1] These catalysts are usually supported on a cordierite honeycomb and for petrol (gasoline) engines they convert carbon monoxide to carbon dioxide, hydrocarbons from unburnt petrol to water and carbon dioxide and nitrogen oxides to nitrogen. The filters are open at both ends and are also known as flow-through filters. Catalysts include platinum and rhodium. For diesel emissions, the honeycombs act as particulate filters for removal of carbon particles and contain catalysts for treating nitrogen oxides.

See: *ceramic honeycomb; diesel particulates*

1. van Vlack, L. H. Automotive materials. In R. J. Brook (ed.), *Concise Encyclopaedia of Advanced Ceramic Materials*, pp. 25–30. Oxford: Pergamon Press, 1991.

toughened materials

Toughness is a property of materials and is a measure of the resistance to the growth of cracks within the material that can lead to failure.[1] The toughness of materials is increased by incorporation of fillers or fibres that act to deflect cracks: for example, glass fibre-reinforced plastics or silicon carbide fibre-reinforced silicon carbide composites and alumina-toughened zirconia components. The shells of molluscs are tough due to their composite nature.

See: *biomimetic materials; nacre; silicon carbide fibres*

1. Askeland, D. K., and W. J. Wright. *The Science and Engineering of Materials*, 7th edn, p. 230. Boston, MA: Cengage Learning, 2011.

trademarks

Trademarks are words and logos with distinctive features that can distinguish goods or services of one business from those of another.[1,2] The features include logos, sounds, colours, gestures, brand names and slogans. A registered trademark is represented by a superscript ® next to the mark while an unregistered mark has a superscript ™ next to the mark. The significance of trademarks to materials arises because when patent protection for a material has lapsed (usually 20 years after the filing date) the trademark can be used in perpetuity. Even though competitors can make the generic product, they are not able (without permission from the original owner) to use the trademark. For example, patent protection has lapsed for Lycra used in clothing but the trademark is widely used to advertise its continuing use in textiles or other applications.

1. McManus, J. *Intellectual Property: From Creation to Commercialisation*. Oxford: Oak Tree Press, 2012.
2. Jackson Knight, H. *Patent Strategy for Researchers and Research Managers*. Chichester: Wiley, 2012.

traditional ceramics

Traditional ceramics are derived from naturally occurring raw materials and include clay-based products such as tableware and sanitaryware as well as structural claywares, for example bricks and pipes.[1] Raw materials include clay minerals such as kaolinite, also known as kaolin, $(Al_2Si_2O_5(OH)_4)$, silica sand and a feldspar [e.g. $(K, Na)_2O;Al_2O_3;6SiO_2$].

See: *ceramic materials*

1. Verdeja, L. F., J. P. Sancho, A. Ballester and R. Gonzalez. *Refractory and Ceramic Materials*. Madrid: Editorial Sintesis, 2014.

transistors

Transistors are a key component of integrated circuits and computers and are based on the electronic properties of a *p–n–p* or *n–p–n* structure in a semiconductor:[1,2] For example, two doped *n*-type regions in a *p*-type silicon substrate. This type of transistor is known as a MOSFET, that is, a metal oxide semiconductor field effect transistor, and contains three terminals (Fig. G.29). One terminal on the *n*-type region is the 'source' and another terminal on the other *n*-type region is the 'drain'. The source and drain are connected by a conducting strip, the gate which represents the third terminal, and the gate is separated from the silicon substrate by a thin insulating layer of silica. The region between the source and drain is known as the 'channel'. If a potential is applied between the source and drain, then electrons cannot flow from the source to drain as the *p*-type region has a low electrical conductivity. When a modulating voltage is applied between the gate and source, electrons move to the region of the gate but cannot enter the gate because of the insulating region of silica. The concentration of electrons below the gate makes the channel more conductive so that a large applied voltage between the source and drain allows electrons to flow from the source to the drain, producing an amplified signal (the 'on' state). When no potential is applied to the gate no electrons move to the region between the source and drain and there is no current flow from the source to drain (the 'off' state). Thus, current flow between source and drain is controlled by the voltage on the gate. Transistors act as amplifiers but also as switches; that is, they recognise on-off instructions. This operation as switches is crucial in their use in logic gates for processing binary notation, that is, in distinguishing 1 from 0 in these circuits. It has been estimated that humans produce more transistors per year than grains of rice.[3] Initial work on transistors was carried out by John Bardeen, Walter Brattain and William Shockley in 1947.

See: p–n *junction diodes*

1. Solymar, L., D. Walsh and R. R. A. Syms. *Electrical Properties of Materials*, 9th edn. Oxford: Oxford University Press, 2014.
2. Sinclair, I. *Electronics Simplified*, 3rd edn. Oxford: Newnes, 2011.
3. Askeland, D. R., and W. J. Wright. *The Science and Engineering of Materials*, 7th edn, p. 702. Boston, MA: Cengage Learning, 2011.

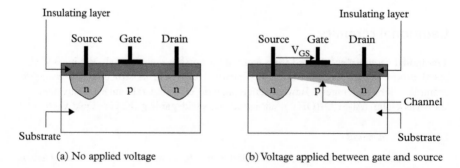

(a) No applied voltage   (b) Voltage applied between gate and source

Fig. G.29 *Schematic diagram of a MOSFET field-effect transistor where the gate acts as a capacitor.*

(M. Nixon. Digital Electronics: A Primer, p. 87. London: Imperial College Press, 2015.)

triacetone triperoxide

There is concern among governments in the 21st century on the availability of explosives that can be made from household chemicals and which are unstable.[1,2] Peroxide-based explosives such as triacetone triperoxide (TATP) is an example of this type of explosive. Another type of homemade explosive is hexamethylene triperoxide diamine (HMTD).

1. Milmo, S. Bomb chemicals face EU regulation. *Chemistry World* (Nov. 2012) 9, 17.
2. Robinson, P. Explosive chemistry. *Chemistry World* (Feb. 2015) 12, 46–9.

twisted nematic liquid crystals

Twisted nematic liquid crystals ('thread-like structure') have a positive dielectric anisotropy where the dielectric constant along the axis of the molecule is greater than the dielectric constant in the perpendicular direction and are particularly useful for liquid crystal displays.[1,2] An example of such a material is *n*-4'-ethoxybenzylidene-4-aminobenzontrile (PEBAB). In use, the crystals are sandwiched between two electrodes and on one electrode the molecules are aligned in a horizontal direction, whereas they are in a perpendicular direction on the other electrode, forming a helical structure. Optical activity of the molecules is controlled by the application of an electrical field that causes the helical structure to untwist and control light transmission that gives rise to the image seen by an observer.

See: *liquid crystal display; nematic liquid crystals*

1. Helfrich, W., and M. Schadt. Optical device. United Kingdom Patent Application, 1,372,868, 1974.
2. Dunmur, D., and T. Sluckin. *Soap, Science, & Flat-Screen TVs: A History of Liquid Crystals.* Oxford: Oxford University Press, 2014.

two-dimensional materials

Graphene is a two-dimensional sheet of carbon that is one atomic layer thick in which the atoms are arranged in a hexagonal array in a honeycomb structure and it is the thinnest known material. There is much activity in using this material for electronic applications due to its high electrical conductivity but graphene is not a semiconductor, properties of which are desirable for electronic applications.[1,2] Attempts continue to prepare other two-dimensional materials with properties comparable to those of graphene. Potential candidates, if they can be prepared, include 'stannene', a sheet of tin atoms, and 'germanene' sheets of germanium atoms while attempt to modify the structure of graphene to impart semiconducting properties, are being explored. For example, triazine-based graphitic nitride is a semiconductor.

See: *graphene*

1. Gross, M. Stannene: the next miracle material. *Chemistry and Industry* (Sept. 2014) 78, 24–7.
2. Evans, J. Beyond graphene. *Chemistry World* (Jan. 2014) 11, 44–7.

tyre compositions

Nowadays, the principal synthetic rubber used for automotive tyres is based on copolymers of styrene ($C_6H_5CH=CH_2$) and butadiene ($CH_2=CHCH=CH_2$), and blends of styrene-butadiene elastomers (SBR) and natural rubber are used in tyres. SBR can be produced in a free-radical polymerisation reaction in an emulsion and these copolymers have good wear and abrasion resistance and strength. Styrene-butadiene copolymers can be prepared with a narrow distribution of molecular weights and are known as living polymers when this criterion is fulfilled. Styrene-butadiene elastomers have been commercialised under the trade name Kraton (Shell Chemical Company).[1]

See: *living polymers*

1. J. A. Tyrell. *Fundamentals of Industrial Chemistry: Pharmaceuticals, Polymers and Business*, p. 96. New York: Wiley, 2014.

Ultrafiltration membranes

Membranes are porous barriers that are selective in the type of species allowed to pass through them.[1] Ultrafiltration membranes have a molecular weight cut-off of around 10,000 and a pore diameter of approximately 4 nm so that species with a molecular weight larger than this value are retained by the membrane. Examples of ultrafiltration membrane are 'Kerasep' membranes that consist of multichannel porous alumina monoliths with a zirconia ultrafiltration separation layer on the inside of the channels. Ultrafiltration membranes can be made from polymers although ceramic membranes have greater mechanical stiffness and superior chemical and thermal durability. Ultrafilters can separate some bacteria and some sugars.

See: *nanofiltration membranes*

1. Basile, A., and S. P. Nunes. *Advanced Membrane Science and Technology for Sustainable Energy and Environmental Applications*. Cambridge: Woodhead, 2011.

Velcro

Inspiration for inventions can come from the natural world. The greater burdoch is a hairy bush whose purple flowers grow on top of small globes, each consisting of many hooked bracts that cover the bush.[1] The bracts carry the seeds and their hooks catch on animal fur or clothes, so that the seeds get widely distributed. George de Mestral, a Swiss inventor, was out walking his dog in the 1940s and noticed the burrs hooking onto the dog's fur. This observation is said to have given him the idea for his invention of Velcro used in fasteners.[2–4] The latter consists of two strips of a support, for example terry- or velvet-type woven textile or metal. One strip contains a series of hooks while the other strip contains a series of loops; hooks and loops can be made out of nylon. The hooks can be pressed into the loops, thus securing the fastener while the strips can be easily pulled apart.

1. May, D. *The Times*, 9 September 2013.
2. de Mestral, G. Separable fastening device. United States Patent. 3,009,235, granted on 21 November 1961.

3. de Mestral, G. Apparatus for manufacturing separable fasteners. United States Patent 3,083,737, granted on 2 April 1963.
4. de Mestral, G. Method for the manufacture of pile fabrics. United States Patent 3,154,837, granted on 3 November 1964.

Viton

Viton is a copolymer of vinylidene fluoride and hexafluoropropylene and can be used as a replacement for natural rubber.[1] Viton is an example of a fluoropolymer and this class of material is thermally stable and relatively inert. Viton is associated with its use as 'O' ring seals in chemical reactors. Another example of a fluoropolymer is Technoflon, a copolymer between vinylidene fluoride and 1-hydropentafluoropropylene.[2] Both Viton and Technoflon have a backbone of carbon–carbon bonds.

See: *elastomers*

1. Smith, D. W. Jr, S. T. Iaccono and S. S. Iyer (eds). *Handbook of Fluoropolymer Science and Technology*. Chichester: Wiley, 2014.
2. Loadman, J. *Tears of the Tree: The Story of Rubber–A Modern Marvel*. Oxford: Oxford University Press, 2014.

vulcanite

The sulphur content used by Charles Goodyear for the vulcanisation of natural rubber was around 3 wt%. As the sulphur content increases, the cross-linking of natural rubber molecules increases and a hard, black, non-elastomeric product known as vulcanite (also referred to as ebonite) was produced when the sulphur concentration approached 30 wt%.[1,2] Applications for this early plastic in the 19th century included materials for fountain pens until it was superseded by other plastics (Fig. G.30).

See: *elastomer*

1. Gratzer, W. *Giant Molecules: From Nylon to Nanotubes*, p. 109. Oxford: Oxford University Press, 2013.
2. Loadman, J. *Tears of the Tree: The Story of Rubber–A Modern Marvel*. Oxford: Oxford University Press, 2014.

Vycor

Vycor is an example of a high-silica-containing glass with a typical composition 96% SiO_2; 3% B_2O_3; 1% Na_2O and has a low coefficient of thermal expansion.[1] It has applications as windows in space vehicles.

See: *phase-separated glasses*

1. Bolton, W., and R. A. Higgins. *Materials for Engineers and Technicians*, 6th edn, p. 342. London: Routledge, 2015.

Fig. G.30 *Vulcanite articles: (a) a jumble of necklace chains, (b) pendants, (c) buttons and brooches, (d) detail of one brooch (top right in (c)), (e) fountain pens, (f) flute, (g) pipe bowl, (h) revolver hand grips, (i) brooch, (j) Queen Victoria Jubilee medal, (k) ornamental comb, (l) combs, (m) medicine holders and (n) cigarette lighter.*

(J. Loadman. Tears of the Tree: The Story of Rubber—A Modern Marvel, p. 236. Oxford: Oxford University Press, 2014.)

(h)

(i)

(j)

(k)

(l)

(m)

(n)

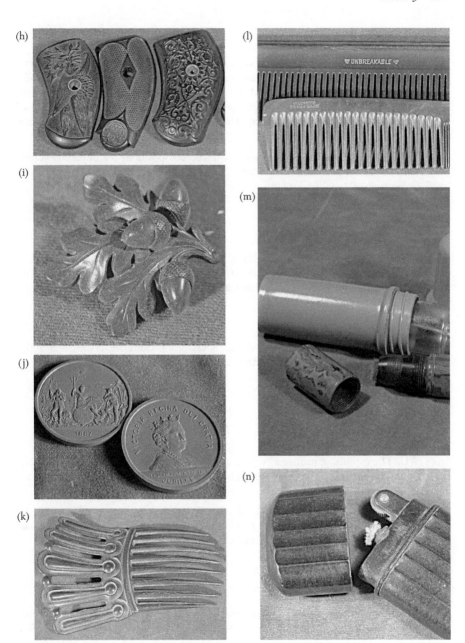

Fig. G.30 *(Continued)*

waterproof cosmetics

Waterproof properties are important for certain cosmetics and personal hygiene products, for example in mascara and lipsticks.[1] Sunscreens that help protect the user from harmful exposure to components of sunlight should be waterproof and have a smooth texture when applied from skin care formulations. Solid polymeric citrate esters made by reaction of alcohols with citric acid where the reactants are obtained from natural products can act as a waterproofing agent for sunscreens.

1. Ranade, R. A., and M. S. Garrison. Cosmetic compositions for imparting super-hydrophobic films. United States Patent Application US 2011/0008401, published on 13 January 2011.

waterproof garments

The word mackintosh is used to describe a waterproof garment such as a raincoat.[1,2] Charles Mackintosh showed in 1824 that a solution of natural latex rubber could be dissolved in naphtha and coated onto sheets of cloth that could be glued together by the rubber and used as waterproof clothing.

See: *elastomer*

1. Walton, D., and P. Lorimer. *Polymers*, p. 15. Oxford: Oxford University Press, 2005.
2. Loadman, J. *Tears of the Tree: The Story of Rubber–A Modern Marvel*. Oxford: Oxford University Press, 2014.

weapons-grade plutonium

The isotope ^{239}Pu is produced in thermal nuclear reactors by neutron bombardment of non-fissile ^{238}U. The capture of neutrons by ^{238}U and formation and decay of ^{239}U and ^{239}Np produces fissionable ^{239}Pu. Weapons-grade plutonium consists of plutonium with approximately 93% or more of ^{239}Pu.[1]

1. Hecker, S. S., M. Englert and M. C. Miller. Nuclear non-proliferation. In D. S. Ginley and D. Cahen (eds), *Fundamentals of Materials for Energy and Environmental Sustainability*, p. 165. Cambridge: Cambridge University Press, 2012.

wearable technology

Running is a popular pastime and it is a common sight to see runners with small monitors strapped to their arms to measure and record physiological data such as heart rate and body temperature. In addition, Google Glass, a wearable pair of glasses with a miniature camera attached for recording the environment, has been evaluated and was issued as an experimental product but did not go on sale. Contact lenses with sensors to detect eye fluids for signs of disease have been developed. These products are examples of wearable technology and the data which they collect can be wirelessly transmitted to other electronic devices for analysis.[1] Other examples of wearable technology include incorporation of devices such as earphones and microphones designed into textile fabric as well as keypads incorporated into fabric and smart watches such as the Sony Smartwatch. Wearable technology utilises miniature sensors fabricated on silicon, often involving microelectromechanical systems.

See: *microelectromechanical systems*

1. Eisberg, N. Wear it well. *Chemistry and Industry* (Mar. 2015) 79, 32–5.

wind energy

Wind energy is having an increasing role worldwide in the generation of carbon-free electrical energy.[1] The blades of the turbine are critical components and lightweight glass fibre and silicon carbide fibre composites are used in their construction.

1. Robinson, M., N. Kelley, P. Moriarty, S. Schreck, D. Simms and A. Wright. Wind energy. In D. S. Ginley and D. Cahen (eds), *Fundamentals of Materials for Energy and Environmental Sustainability*, p. 396. Cambridge: Cambridge University Press, 2012.

Wint-O-Green mints

Wint-O-Green mints[1] consist of tablets of compressed sugar and when broken apart in the mouth blue sparks are produced. The sparks are produced by triboluminescence in which some crystalline materials develop static electrical charges as a result of frictional forces generated on crushing. In the case of these mints, sugar crystals are broken and become electrically charged. The charged crystals interact with nitrogen molecules in the atmosphere to produce charged gaseous species that emit ultraviolet radiation. The latter is absorbed by a component of the sweets, oil of wintergreen (used as a flavouring), that contains methyl salicylate, which re-radiates light with a blue wavelength. Hence, sweets can be considered an unusual combination of materials.

1. Hartel, R. W., and A. Hartel. *Candy Bites: The Science of Sweets*. London: Springer, 2014.

wound dressings

Hydrogels can absorb and retain large quantities of fluid relative to their weight and are ideal candidate materials for bandages or wound dressings.[1] The hydrogel, for example, a polyurethane gel or a hydroxyethyl cellulose gel, helps to keep the wound moist.

See: *hydrogels*

1. Segal, D. *Exploring Materials through Patent Information*. Cambridge: Royal Society of Chemistry, 2014.

yeast

Yeasts are eukaryotes, that is, single-celled organisms and are fungi.[1] Yeasts can enzymatically catalyse the conversion of sugar to alcohol, a process used in brewing beer, and this process supplies energy to the organism for growth. Both baker's yeast, *Saccharomyces cerevisiae*, and the non-fermenting yeast *Pichia pastoris* have important roles in biotechnology where recombinant DNA technology is used to engineer their expression of valuable recombinant proteins. Examples of recombinant proteins are insulin, the anticoagulant hirudan, liraglutide for treatment of Type 2 diabetes, glucagon for treatment of hypoglycemia and elastase inhibitor for treatment of cystic fibrosis. The word expression that is widely used in biotechnology can be interpreted in everyday language as equivalent to synthesis *in vivo*.

See: *insulin; recombinant DNA technology*

1. Shi, J., Q. Qing, T. Zhang, C. E. Wyman and T. A. Lloyd. Biofuels from cellulosic biomass via aqueous processing. In D. S. Ginley and D. Cahen (eds), *Fundamentals of Materials for Energy and Environmental Sustainability*, p.342. Cambridge: Cambridge University Press, 2012.

zeolites

Zeolites are hydrated crystalline aluminosilicates.[1–3] They are both naturally occurring (e.g. clinoptilolite) and are prepared by synthetic means often involving a hydrothermal process. They have a well-defined pore size on the order of a nanometre or less. Zeolite A has a pore size between 0.35 and 0.45 nm while zeolite Y has a pore size of 0.6–0.8 nm. They are used as ion-exchange media to remove calcium ions from hard water, as replacements for phosphates in detergents to bind calcium as phosphates can contribute to blooming of algae in marine environments and in gas separations, for example removal of hydrogen sulphide from natural gas. They are used extensively as catalysts particularly in petroleum refining as fluid cracking catalysts. Zeolites are often referred to as molecular sieves when used for gas separations.

See: *classification of porosity*

1. Tyrell, J. A. *Fundamentals of Industrial Chemistry: Pharmaceuticals, Polymers and Business*, pp. 22–33. John Wiley & Sons Inc., 2014.
2. B Trewyn (editor). Heterogeneous catalysis for today's challenges, Royal Society of Chemistry, 2015.
3. Ploszajski, A. Zeolites. *Materials World* (Oct. 2015) 23, 62–4.

Ziegler–Natta catalysts

Until the early 1950s, polyethylene was manufactured in a high-pressure, high-temperature radical process that produced what is referred to as low-density polyethylene (LDPE). It is low density as the polymer is not linear and has some branching so that polymer chains do not all pack tightly together. Ziegler–Natta catalysts,[1,2] named after Karl Ziegler and Guilio Natta, enables polyethylene to be made under more moderate conditions, namely at around 323 K and 1 atm; the resulting polymer is known as high-density polyethylene (HDPE). A representative Ziegler–Natta catalyst consists of titanium tetrachloride and an aluminium alkyl (e.g. aluminium triethyl Al$(C_2H_5)_3$) supported on magnesium chloride. Ziegler–Natta catalysts are also used for preparation of polypropylene.

See: *polyethylene*

1. Sutton, M. Paving the way to polythene. *Chemistry World* (Nov. 2013) 10, 50–3.
2. Walton, D., and P. Lorimer. *Polymers*, p. 78. Oxford: Oxford University Press, 2005.

Zircaloy

Zircaloy[1] is an alloy of zirconium and compositions contain a majority of this metal with minor components of other components such as tin, niobium and iron, depending on its application. Fuel rods in thermal nuclear reactors consist of Zircaloy tubes that contain cylindrical pellets of nuclear fuel. The rods, which are between 1 and 4 m in length, are grouped together into fuel assemblies that are inserted into the reactor. The inside of the fuel rod operates at around 400°C while the outside can be up to temperatures around 1500°C. In pressurised water reactors the coolant is water while carbon dioxide is the coolant in high-temperature gas-cooled reactors.

See: *nuclear fuel*

1. Stanek, C. R., R. W. Grimes, C. Unal, S. A. Maloy and S. C. Scott. Nuclear energy: current and future schemes. In D. S. Ginley and D. Cahen (eds), *Fundamentals of Materials for Energy and Environmental Sustainability*, pp. 147–161. Cambridge: Cambridge University Press, 2012.

ZMapp

Pharmaceutical compounds such as acetylsalicylic acid are small molecules that have been traditionally prepared by chemical synthesis. There is increasing interest in the use of biological molecules such as peptides and antibodies for the treatment of diseases. An example is the experimental drug ZMapp that contains three monoclonal antibodies for the treatment of people infected with the Ebola virus during an outbreak of Ebola in West Africa in 2014.[1] ZMapp has been obtained by using recombinant DNA technology on genetically engineered tobacco plants.

See: *acetylsalicylic acid; recombinant DNA technology*

1. Burke, M. Ebola drug verdict. *Chemistry and Industry* (Sept. 2014) 78, 7.

Appendix 1
Suggestions for Further Reading

Ashby, M. F., D. F. Balas and J. S. Coral. *Materials and Sustainable Development*. Oxford: Butterworth, 2016. (*A very readable account of a range of materials, their uses and in particular the role of materials in sustainable development in a world of increasing population and finite resources.*)

Askeland, D. R., and W. J. Wright. *The Science and Engineering of Materials*, 7th edn. Boston, MA: Cengage Learning, 2011. (*Comprehensive coverage of a wide range of materials for undergraduates and the interested general reader.*)

Atkins, P. *Atkins' Molecules*, 2nd edn, p. 94. Cambridge: Cambridge University Press, 2003. (*Very well-written account of a wide range of molecules and their properties for the general reader and specialist.*)

Bansal, N. P., and J. Lamon (eds). *Ceramic Matrix Composites: Materials, Modeling and Technology*. Chichester: Wiley, 2015. (*A detailed and comprehensive account of composites.*)

Bolton, W., and R.A. Higgins. *Materials for Engineers and Technicians*, 6th edn. London: Routledge, 2015. (*For both the specialist and the interested reader.*)

Boysen, E., and N. Boysen. *Nanotechnology for Dummies*, 2nd edn. New York: Wiley, 2011. (*Suitable for the general reader and specialist.*)

Brook, R. J. (ed.). *Concise Encyclopaedia of Advanced Ceramic Materials*. Oxford: Pergamon Press, 1991. (*A reference book with a comprehensive account of a wide range of ceramic systems.*)

Buzak, S., and D. Rende. *Colloid and Surface Chemistry: A Laboratory Guide for Exploration of the Nano World*. Boca Raton, FL: CRC Press, 2014. (*Contains instructions on how to prepare a range of colloidal systems.*)

Christie, R. M. *Colour Chemistry*, 2nd edn. Cambridge: Royal Society of Chemistry, 2015. (*This book covers a wide range of materials used for producing coloured products.*)

Currano, J. N., and D. L. Roth (eds). *Chemical Information for Chemists: A Primer*. Cambridge: Royal Society of Chemistry, 2014. (*Chapter 3 gives a comprehensive account of the patenting process.*)

Daintith, J. (ed.). *A Dictionary of Chemistry*, 6th edn. Oxford: Oxford University Press, 2008. (*A very useful reference book with nearly 5000 entries.*)

Daintith, J. (ed.). *A Dictionary of Physics*, 6th edn. Oxford: Oxford University Press, 2009. (*A very useful reference book with nearly 4000 entries.*)

DeRosa, T. F. *Engineering Green Chemical Processes: Renewable and Sustainable Design*. New York: McGraw-Hill, 2015. (*The majority of references are in the patent literature and these are used with great effect to describe industrial processes.*)

Dodgson, M., and D. Gann. *Innovation: A Very Short Introduction*. Oxford: Oxford University Press, 2010. (*A useful description for the development of Kevlar.*)

Donald, G. *The Accidental Scientist: The Role of Chance and Luck in Scientific Discovery*. London: Michael O'Mara Books, 2013. (*A very useful account of the development of Post-it notes and cellulose nitrate.*)

Downie, N. A. *The Ultimate Book of Saturday Science*. Princeton, NJ: Princeton University Press, 2012. (*Excellent technical descriptions of experiments for the amateur and professional scientist.*)

van Dulken, S. *Inventing the 21st Century*. London: British Library, 2014. (*Informative text for a general and specialist audience.*)

Dunmur, D., and T. Sluckin. *Soap, Science, and Flat-Screen TVs: A History of Liquid Crystals*. Oxford: Oxford University Press, 2014. (*Highly readable account of the development of liquid crystals and their applications.*)

Emsley, J. *Molecules at an Exhibition*. Oxford: Oxford University Press, 1998. (*Well-written account of a wide range of materials for the general reader.*)

Espacenet, http://worldwide.espacenet.com. (*For anyone interested in patent searching.*)

Freemont, P. S., and R. I. Kitney (eds). *Synthetic Biology: A Primer*. London: Imperial College Press, 2012. (*An introductory overview of synthetic biology.*)

Fried, J. R. *Polymer Science and Technology*, 3rd edn. Upper Saddle River, NJ: Prentice Hall, 2014. (*Comprehensive and detailed account of polymer systems.*)

Fu, Jie (ed.). *Dyeing: Processes and Applications*. Hauppauge, NY: Nova Science Publishers, 2014. (*Very informative coverage of different types of dyeing processes used for textiles.*)

Ginley, D. S., and D. Cahen (eds). *Fundamentals of Materials for Energy and Environmental Sustainability*. Cambridge: Cambridge University Press, 2012. (*Covers a very wide range of materials and applications.*)

Giustino, F. *Materials Modeling Using Density Functional Theory*. Oxford: Oxford University Press, 2014. (*A very advanced text suitable for the specialist.*)

Glynn, J. *My Sister Rosalind Franklin*. Oxford: Oxford University Press, 2012. (*A very readable account of the DNA story with reference to the work of Rosalind Franklin.*)

Goldstein, D. (ed.). *The Oxford Companion to Sugar and Sweets*, pp. 190–1. Oxford: Oxford University Press, 2015. (*Refers to candy floss and contains excellent descriptions of sugars, starches and other food ingredients.*)

Gordon, J. E. *The New Science of Strong Materials*, 2nd edn. London: Penguin, 1991. (*Informative for both the general reader and specialist.*)

Gratzer, W. *Giant Molecules: From Nylon to Nanotubes*. Oxford: Oxford University Press, 2013. (*Informative for both the general reader and specialist.*)

Hall, C. *Materials: A Very Short Introduction*. Oxford: Oxford University Press, 2014. (*A useful description for the development of integrated circuits.*)

Hartel, R. W., and A. Hartel. *Candy Bites: The Science of Sweets*. New York: Springer, 2014. (*Informative for the general reader.*)

Hester, R. E., and R. M. Harrison (eds). *Fracking, Issues in Environmental Science and Technology*, vol. 39. Cambridge: Royal Society of Chemistry, 2015. (*A comprehensive text for the specialist.*)

Hosford, W. F. *Iron and Steel*. Cambridge: Cambridge University Press, 2012. (*A very clear account for the general reader and students of the processes involved in iron and steel production and the properties of these materials.*)

Johnstone, B. *Brilliant!: Updated edition*. Amherst, NY: Prometheus Books, 2015. (*A very readable account of the development of blue light-emitting diodes by Shuji Nakamura.*)

Kerton, F., and R. Marriot. *Alternative Solvents for Green Chemistry*, 2nd edn. Cambridge: Royal Society of Chemistry, 2013. (*Overview of ionic liquids for the specialist.*)

Lecce, L., and A. Concilio (eds). Shape Memory Alloy Engineering for Aerospace, Structural and Biomedical Applications. Oxford: Butterworth-Heinemann, 2015. (*A specialist text.*)

Lee, M. (ed.). *Remarkable Natural Material Surfaces and Their Engineering Potential*. New York: Springer, 2014. (*Outstanding illustrations and text for a range of materials in the natural world.*)

Liu, P. S., and G. F. Chen. *Porous Materials*. London: Elsevier, 2014. (*Comprehensive account for students in the physical sciences.*)

Lucas, J., P. Lucas, T. Le Mercier, A. Rollat and W. Davenport. *Rare Earths: Science, Technology, Production and Use*. London: Elsevier, 2015. (*Comprehensive and detailed reference book.*)

May, P., and S. Cotton. *Molecules That Amaze Us*. Boca Raton, FL: CRC Press, 2015. (*A very readable account of 67 compounds in 67 chapters for the general public and specialist. Contains a comprehensive set of structural chemical formulae.*)

Miller J. (Gen. Ed.). *Miller's Antiques Encyclopedia*. London: Octopus Publishing, 2013. (*Comprehensive account of the materials used in a wide range of antiques and collectable items such as porcelain.*)

Mitchell, G. R. (ed.). *Electrospinning: Principles, Practice and Possibilities*. London: Royal Society of Chemistry, 2015. (*Comprehensive account of electrospinning for the specialist.*)

Monk, P. M. S., R. J. Mortimer and D. R. Rosseinsky. *Electrochromism and Electrochromic Devices*. Cambridge: Cambridge University Press, 2007. (*Comprehensive account of electrochromism.*)

Ngo, C., and M. van de Voorde. *Nanotechnology in a Nutshell: From Simple to Complex Systems*. Amsterdam: Atlantis Press, 2014. (*Wide-ranging descriptions for the application of nanotechnology for the general reader and specialist.*)

Novotny, L., and B. Hecht. *Principles of Nano-optics*, 2nd edn. Cambridge: Cambridge University Press, 2006. (*For serious students of chemistry and physics.*)

Ozin, G. A., A. C. Arsenault and L. Cademartiri (eds). *Nanochemistry: A Chemical Approach to Nanomaterials*. Cambridge: Royal Society of Chemistry, 2009. (*A detailed text for students in the physical sciences.*)

Platt, C., and F. Jansson. *Encyclopaedia of Electronic Components*, vol. 2. San Francisco, CA: Maker Media, 2014. (*Clear account for both the interested reader and specialist.*)

Rapley, R., and D. Whitehouse (eds). *Molecular Biology and Biotechnology*. Cambridge: Royal Society of Chemistry, 2015. (*For undergraduates and postgraduate students as well as the interested non-specialist.*)

Ross, D., C. Shamieh and G. McComb. *Electronics for Dummies*. Chichester: Wiley, 2010. (*Suitable for the general reader and specialist.*)

Segal, D. *Chemical Synthesis of Advanced Ceramic Materials*. Cambridge: Cambridge University Press, 2014. (*Describes the preparation of ceramic materials in the form of powders, fibres, coatings and monoliths by chemical syntheses such as sol-gel processing and hydrothermal synthesis.*)

Segal, D. *Exploring Materials through Patent Information*. Cambridge: Royal Society of Chemistry, 2014. (*Describes how development of materials, for example graphene and hydrogels, can be followed by a technical analysis of the patent literature.*)

Sinclair, I. *Electronics Simplified*, 3rd ed. Oxford: Newnes, 2011. (*Comprehensive and clear descriptions for the general public and specialist.*)

Solymar, L., D. Walsh and R. R. A. Syms. *Electrical Properties of Materials*, 9th edn. Oxford: Oxford University Press, 2014. (*For students in the physical sciences.*)

Smith, Jr, D. W., S. T. Iaccono and S. S. Iyer (eds). *Handbook of Fluoropolymer Science and Technology*. Chichester: Wiley, 2014. (*A detailed reference work.*)

Stewart, I. C., and J. P. Lomont. *The Handy Chemistry Answer Book*. Canton, MI: Visible Ink Press, 2014. (*Recommended to readers of all ages and educational background who want to know clear answers to a wide range of questions on chemistry.*)

Thackray, A., D. C. Brock and R. Jones. *Moore's Law*. New York: Basic Books, 2015. (*Refers to the work of Gordon Moore who proposed Moore's Law. Recommended for the general public and anyone interested in the development of the electronic age of integrated circuits.*)

Tilley, R. J. D. *Understanding Solids: The Science of Materials*. Chichester: Wiley, 2013. (*For students in the physical sciences.*)

Tindell, H. *Engineering Materials*. Marlborough, UK: Crowood Press, 2014. (*Comprehensive account of metals and alloys for the enthusiastic hobbyist and specialist.*)

Tyrell, J. A. *Fundamentals of Industrial Chemistry: Pharmaceuticals. Polymers and Business*. New York: Wiley, 2014. (*The majority of references are in the patent literature and these are used with great effect to describe industrial processes.*)

United Kingdom Intellectual Property Office, http://www.ipo.gov.uk. (*For anyone interested in intellectual property.*)

United States Patent and Trademark Office, http://www.uspto.gov. (*For anyone interested in patent searching.*)

Walton, D., and P. Lorimer. *Polymers*, p. 97. Oxford: Oxford University Press, 2005. (*Refers to aliphatic polyamides.*)

Weightman, G. *Eureka: How Invention Happens*. New Haven, CT: Yale University Press, 2015. (*An account of the use of gutta percha in underwater cables.*)

Woolfson, M. M. *Resonance: Applications in Physical Science*. London: Imperial College Press, 2014. (*Includes a concise account of stimulated and spontaneous emission of light that is relevant to lasers.*)

World Intellectual Property Office, http://www.wipo.int/portal/index.html.en. (*For anyone interested in patent searching.*)

Appendix 2
Selected Patent Documents Referred to in the Text

Andrews, B. M., N. Carr, G.W. Gray and C. Hogg. Liquid crystal compounds. European Patent 0097033 B, 1990. (*Refers to cyanobiphenyl liquid crystals.*)

Baekeland, L. H. Method of making insoluble products of phenol and formaldehyde. United States Patent 942,699, granted on 7 December 1909. (*Refers to phenol-formaldehyde resins.*)

Banting, F. G., and C. H. Best. Extract obtainable from the mammalian pancreas or from related glands of fishes. Canadian Patent Application 234336 A, published 18 September 1923. (*Refers to insulin.*)

Baynes, B. M., et al. Biological synthesis of 6-aminocaproic acid from carbohydrate feedstocks. United States Patent 8,404,465, granted on 26 March 2013. (*Refers to recombinant 6-aminocaproic acid.*)

Benavides, J. M. Method for manufacturing high quality carbon nanotubes. United States Patent 7,008,605, granted on 7 March 2006. (*Refers to fullerenes.*)

Borbely, J. Additives for cosmetic products and the like. United States Patent Application 2008/0070993, published on 20 March 2008. (*Refers to chitosan and hyaluronic acid.*)

Bowe, M. J., D. L. Segal, C. D. Lee-Tuffnell, D. C. W. Blaikley, J. A. Maude, J. W Stairmand and I. F. Zimmerman. Catalytic reactor. United States Patent 8,118,889, granted on 21 February 2012. (*Refers to synthetic diesel.*)

Brandenberger, J. E. Composite cellulose film. United States Patent 1,266,766, granted on 21 May 1918. (*Refers to cellophane.*)

Buchanan, C. M., and N. L. Buchanan. Cellulose esters and their production in halogenated ionic liquids. United States Patent 8,273,872, granted on 25 September 2012. (*Refers to cellulose.*)

Buehler, W. J., and R. C. Wiley. Nickel-based alloys. United States Patent 3,174,851, granted in March 1965. (*Refers to shape memory alloys.*)

Carlick, D. J., R. W. Bassemir, R. Krishnan and R. R. Durand. Low rub printing ink. United States Patent 5,158,606, granted on 27 October 1992. (*Refers to polytetrafluoroethylene.*)

Carothers, W. H. Diammine-dicarboxylic acid salts and process of preparing same. United States Patent 2,130,947, granted on 20 September 1938. (*Refers to nylon.*)

Chen, L.-Z., C.-H. Liu, J.-P. Wang and S.-S. Fan. Method for using a Poisson ratio material. United States Patent 8,545,745, granted on 1 October 2013. (*Refers to auxetic materials.*)

Choi, S. J., J. Palgunadi, J. E. Kang, H. S. Kim and S. Y. Chung. Amidium-based ionic liquids for carbon dioxide absorption. United States Patent 8,282,710, 2012. (*Refers to industrial effluents: carbon dioxide.*)

Cohen, S. N., and H. W. Boyer. Biological functional molecular chimeras. United States Patent 4,740,470, granted on 26 April 1988. (*Refers to recombinant DNA technology.*)

Cohen, S. N., and H. W. Boyer. Process for producing biologically functional molecular chimeras. United States Patent 4,237,224, granted on 2 December 1980. (*Refers to recombinant DNA technology.*)

Dicosimo, R. Production of peracids using an enzyme having perhydrolysis activity. United States Patent 8,329,441, granted on 11 December 2012. (*Refers to bleaching agents.*)

Earle, M. J., K. R. Seddon and N. V. Plechkova. Production of bio-diesel. United States Patent Application 2009/0235574, published on 24 September 2009. (*Refers to bio-diesel.*)

Edmonds, J. T., and H. W. Hill. United States Patent 3,354,129, 1967. (*Refers to poly (p-phenylene sulphide).*)

Feher, F. J., et al. Polymers of isoprene from renewable resources. United States Patent 8,420,759, granted on 16 April 2013. (*Refers to recombinant cis-isoprene.*)

Fenyvesi, G., et al. Personal care and cosmetics compositions comprising biologically-based mono- and di-esters. United States Patent 8,309,116, granted on 13 November 2012. (*Refers to recombinant face cream.*)

Friend, R. H., J. H. Burroughes and D. D. Bradley. Electroluminescent devices. United States Patent 5,247,190, 1993. (*Refers to poly-(p-phenylenevinylene).*)

Goodyear, C. Improvement in India-rubber fabrics. United States Patent 3633, granted on 15 June 1844. (*Refers to vulcanisation of latex gum.*)

Gore, R. W. United States Patent 3,953,566, 1976. (*Refers to Gore-Tex.*)

Gore, R. W. United States Patent 4,187,390, 1980. (*Refers to Gore-Tex.*)

Gore, R. W., and S. B. Allen. United States Patent 4,194,041, 1980. (*Refers to Gore-Tex.*)

Haber, F., and R. Le Rossignol. Production of ammonia. United States Patent 971,501, granted on 27 September 1910. (*Refers to artificial fertilizer.*)

Hack, M., M.-Hao, M. Lu and M. S. Weaver. Organic light-emitting devices for illumination. United States Patent Application 2011/0284899, 2011. (*Refers to solid-state lighting.*)

Hall, S. I., J. T. Shawcross, R. McAdams, D. L. Segal, M. Inman, K. Morris, D. Raybone and J. Stedman. Non-thermal plasma reactor. United States Patent Application 2004/0208804, published on 21 October 2004. (*Refers to diesel particulates.*)

Hasan, S. K. Three-dimensional tissue generation. United States Patent Application 2011/0250688, published on 13 October 2011. (*Refers to bioprinting.*)

Hay, A. S. United States Patent 3,306,874, 1967. (*Refers to polyphenylene oxide.*)

Hay, A. S. United States patent 3,306,875, 1967. (*Refers to polyphenylene oxide.*)

Hay, A. S. United States Patent 4,028,341, 1975. (*Refers to polyphenylene oxide.*)

Heeger, A. J., A. G. MacDiarmid, C. K. Chiang and H. Shirakawa. P-type electrically conducting doped polyacetylene film and method of preparing the same. United States Patent 4,222,903, 1980. (*Refers to polyacetylene.*)

Helfrich, W., and M. Schadt. Optical device. United Kingdom Patent Application, 1,372 868, 1974. (*Refers to twisted nematic liquid crystals.*)

Hermeling, D., U. Stuven and U. Hoss. Super-absorbing hydrogel with specific particle size distribution. United States Patent Application 2004/0265387, published on 30 December 2004. (*Refers to superabsorbent polymers.*)

Hill, H. W. Jr, S. L. Kwolek and W. Sweeny. Aromatic polyamides. United States Patent 3,380,969, granted on 30 April 1968. (*Refers to Kevlar.*)

Hockessin, H. B. United States Patent US 3,767,756, 1973. (*Refers to Nomex.*)

Hoffmann, F. Acetyl salicylic acid. United States Patent 644,077, granted 27 February 1900. (*Refers to acetylsalicylic acid.*)

Hogan, J. P., and R. L. Banks. Polymers and production thereof. United States Patent 2,825,721, granted on 4 March 1958. (*Refers to polyethylene.*)

Hubbell, J. A., C. P. Pathak, A. S. Sawhney, N. P. Desai and J. L. Hill-West. Photopolymerizable biodegradable hydrogels as tissue contacting materials and controlled-release carriers. United States Patent 5,567,435, granted on 22 October 1996. (*Refers to biodegradable hydrogels.*)

Hull, C. W. Apparatus for production of three-dimensional objects by stereo-lithography. United States Patent 4,575,330, 1986. (*Refers to stereolithography.*)

Huver, T., C. Nicolaison and S. Camp. United States Patent 5,561,198, 1995. (*Refers to Super Glue.*)

Hyatt, J. W. Jr. Improved molding composition to imitate ivory and other substances. United States Patent 88,633, granted on 6 April 1869. (*Refers to celluloid.*)

Hudgins, R. G., and J. M. Criss Jr. Phase separated branched copolymer hydrogel. United States Patent 7,919,542, granted on 5 April 2011. (*Phase separated polymers.*)

Kazakov, S., M. Kaholek and K. Levon. Lipobeads and their production. United States Patent 7,618,565, granted on 17 November 2009. (*Refers to liposomes.*)

Kistler, S. S. Method of making aerogels. United States Patent 2,249,767, granted on 22 July 1942. (*Refers to aerogels.*)

Kwolek, S. L. Optically anisotropic aromatic polyamide dopes. United States Patent 3,671,542, granted on 20 June 1972. (*Refers to Kevlar.*)

Langer, R. S., et al. Biodegradable poly (β-amino esters) and uses thereof. United States Patent 8,287,849, granted on 16 October 2012. (*Refers to biodegradable nanoparticles.*)

Lee, J.-H., J.-H. Cho, J.-S. Choi and D.-J. Lee. Spin-on glass composition and method of forming silicon oxide layer in semiconductor manufacturing process using the same. United States Patent 7,270,886, granted on 18 September 2007. (*Refers to polysilazanes and spin-on-glass.*)

Lee, W. K., B. D. Kim, C. D. Jung and S. H. Jin. Flame resistant thermoplastic resin composition. United States Patent Application 2012/0172502, published on 5 July 2012. (*Refers to styrene-acrylonitrile-butadiene.*)

Lim, H. J., E. C. Cho, J. H. Lee and J. Kim. Chemically cross-linked hyaluronic acid hydrogel nanoparticles and the method for preparing thereof. International Patent Application WO 2008/100044, published on 21 August 2008. (*Refers to hyaluronic acid.*)

Maase, I. M., K. Massonne, K. Halbritter, R. Noe, M. Bartsch, W Siegel, V. Stegmann, M. Flores, O. Huttenloch and M. Becker. International Patent Application WO 03/062171, published on 31 July 2003. (*Refers to industrial effluents: acids.*)

McGee, T., and R. P. Sgaramela. Substrate care products. United States Patent Application, 2008/0234172, published on 25 September 2008. (*Refers to fabric conditioners.*)

Maiman. T. United States Patent 3,353,115, 1960. (*Refers to the ruby laser.*)

Medoff, M., T. C. Masterman, C. Cooper and J. Khan. Conversion of biomass. United States Patent Application 2014/0011248, published on 9 January 2014. (*Refers to saccharification.*)

de Mestral, G. Separable fastening device. United States Patent. 3,009,235, granted on 21 November 1961. (*Refers to Velcro.*)

de Mestral, G. Apparatus for manufacturing separable fasteners. United States Patent 3,083,737, granted on 2 April 1963. (*Refers to Velcro.*)

de Mestral, G. Method for the manufacture of pile fabrics. United States Patent 3,154,837, granted on 3 November 1964. (*Refers to Velcro.*)

Morrison, W. J., and J. C. Wharton. Candy machine. United States Patent 618428, granted on 31 January 1899. (*Refers to candy floss.*)

Mullis, K. B. Process for amplifying nucleic acid sequences. United States Patent 4,683,202, granted on 28 July 1987. (*Refers to polymerase chain reaction.*)

Nakamura, S., N. Iwasa and M. Senoh. Method of manufacturing p-type compound semiconductor. United States Patent 5,468,678, 1995. (*Refers to gallium nitride.*)

Nieuwland, J. A. Vinyl derivatives of acetylene and method of preparing the same. United States Patent 1,811,959, granted on 30 June 1931. (*Refers to neoprene.*)

Noebe, R. D., S. L. Draper, M. V. Nathal and E. A. Crombie. Precipitation hardenable high temperature shape memory alloy. United States Patent 7,749,341, granted on 6 July 2010. (*Refers to shape memory alloys.*)

Paley, M. S., R. S. Libb, R. N. Grugel and R. E. Boothe. Ionic liquid epoxy resins. United States Patent 8,193,280, 2012. (*Refers to ionic liquid adhesives.*)

Palmer, P., T. Schreck, L. Hamel, S. Tzannis and A. Poutiatine. Small volume oral transmucosal dosage forms containing sufentanil for treatment of pain. United States Patent, 8,535,714, 2013. (*Refers to drug delivery.*)

Palmese, G. R., and A. L. Watters. Room temperature ionic liquid-epoxy systems as dispersants and matrix materials for nanocomposites. United States Patent Application 2012/0296012, published on 22 November 2012. (*Refers to nanomaterials.*)

Partanan, J. P., and D. R. Smalley. Apparatus and method for controlling exposure of a solidifiable medium using a pulsed radiation source in building a three-dimensional object using stereolithography. United States Patent 6,215,095, granted on 10 April 2001. (*Refers to stereolithography.*)

Perkin, W. H. Producing a new colouring matter for the dyeing with a new lilac or purple color stuff of silk, cotton, wool or other materials. United Kingdom Patent GB 1984, granted on 24 February 1857. (*Refers to mauveine dye.*)

Plunkett, R. J. Tetrafluoroethylene polymers. United States Patent 2,230,654, granted on 4 February 1941. (*Refers to polytetrafluoroethylene.*)

Predtechensky, M. R., O. M. Tukhto and I. Y. Koval. United States Patent 8,137, 653, granted on 20 March 2012. (*Refers to carbon nanotubes.*)

Prentice, T. C., K. T. Scott and D. L. Segal. Preparation of hydroxyapatite. United Kingdom Patent Application GB 2433257 A, published on 20 June 2007. (*Refers to hydroxyapatite.*)

Ranade, R. A., and M. S. Garrison. Cosmetic compositions for imparting superhydrophobic films. United States Patent Application US 2011/0008401, published on 13 January 2011. (*Refers to waterproof cosmetics.*)

Rohm, O., and E. Trommadorff. Process for the polymerization of methyl methacrylate. United States Patent 2,171,765, granted on 5 September 1939. (*Refers to poly methylmethacrylate.*)

Semon, W. L. Synthetic rubber-like composition and method of making the same. United States Patent 1,929,453, granted on 10 October 1933. (*Refers to polyvinylchloride.*)

Semon, W. L. Method of preparing polyvinyl halide products. United States Patent 2,188,396, granted on 30 January 1940. (*Refers to polyvinylchloride.*)

Shaikh, J. Self-heating fluid connector and self-heating fluid container. International application WO 2006/109098, published 19 October 2006. (*Refers to phase-change materials.*)

Shi, B. Biodegradable and renewable films. United States Patent 8,329,601, granted on 11 December 2012. (*Refers to biodegradable hygiene products.*)

Silva, S. R., S. Haq and B. O. Boskovic. Production of carbon nanotubes. United States Patent 8,715,790, granted on 6 May 2014. (*Refers to carbon nanotubes.*)

Silver, S. F. Acrylate copolymer microspheres. United States Patent 3,691,140, granted 12 September 1972. (*Refers to Post-it notes.*)

Simmons, M., and J. Cawse. Composite materials. United States Patent Application 2014/ 0047710, published on 20 February 2014. (*Refers to nanocomposites.*)

Singh, B., and F. C. Schaefer. United States Patent 4,374,987, 1983. (*Refers to methotrexate.*)

Slayter, G. Method and apparatus for making glass wool. United States Patent 2,133,235, 1938. (*Refers to section on candy floss.*)

Smith, C., et al. Biodegradable, biocompatible and non-toxic material, sheets consisting of said material and the use thereof in food, pharmaceutical, cosmetic and cleaning products. United States Patent Application 2013/0034638, published on 7 February 2013. (*Refers to starch nanoparticles.*)

Talapatra, S., S. Kar, S. Pal, R. Vajtai and P. Ajayan. Carbon nanotube growth on metallic substrate using vapour phase catalyst delivery. United States Patent 8,207,658, granted on 26 June 2012. (*Refers to carbon nanotubes.*)

Yamauchi, K., K. Mori and S. Yamashita. Balloon expandable superelastic stent. United States Patent 7,658,761, granted on 9 February 2010. (*Refers to shape memory alloys.*)

Zhang, J., and C. Shachaf. Methods of making silver nanoparticles and their applications. International Patent Applications WO 2014/052973, published on 3 April 2014. (*Refers to silver nanoparticles.*)

Index

Page numbers in italics refer to figures and tables.
Page numbers in bold type refer to items in the glossary.